D1029460

Applications and Systematics of *Bacillus* and Relatives

Dedicated to
Ruth Gordon

Applications and Systematics of *Bacillus* and Relatives

Edited by

Roger Berkeley
University of Bristol
Bristol
England

Marc Heyndrickx
Centrum voor Landbouwkundig
Melle
Belgium

Niall Logan
Glasgow Caledonian University
Glasgow
Scotland

Paul De Vos
Laboratorium voor Microbiologie
Gent
Belgium

OCT 2 3 2002

REVIEW COPY

from

Blackwell
Publishing

© 2002 by Blackwell Science Ltd
a Blackwell Publishing company

350 Main Street, Malden, MA 02148-5018, USA
108 Cowley Road, Oxford OX4 1JF, UK
550 Swanston Street, Carlton, Victoria 3053, Australia
Kurfürstendamm 57, 10707 Berlin, Germany

The right of the Editors to be identified as the Authors of the Editorial Material
in this Work has been asserted in accordance with the UK Copyright, Designs,
and Patents Act 1988.

All rights reserved. No part of this publication may be reproduced, stored
in a retrieval system, or transmitted, in any form or by any means, electronic,
mechanical, photocopying, recording or otherwise, except as permitted by
the UK Copyright, Designs and Patents Act 1988, without the prior
permission of the publisher.

First published 2002

Library of Congress Cataloging-in-Publication Data

Applications and systems of bacillus and relatives/
Roger Berkeley . . . [et al.].
p.   cm.
Includes bibliographical references and index.
ISBN 0-632-05758-0 (hb: alk. paper)
1. Bacillus (Bacteria)   I. Berkeley, Roger.
QR82.B3 A66 2002
579.3′62—dc21
2002022135

A catalogue record for this title is available from the British Library.

Set in 9.5/12pt Sabon
by Graphicraft Limited, Hong Kong
Printed and bound in the United Kingdom
by MPG Books Ltd, Bodmin, Cornwall

For further information on
Blackwell Publishing, visit our website:
http://www.blackwellpublishing.com

Contents

List of Contributors

D.R. Arahal, *Department of Microbiology and Parasitology, Faculty of Pharmacy, University of Seville, Seville, Spain*

D.J. Beecher, *Hazardous Materials Response Unit, FI Academy, Quantico, VA 22135, USA*

R.C.W. Berkeley, *Badock Hall, University of Bristol, Stoke Park Road, Bristol BS9 1JQ, England*

A. Bishop, *School of Chemical and Life Sciences, University of Greenwich, Wellington Street, Woolwich, London SE18 6PF, England*

C.P. Chanway, *Faculties of Forestry and Agricultural Sciences, University of British Columbia, Vancouver, British Columbia, Canada V6T 1Z4*

P. De Vos, *Laboratorium voor Microbiologie, Universiteit Gent, K L Ledgeganckstraat 35, B-9000 Gent, Belgium*

D. Fritze, *DSMZ-Deutsche Sammlung von Mikroorganismen und Zellkulturen GmbH, Mascheroder Weg 1b, D-38124 Braunschweig, Germany*

A. Gambacorta, *Institute of Chimica Biomolecolare of CNR, Comprensorio Olivetti isol 70, Via Campi Flegrei 34, 80078 Pozzuoli, Napoli, Italy*

R. Goodacre, *Institute of Biological Sciences, University of Wales, Aberystwyth, Ceredigion SY23 3DD, Wales*

P.E. Granum, *Department of Pharmacology, Microbiology and Food Hygiene, Norwegian Veterinary College of Veterinary Medicine, PO Box 8146 Dep, N-0033 Oslo, Norway*

M. Heyndrickx, *Centrum voor Landbouwkundig Ondorzoek Gent, DVK, Bursselsesteenweg 370, B-9090 Melle, Belgium*

K.K. Hill, *Bioscience Division, Los Alamos National Laboratory, Los Alamos, NM 87545, USA*

P.J. Jackson, *Bioscience Division, Los Alamos National Laboratory, Los Alamos, NM 87545, USA*

S.T. Jørgensen, *Novozymes A/S, Novo Alle, DK-2880 Bagsvaerd, Denmark*

P. Kämpfer, *Institut fur Angewandte Mikrobiologie, Fachbereig Agrarwissenschaften und Unweltsicherung, Uustus-Liebig-Universiteit Giessen, Heinrich-Buff-Ring 26–32, D-35392 Giessen, Germany*

P. Keim, *Department of Biological Sciences, Northern Arizona University, Flagstaff, AZ 86011, USA*

A.-B. Kolstø, *Biotechnology Centre of Oslo and Institute of Pharmacy, University of Oslo, PB 1125 Blindern, 0316 Oslo, Norway*

L. Lama, *Institute of Chimica Biomolecolare of CNR, Comprensorio Olivetti isol 70, Via Campi Flegrei 34, 80078 Pozzuoli, Napoli, Italy*

N.A. Logan, *School of Biological and Biomedical Sciences, Glasgow Caledonian University, City Campus, Cowcaddens Road, Glasgow G4 0BA, Scotland*

J.T. Magee, *Department of Medical Microbiology and Public Health Laboratory, University Hospital of Wales, Heath Park, Cardiff CF4 4XW, Wales*

J. Mahillon, *Laboratorie de Genetique Microbienne UCL, Place Croix du Sud 2/12, B-1348 Louvain-la-Neuve, Belgium*

B. Nicolaus, *Institute of Chimica Biomolecolare of CNR, Comprensorio Olivetti isol 70, Via Campi Flegrei 34, 80078 Pozzuoli, Napoli, Italy*

J.R. Norris, *Langlands, 10 Langley Road, Bingley, West Yorkshire BD16 4AB, England*

H. Outtrup, *Skovhaven 89, DK-3500 Vaerloese, Denmark*

F.G. Priest, *Department of Biological Sciences, Heriot-Watt University, Riccarton, Edinburgh EH14 4AS, Scotland*

P. Scheldeman, *Centrum voor Landbouwkundig Onderzoek Gent, DVK, Brusselsesteenweg 370, B-9090 Melle, Belgium*

E. Stackebrandt, *DSMZ-Deutsche Sammlung von Mikroorganismen und Zellkulturen GmbH, Mascheroder Weg 1b, D-38124 Braunschweig, Germany*

J. Swiderski, *DSMZ-Deutsche Sammlung von Mikroorganismen und Zellkulturen GmbH, Mascheroder Weg 1b, D-38124 Braunschweig, Germany*

P.C.B. Turnbull, *Arjemptur Technology Ltd, Science Park, DSTL, Porton Down, Salisbury SP4 0JQ, England*

J. Van Rie, *Aventis Crop Science, Plateaustraat 22, B-9000 Gent, Belgium*

A. Ventosa, *Department of Microbiology and Parasitology, Faculty of Pharmacy, University of Seville, Seville, Spain*

Foreword

In 1979, in Cambridge, the Systematics Group of the Society for General Microbiology held a meeting on the systematics of the aerobic, endospore-forming bacteria, and in 1981 a book based upon it was published by the SGM in its Special Publications series. That book, *The Aerobic Endospore-forming Bacteria: Classification and Identification*, was edited by Roger Berkeley and Mike Goodfellow, and for a number of years it served as a valuable reference work in the field, and was widely cited in publications dealing with *Bacillus* species.

All the contributors to the Cambridge meeting were well aware that the *Approved Lists of Bacterial Names* were soon to be published, and indeed these appeared in the following year. In it, the number of valid *Bacillus* species was reduced to 31, reflecting a considerable tidying up of the genus.

By 1997, when planning for the meeting on which the present volume is based began, members of the genus *Bacillus* had been allocated to six genera, with a total of about 140 species. This progress was largely driven by the application of sophisticated chemotaxonomic and genetic characterization methods, and the use of powerful computers to analyse the resulting data. Indeed, it was this explosion in species numbers, the pace of change in the taxonomy of the aerobic endospore-forming bacteria, and the absence of any comprehensive and up-to-date treatment of the systematics of the group, that suggested the idea for the 'Bacillus 2000' meeting.

The background of the meeting was thus a desire to bring taxonomists interested in *Bacillus* and its relatives together with those who use or combat these organisms in medicine, agriculture, food and industry. The meeting was held in Bruges (Belgium) in August 2000, and one measure of its success was the number of people who at its end agreed that they found it difficult to remember when they last enjoyed hearing every paper in each session from its brief introduction to the concluding remarks. We therefore warmly thank all those who contributed to the meeting, the poster display and this book.

The organizers of the meeting and editors of this book also acknowledge with gratitude the financial support from FEMS, without which the whole enterprise would have been impossible. Important financial contributions were also made by the Belgian Society for Microbiology and by a number of commercial organizations (bioMérieux, Applied Maths, Belgian Coordinated Collection of Micro-organisms, B. Braun Biotech International, Van Hopplynus, Bio-Rad Laboratories, *P.E.* Biosystems and MERCK Eurolab), and we are most grateful to all of them too. We also acknowledge the University of Ghent for its practical and material support in the organization of the poster session. Finally, we wish to thank our secretarial colleagues for their invaluable assistance.

Another, and most startling, feature of the meeting was that, on the evening of the day before it began, one conference member asked 'When is the next meeting like this going to take place?'! We strongly believe that the quality of the meeting lived up to the expectation implied in that question, and initial arrangements are being made for the next meeting, probably to be held in Slovenia in the summer of 2003.

Roger Berkeley
Marc Heyndrickx
Niall Logan
Paul De Vos

Chapter 1

Whither *Bacillus*?

Roger C.W. Berkeley

Introduction

'Whither . . . – To what place, position; what is the future of.' (*The Concise Oxford Dictionary*, Sixth edition.)

The beginning of a new millennium is a major historical milestone and it is appropriate to look at what has happened recently to the genus *Bacillus* and what might happen to it in the future. But first, to give a proper perspective, this should be preceded by a glance into the past.

The history of the genus *Bacillus* is long, and interwoven with the early history of bacteriology. '*Vibrio subtilis*' – now *Bacillus subtilis* – was described in 1835 by Ehrenberg and in 1864 Davaine allocated the name '*Bacteridium*' to the organism associated with anthrax. But it was Cohn, in 1872, who proposed the genus *Bacillus*. All this happened before the final resolution of the debate about whether spontaneous generation occurred or not!

In the 130 years since the creation of this genus its systematics have, unsurprisingly, undergone massive changes. Those up to 1979 were reviewed by Gordon (1981), and a numerical summary from her chapter, listing the number of species assigned to *Bacillus* in each of the first eight editions of *Bergey's Manual of Determinative Bacteriology* (table 1.1), gives a flavour of changes in the 50 years or so spanning the middle of the last century.

In her review, Ruth Gordon remarked that the (large) number of species assigned to the genus in the first to the fifth editions of the *Manual* make it obvious that many new species were named and described without, using the words of Cowan and Steel (1974), '. . . the comparative work necessary to put an organism into its rightful place in an existing genus or species'. Ruth's standards of comparative work were high. The work in which she was involved was based initially on a collection of 621 strains, later expanded to 1134 strains. Furthermore, as I heard her explain in her characteristically simple, quiet and modest way, at the very beginning of my scientific career, to a meeting of the Society for General Microbiology in London: if a colony of appearance different to the majority appeared on a plate, it was not assumed that the culture was contaminated. Instead, attempts would be made to isolate the organism with the different colonial morphology and to study it until it was certainly established that it really was a contaminant and not a variant. In this way, and by studying the limits of variability for some species, she and her colleagues were able to eliminate

Table 1.1 Numbers of species assigned to the genus *Bacillus* in different editions of *Bergey's Manual* up to 1974 (modified from Gordon 1981).

Bergey's Manual	Year	Number of species
1st edition	1923	75
2nd edition	1925	75
3rd edition	1930	93
4th edition	1934	95
5th edition	1938	146
6th edition	1948	33
7th edition	1954	25
8th edition	1974	Group I: 22
		Group II: 26

some species and to demote others to lesser rank. Thus, the 146 species in the 5th edition of *Bergey's Manual* were reduced to 33 in the 6th edition and to 25 in the 7th.

In the 8th edition there was a further reduction in the number of species. These fell into two groups. In Group I there were 22 which were widely accepted as distinct entities, whereas the 26 in Group II had received less widespread recognition (table 1.1). An editorial note records that there was considerable correspondence between the *Manual*'s editors and the authors of the section on *Bacillus*, about the status of the species in Group II. This states that 'In many genera, most, if not all the species in this Group would have been listed as species *incertae sedis*, and one author agrees'. That person was certainly Ruth Gordon whose thorough, painstaking work is a model for us all. It resulted in a taxonomic arrangement of the genus *Bacillus* which largely 'worked' – although not without problems – for most of the last half of the last century and which still, essentially, forms the foundation of the current taxonomy of these organisms.

The next milestone in the development of *Bacillus* systematics was the publication of the *Approved List of Bacterial Names* (Skerman *et al.* 1980). In this, the number of species recognized increased to 31. Six years later the number in *Bergey's Manual of Systematic Bacteriology* had climbed further to 40 validated species, with another 27 *incertae sedis* (Claus & Berkeley 1986). This was still a relatively small number compared to that in the 5th edition of *Bergey's Manual*, but this was not a reflection of a satisfactory taxonomic arrangement.

One of the major problems with the genus *Bacillus* was that it was clearly heterogeneous. As noted in the 8th edition of the *Manual* (Gibson & Gordon 1974), it embraced, as compared with other genera, organisms with a great diversity of properties. Confirming its lack of homogeneity was a range of DNA base composition of over 30% (Claus & Berkeley 1986), as opposed to the agreed upper limit for a homogenous genus of 10% (Bull *et al.* 1992) and there were numerical studies such as that by Logan and Berkeley (1984) suggesting that the genus should be separated into five or six genera. Not inconsistent with all this were a number of proposals made between 1889 and 1952 for at least five new genera that included species usually regarded as belonging to *Bacillus* (see Gibson & Gordon 1974); none of these, however, became established.

Table 1.2 Aerobic endospore-forming genera included in the 8th edition of *Bergey's Manual*, and closely related to *Bacillus* but morphologically or physiologically different from it.

Sporolactobacillus	Kitahara & Suzuki (1963)
Sporosarcina	Kluuyver & van Niel (1936)
Thermoactinomyces	Tsiklinsky (1899)

Table 1.3 Trichome-forming bacteria from the guts of animals, and said to form endospores.

'*Anisomitus*'	Grassé (1925)
'*Arthromitis*' (= '*Entomitus*')	Leidy (1850) Grassé (1924)
'*Bacillospira*' (= '*Sporospirillum*')	Hollande (1933) Delaporte (1964)
'*Coleomitus*'	Duboscq & Grassé (1930)
'*Metabacterium*'	Chatton & Pérard (1913)

Another problem area concerned related genera. There were three genera of aerobic endospore-formers, *Sporolactobacillus*, *Sporosarcina* and *Thermoactinomyces* (table 1.2), which, although morphologically or physiologically very different from *Bacillus*, had been established by molecular studies to be close relatives of species belonging to this genus (Herndon & Bott 1969; Pechman *et al.* 1976; Fox *et al.* 1977; Stackebrandt & Woese 1981; Stackebrandt *et al.* 1987).

Pasteuria too was described as endospore-forming although it was very different from *Bacillus*. The taxonomy of *Pasteuria* was confused (see Sayr & Starr 1989), but some molecular evidence indicated that *Pasteuria penetrans* is a deeply rooted member of the *Bacillus/Clostridium* line of descent. It is, however, not related closely either to the true endospore-formers or to the actinomycetes (E. Stackebrandt, pers. comm.).

In addition, there were reports of several trichome-forming bacteria, said to form endospores, isolated from the gut of animals. These organisms have not been obtained in pure culture and their oxygen relationships have not been established (table 1.3). The spores of one of these, *Metabacterium polyspora*, show some cytological similarities to the endospores of *Bacillus* (Robinow 1951).

Finally, affinity between one *Bacillus* species and nonspore-forming organisms such as *Caryophanon latum*, *Filibacter limicola* and *Planococcus citreus* was demonstrated by Stackebrandt and his colleagues (1987).

In short, *Bacillus* was part of a very large and very diverse group of organisms and by 1986 Claus and Berkeley, recognizing the almost irresistible temptation to publish proposals to split the genus on the basis of the existing evidence, suggested that, as some areas of the genus were as yet inadequately studied, premature division of the genus should be avoided as it could cause difficulties for practitioners working with *Bacillus*.

Table 1.4 Recently described genera which include species once assigned to the genus *Bacillus*.

Alicyclobacillus	Wisotzkey *et al.* (1992)
Aneurinibacillus	Shida *et al.* (1996)
Brevibacillus	Shida *et al.* (1996)
Gracilibacillus	Wainö *et al.* (1999)
Paenibacillus	Ash *et al.* (1994)
Salibacillus	Wainö *et al.* (1999)
Virgibacillus	Heyndrickx *et al.* (1998)

Table 1.5 Genera containing endospore-forming species not transferred from *Bacillus*.

Ammoniphilus	Zaitsev *et al.* (1998)
Amphibacillus	Niimura *et al.* (1990)
Halobacillus	Spring *et al.* (1996)
Sulfobacillus	Golovacheva & Karavaiko (1991)
Thermobacillus	Touzel *et al.* (2000)

Recent changes

Evidence relating to the phylogeny of *Bacillus* has been accumulating since the early days of the application of molecular techniques to bacterial systematics, and division of the genus might have started much earlier than it did. Any such attempt, however, would probably have been unsatisfactory as until 1991 there was 16S rRNA oligonucleotide cataloguing information for only nine *Bacillus* species. Whether or not the plea for restraint in relation to division of the genus (Claus & Berkeley 1986) had any influence or not, no attempt at division occurred until 1991. In that year, Ash and her colleagues published results of studies on rRNA sequences of single strains of 51 *Bacillus* species. These showed the existence of at least five phylogenetically distinct clusters which, these authors suggested, would provide the basis for the division of *Bacillus* into several phylogenetically distinct genera.

In the next eight years, seven new genera were established, some of them based on these clusters of Ash and her co-workers (table 1.4). Also, both before and during this period, five other new genera were described for aerobic endospore-forming species not previously classified as members of the genus *Bacillus* (table 1.5). Added to *Bacillus* itself, this gives a total of 13 genera containing organisms that would probably once have been included in this genus. Leaving aside both the genera based on organisms which apparently produce endospores but whose relationships with *Bacillus* are currently completely unknown, and the nonendospore-forming species *Caryophanon latum*, *Filibacter limicola* and *Planococcus citreus*, but adding *Sporolactobacillus*, *Sporosarcina* and *Thermoactinomyces*, this brings the number of genera to 16.

Regrettably, not all of these genera meet the standards suggested by Ruth Gordon; indeed, some of them (see Chapters 8 and 9) are based on single strain species and others on so few strains that there is no possibility of assessing the limits of variability of the taxa. This points to the need to try to isolate, from a variety of environments, additional representatives of the taxa that are poorly represented in culture collections. In doing this, though, it is hoped that such bacteriology as practiced by Leidy (1850) can be avoided. When studying 'Arthromitis,' he wrote: 'Whilst the legs of fragments of the animals were yet moving upon my table, or one half the body even walking, I have frequently been examining the plants growing upon the intestinal canal of the same individual'!

Applied aspects

Challenging though their work is in itself, systematists must not lose sight of the needs of practitioners in applied areas, for whom classification and nomenclature are both means to an end, and not ends in themselves.

The chapters which follow contain accounts by practitioners or practitioner/systematists working with, on the one hand, some of the more important beneficial uses of *Bacillus* species as sources of insecticides (Chapters 11 and 13) or genes used to produce insect resistant plants (Chapter 12), as sources of enzymes for a variety of uses (Chapter 14) and as growth promoters for plants (Chapter 15), and on the other, some of the main but less desirable activities of aerobic endospore-formers, as causes of disease in humans (Chapter 4) and in causing problems in the dairy and food industries (Chapter 6).

The future

Division of the genus *Bacillus sensu lato* is now so substantial that in a sense *Bacillus* has withered! The number of genera derived from it, however, give perhaps a misleading impression of a reduction in its size. In fact, there are actually still more species in *Bacillus* than in all the genera containing close relatives, some transferred from *Bacillus* and some newly described, put together.

As suggested in Chapter 2, partition of some clearly heterogeneous species is likely to be a next area of activity.

Further ahead, one (probably safe) prediction is that, given the number of taxa with very small numbers of representative strains or species, there will be retrenchment, and perhaps publication of a new version of the *Approved List of Bacterial Names*. This would follow logically from a consensus being arrived at concerning the concept of the bacterial species.

It is equally certain that this process of cutting back the number of taxa will be informed by phylogenetic information and that the species of aerobic endospore-forming bacteria, let alone their close relatives, will probably never, as hoped by Ruth Gordon (1981), number fewer than the 146 listed in the fifth edition of *Bergey's Manual*.

As is implied in, for example, Chapters 8 and 17, it is desirable, if this is indeed possible (see Chapter 19), that there be agreement about the concept of the bacterial species, reconciling that based on molecular approaches to systematics with those depending on other approaches. There will be a need to address questions such as: 'Should taxa which are based on phylogenetic information, and which cannot be identified by any known phenotypic tests, be validly described?' and 'Is there a desirable minimum number of strains on which to base valid species?'. The discussion will be fascinating and important if the tensions (see Chapter 2) caused by the two approaches are to be reduced or eliminated.

Whatever the outcome of work and debates to that end, I am sure, given the desirability of the predictive value of nomenclature, and, not overlooking the existence of genera such as *Lactobacillus, Streptobacillus, Thiobacillus*, etc., that generic epithets for aerobic endosporers should contain the root *-bacillus* (cf. *Ammoniphilus*) continuing its association with this important group of bacteria.

References

Ash, C., Farrow, J.A.E., Wallbanks, S. & Collins, M. (1991) Phylogenetic heterogeneity of the genus *Bacillus* revealed by comparative analysis of small-subunit-ribosomal RNA sequences. *Letters in Applied Microbiology* 13, 202–206.

Ash, C., Priest, F.G. & Collins, M.D. (1994) *Paenibacillus polymyxa* comb. nov. In: *Validation of the Publication of New Names and New Combinations Previously Effectively Published Outside the IJSB. List No. 51. International Journal of Systematic Bacteriology* 44, 852.

Bull, A.T., Goodfellow, M. & Slater, J.H. (1992) Biodiversity as a source of innovation in biotechnology. *Annual Review of Microbiology* 46, 219–252.

Chatton, E. & Pérard, C. (1913) Schizophytes du ceacum du cobaye. II. *Metabcaterium polyspora* n.g., n. sp.. *Comptes Rendus de la Société de Biologie, Paris* 74, 1232–1234.

Claus, D. & Berkeley, R.C.W. (1986) Genus *Bacillus* 1872. In: *Bergey's Manual of Systematic Bacteriology* (eds P.H.A. Sneath *et al.*), Vol. 2, pp. 1105–1139. Williams & Wilkins, Baltimore, MD.

Cohn, F. (1872) Untersuchungen über Bakterien. *Beitrage zur Biologie der Pflanzen* 1, 127–224.

Cowan, S.T. & Steel, K.J. (1974) *Identification of Medical Bacteria*. Cambridge University Press, Cambridge.

Davaine, M.C. (1864) Nouvelles recherches sur la nature de la maladie charbonneuse connue sous le nom de sang de rate. *Comptes Rendus de l'Académie des Sciences, Paris* 59, 393–396.

Delaporte, B. (1964) Étude comparée de grands spirilles formant des spores: *Sporospirillum (Spirillum) praeclarum* (Collin) n.g., *Sporospirillum gyrini* n.sp. et *Sporospirillum bisporum* n.g.. *Annales de l'Institut Pasteur, Paris* 107, 246–262.

Duboscq, R. & Grassé, P-P. (1930) Protistologica XXI. *Coleomitus* n.g. au lieu de *Coleonema* pour la

schizophyte *C. pruvoti* Duboscq et Grassé, parasite d'un *Calortermes* des Iles Loyalty. *Archives de Zoologie, Expérimental et Générale* 70, 28.

Eherenberg, C.G. (1835) Dritter Beitrag zur Erkenntnis grosser Organisation in der Richtung des kleinsten Raumes. *Abhandlung der Königlichen Akademie der Wissenschaften zu Berlin aus der Jahre 1833–1835*, 145–336.

Fox, G.E., Pechman, K.R. & Woese, C.R. (1977) Comparative cataloguing of 16 S ribosomal ribonucleic acid: molecular approaches to prokaryotic systematics. *International Journal of Systematic Microbiology* 27, 44–57.

Gibson, T. & Gordon, R.E. (1974) *Bacillus*. In: *Bergey's Manual of Determinative Bacteriology* (eds R.E. Buchanan & N.E. Gibbons), 8th edn, pp. 529–550. Williams & Wilkins, Baltimore, MD.

Golovacheva, R.S.S. & Karavaiko, G.I. (1991) *Sulfobacillus* new genus, *Sulfobacillus thermosulfidooxidans* new species. In: *Validation of the Publication of New Names and New Combinations Previously Effectively Published Outside the IJSB. List No. 36. International Journal of Systematic Bacteriology* 41, 179.

Gordon, R.E. (1981) One hundred and seven years of the genus *Bacillus*. In: *The Aerobic Endospore-forming bacteria* (eds R.C.W. Berkeley & M. Goodfellow), pp. 1–15. Academic Press, London.

Grassé, P-P. (1924) Notes protistologiques. I. La sporulation des *Oscillospiraceae*; II Le genere *Alysiella* Langeron 1923. *Archives de Zoologie, Expérimental et Générale* 62, 25–34.

Grassé, P-P. (1925) *Anisomitus denisi* n.g., n.sp. Schizophyte de l'intestine du canard domestique. *Annales de Parasitologie Humaine et Comparée* 3, 343–348.

Herndon, S.E. & Bott, K.F. (1969) Genetic relationship between *Sarcina urea* and members of the genus *Bacillus*. *Journal of Bacteriology* 97, 6–12.

Heyndrickx, M., Lebbe, L., Kersters, K., De Vos, P., Forsyth, G. & Logan, N.A. (1998) *Virgibacillus*: a new genus to accommodate *Bacillus pantothenticus* (Proom and Knight 1950). Emended description of *Virgibacillus pantothenticus. International Journal of Systematic Bacteriology* 48, 99–106.

Hollande, A.C. (1933) La structure cytologique des *Bacillus enterothrix, camptospora,* Collin et de *Bacillospira (Spirillium) praeclarum,* Collin. *Comptes Rendus Hebdominaires des Séances de l'Académie des Sciences. Série D, Sciences Naturelles* 196, 1830–1832.

Kitahara, K. & Suzuki, J. (1963) *Sporolactobacillus* nov. subgen. *Journal of General and Applied Microbiology* 9, 59–71.

Kluuyver, A.J. & van Neil, C.B. (1936) Prospects for a natural classification of bacteria. *Zentralblatt für Bakteriologie, Parasitenkunde, Infektionskrankheiten und Hygiene, II Abt.* 94, 369–403.

Leidy, J. (1850) On the existence of endophyta in healthy animals, as a natural condition. *Proceedings of the Academy of Natural Sciences, Philadelphia* 4, 225–229.

Logan, N.A. & Berkeley, R.C.W. (1984) Identification of *Bacillus* strains using the API system. *Journal of General Microbiology* 130, 1871–1882.

Niimura, Y., Koh, E., Yanagida, F., Suzuki, K.-I., Komagata, K. & Kozaki, M. (1990) *Amphibacillus xylanus* gen. nov., sp. nov., a facultatively anaerobic sporeforming xylan-digesting bacterium which lacks cytochrome, quinone, and catalase. *International Journal of Systematic Bacteriology* 40, 297–301.

Pechman, K.J., Lewis, B.J. & Woese, C.R. (1976) Phylogentic status of *Sporosarcina urea. International Journal of Systematic Bacteriology* 26, 305–310.

Robinow, C.F. (1951) Observations on the structure of *Bacillus* spores. *Journal of General Microbiology* 5, 439–457.

Sayr, R.M. & Starr, M.P. (1989) Genus *Pasteuria* Metchnikoff 1888, 166[AL]. In: *Bergey's Manual of Systematic Bacteriology* (eds P.H.A. Sneath *et al.*), Vol. 2, pp. 2601–2614. Williams & Wilkins, Baltimore, MD.

Shida, O., Takagi, H., Kadowaki, K. & Komagata, K. (1996) Proposal for two new genera, *Brevibacillus* gen. nov. and *Aneurinibacillus* gen. nov. *International Journal of Systematic Bacteriology* 46, 939–946.

Skerman, V.B.D., McGowen, V. & Sneath, P.H.A. (1980) Approved lists of bacterial names. *International Journal of Systematic Bacteriology* 30, 225–420.

Spring, S., Ludwig, W., Marquez, M.C., Ventosa, A. & Schleifer, K.-H. (1996) *Halobacillus* gen. nov., with descriptions of *Halobacillus litoralis* sp. nov. and *Halobacillus trueperi* sp. nov., and transfer of *Sporosarcin halophila* to *Halobacillus halophilus* comb. nov. *International Journal of Systematic Bacteriology* 46, 492–496.

Stakebrandt, E., Ludwig, W., Weizenegger, M. *et al.* (1987) Comparative 16S RNA oligonucleotide analyses and murein types of round-spore-forming bacilli and non-spore-forming relatives. *Journal of General Microbiology* 133, 2523–2529.

Stakebrandt, E. & Woese, C.R. (1981) The evolution of prokaryotes. In: *Molecular and Cellular Aspects of Microbial Evolution* (eds M.J. Carlile *et al.*), pp. 1–31. Cambridge University Press, Cambridge.

Touzel, J.P., O'Donohue, M., Debeire, P., Samain, E. & Breton, C. (2000) *Thermobacillus xylanilyticus* gen. nov., sp. nov., a new aerobic thermophilic xylan-degrading bacterium isolated from farm soil. *International Journal of Systematic and Evolutionary Microbiology* 50, 315–320.

Tsiklinsky, P. (1898) Sur les microbes thermophiles. *Annales de l'Institut Pasteur* 13, 286–288.

Wainö, M., Tindall, B.J., Schumann, P. & Ingvorsen, K. (1999) *Gracilibacillus* gen. nov., with description of *Gracilibacillus halotolerans* gen. nov., sp. nov.; transfer of *Bacillus dipsosauri* to *Gracilibacillus dipsosauri* comb. nov., and *Bacillus salexigens* to the genus *Salibacillus* gen. nov., as *Salibacillus salexigens* comb. nov. *International Journal of Systematic Bacteriology* 49, 821–831.

Wisotzkey, J.D., Jurtshuk, P., Jr, Fox, G.E., Deinhard, G. & Poralla, K. (1992) Comparative sequence analyses on the 16S rRNA (rDNA) of *Bacillus acidocaldarius, Bacillus acidoterrestris,* and *Bacillus cycloheptanicus* and proposal for creation of a new genus, *Alicyclobacillus* gen. nov. *International Journal of Systematic Bacteriology* 42, 263–269.

Zaitsev, G., Tsitko, I.V., Rainey, F.A. *et al.* (1998) New aerobic ammonium-dependent obligately oxalotrophic bacteria: description of *Ammoniphilus oxalaticus* gen. nov., sp. nov. and *Ammoniphilus oxalivorans* gen. nov., sp. nov. *International Journal of Systematic Bacteriology* 48, 151–163.

Chapter 2

From Phylogeny to Systematics: the dissection of the genus *Bacillus*

Erko Stackebrandt and Jolantha Swiderski

Introduction

Unexpected relationships between members of *Bacillus* and other genera such as *Sporosarcina* were first revealed by Fox *et al.* (1977) on the basis of comparative 16S rDNA cataloguing. Over the following decade the main outlines of prokaryotic phylogeny became available, in which *Bacillus* species were shown to cluster with other taxa of Gram-positive bacteria which exhibit a low DNA base composition (<50 mol% G+C) (Ludwig & Schleifer 1994; Olsen *et al.* 1994). Despite this tremendous progress in our understanding of bacterial phylogeny, the interpretation of molecular data is not straightforward, depending as it does on mathematical algorithms, selection and number of reference sequences and selection of sequence positions. For example, the phylogenetic coherency of the Gram-positive bacteria – the so-called *Clostridium/Bacillus* subline and the Actinobacteria subline – have still to be shown convincingly (Van de Peer *et al.* 1994). Studies on genes and gene products other than ribosomal RNA genes have been performed mainly on *B. subtilis*. Analyses of genomic properties do not, therefore, contribute significantly to the phylogeny of the genus *Bacillus*, but this will most likely change as more species are included in genome sequencing projects. With one exception, *B. subtilis* does indeed group with Gram-positive reference organisms which exhibit low G+C contents; e.g. in studies on RNases H (Ohtani *et al.* 1999), family C DNA polymerases (Huang & Ito 1998), DNAK (heat shock protein) (Gupta *et al.* 1997), GroEL (chaperonin) (Viale *et al.* 1994; Dale *et al.* 1998) and σ^{70}-type sigma factors (Gruber & Bryant 1997). In contrast, analysis of *nifH* genes (Achouak *et al.* 1999) showed members of *Paenibacillus* to cluster next to cyanobacteria, while members of *Clostridium*, their relatives according to 16S rDNA and other genes and proteins (Olsen *et al.* 1994; Van de Peer *et al.* 1994), grouped only distantly.

The order in which branches diverge from each other is a matter for discussion, ranging from the most remotely related lineages to the fine details of taxa that have evolved recently. While the influences of certain factors on tree topologies are known to experienced taxonomists, the neophyte is often puzzled by changes in the positions of taxa within phylogenetic dendrograms. This chapter shows the effect of commonly used algorithms on tree topologies within the *Bacillus* cluster, and tries to explain that despite certain uncertainties in the position of most deeper-branching lineages, recent changes in the systematics of these organisms are, by and large, justified from a phylogenetic point of view.

Reclassification based upon phylogenetic diversity

Despite major revisions in the taxonomy of *Bacillus*, the taxonomic entity of this genus as defined at the time of its original description (Cohn 1872) still exists in the description of the type species *B. subtilis* and phylogenetically affiliated species. Actually, the vast majority (88%) of the 114 species described as members of this genus up to the year 2000 are still members of the genus *Bacillus*. Following the pioneering study of Ash *et al.* (1991) – which itself was a continuation of earlier studies by Fox *et al.* (1977), Clausen *et al.* (1985), Stackebrandt *et al.* (1987) and others – some of the phyogenetically distinct entities were later reclassified as new genera (present number of species in brackets), i.e. *Alicyclobacillus* (3), *Aneurinibacillus* (3), *Brevibacillus* (10), *Gracilibacillus* (2), *Halobacillus* (3), *Paenibacillus* (24), *Salibacillus* (1) and *Virgibacillus* (2). Other organisms, which formerly would have been placed in *Bacillus*, were described as members of novel genera, such as *Thermobacillus* (1) and *Amphibacillus* (1). The dissection of *Bacillus* followed a trend that brought taxonomy on a par with phylogeny. As a consequence, traditional key characters, such as rod-shaped morphology, aerobic metabolism and spore formation lost their significance in circumscribing the genus. Some taxonomists may disagree with this change in dealing with taxonomy, and a look through the microscope and determination of growth properties may still be faster than determination of the primary structure of 16S rDNA and subsequent phylogenetic analyses. However, today, the main concern is directed less towards the dissection of the genus and more towards the splitting of strain-rich species, and species clusters, in which either DNA–DNA reassociation similarities or the presence of subspecific, genomically coherent traits guide the splitting process. These decisions are often not accompanied by the description of sufficient phenotypic properties for a diagnostic laboratory to identify a strain or recognize it as a representative of a novel species. It is in this field of tension, also seen with other groups of organisms, that microbiologists are currently asked to do taxonomic work: on the one hand, using the potential of doing detailed molecular analyses, down to the level of strains, even clones; and on the other, knowing that the tools for unravelling these genomic properties will for a long time be unavailable to the majority of users worldwide.

Assessing the taxonomic boundaries of the genus *Bacillus* and related taxa

The vast majority of 16S rDNA dendrograms including members of *Bacillus* have been based upon distance analyses, using the Jukes and Cantor (1969) correction of similarity values. It is not surprising, therefore, that the topologies of the phylogenetic patterns, generated by basically the same method, differ from each other only in detail in different publications. *Bacillus* species were found to form clusters that have been named RNA groups 1 to 5 (Ash *et al.* 1991). Later, the presence of an additional RNA group – named group 6 – has been described for alkaliphilic and alkalitolerant species (Nielsen *et al.* 1994). Some of these groups,

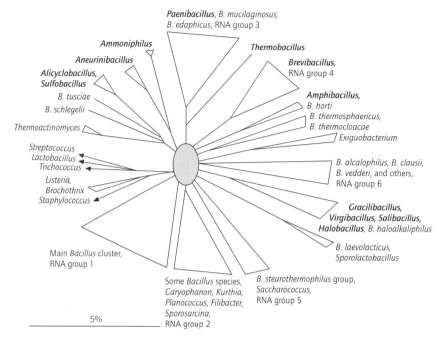

Figure 2.1 Schematic outline of the phylogenetic diversity of 16S rDNA of aerobic, rod-shaped and spore-forming, Gram-positive bacteria, classified as species of *Bacillus*, genera that originated from the dissection of *Bacillus*, and species that were affiliated to novel genera because of their distinct phylogenetic positions. The areas of the triangles represent approximations of the number of species included in the taxa covered by the triangle. The circle indicates the uncertainty of the order at which the lineages diverge from each other. *B, Bacillus*.

such as *Paenibacillus* (group 3), and *Brevibacillus* (group 4), have been reclassified since 1991, while other separate lineages have been reclassified as *Aneurinibacillus*, *Alicyclobacillus*, *Halobacillus*, *Gracilibacillus*, *Salibacillus* and *Virgibacillus*. The main radiation of these organisms, based upon neighbour-joining analysis (Felsenstein 1993) is shown schematically in figure 2.1. This dendrogram also depicts the presence of non-*Bacillus* genera among *Bacillus* groups and clusters, and identifies potential new genera for those species which are not related to those already reclassified, e.g. *B. tusciae*, *B. schlegelii*, *B. horti*, *B. laevolacticus*, *B. thermocloacae*, and members of RNA group 6.

Although the number of sequences of type strains has been increased significantly during recent years (from about 50 by Ash *et al.* 1991 to >120 in 2000), the groups defined in 1991 still, by and large, emerge in any of the phylogenetic analyses published. However, as the topology of a dendrogram depends strongly upon the overall number of sequences and the number of sequences in any particular group, the order in which they emerge in the dendrogram may differ significantly between studies. Rather than showing the precise branching order at deeper phylogenetic levels as unravelled by the phylogenetic analysis, this region is not further resolved in figure 2.1. Figures 2.5 and 2.6 are more detailed phylogenetic analyses based upon the neighbour-joining method (see below).

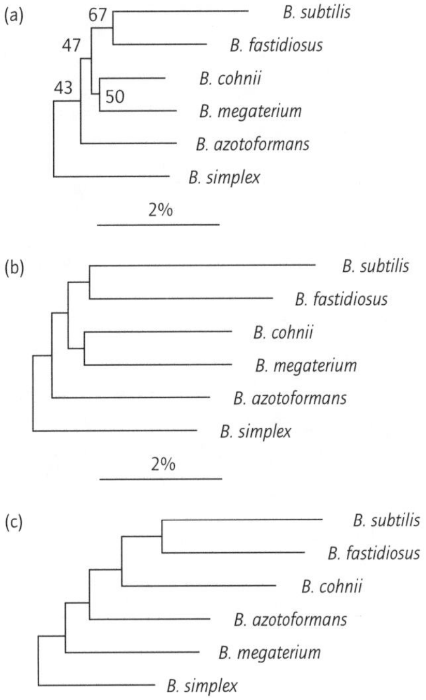

(a)

67　B. subtilis
47　B. fastidiosus
43　50　B. cohnii
B. megaterium
B. azotoformans
B. simplex

2%

Figure 2.2 Comparative analysis of 16S rDNA of six type strains of *Bacillus* species of RNA group 1. Bar indicates 2% nucleotide substitutions. (a) Distance-matrix analysis using the least squares algorithm of DeSoete (1983) and the Jukes and Cantor (1969) correction to compensate for different evolutionary rates. The four highest bootstrap values are indicated. (b) Distance-matrix analysis using the neighbour-joining method (Felsenstein 1993) and the Jukes and Cantor (1969) correction to compensate for different evolutionary rates. (c) Maximum-likelihood analysis, using the program DNAML (transition–transversion rate 2.000) (Felsenstein 1993). *B, Bacillus.*

(b)

B. subtilis
B. fastidiosus
B. cohnii
B. megaterium
B. azotoformans
B. simplex

2%

(c)

B. subtilis
B. fastidiosus
B. cohnii
B. azotoformans
B. megaterium
B. simplex

Different treeing algorithms generate different topologies

Algorithms such as DNAMI, included in PHYLIP (Felsenstein 1993), have been used only rarely to determine the phylogenetic relatedness of *Bacillus* species and related taxa, probably owing to the long computing time required to analyse dozens of sequences by the maximum-likelihood method. Distance-matrix programs such as NEIGHBOR use dissimilarity values to correct for rate variation, while maximum-likelihood methods estimate phylogenies from nucleotide sequences. The latter model allows for unequal expected frequencies of the four nucleotides and for different rates of change in different categories of sites. Figure 2.2 compares the branching patterns of a small set of species from RNA group 1, which present phylogenetically well-separated taxa (93.8–96.3% 16S rDNA sequence similarity). Figures 2.2a and 2.2b have been generated by distance-matrix analyses using the Jukes and Cantor (1969) correction to compensate for different evolutionary rates; figure 2.2a is based on the algorithm of DeSoete (1983), while figure 2.2b is a neighbour-joining (NJ) dendrogram (Felsenstein 1993). The bootstrap values presented in figure 2.2a are low, indicating a low degree of statistical significance in the branching order. Figure 2.2c is a maximum-likelihood (ML) dendrogram (Felsenstein 1993). Quite obviously, the two distance-matrix trees have similar topologies, while the ML dendrogram differs in the branching of *B. megaterium* and *B. azotoformans*.

One can assume that differences in the topologies of dendrograms are further increased when less-related species are included in the analyses. Figure 2.3

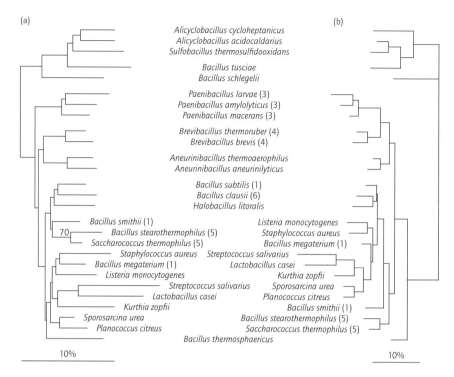

Figure 2.3 Comparative analysis of 16S rDNA of a broad selection of type strains of *Bacillus* species and non-*Bacillus* reference strains. The sequence of *Paenibacillus amylolyticus* has been generated from the nontype strain NCIMB 8144. The bar indicates 10% nucleotide substitutions. (a) Distance-matrix analysis using the neighbour-joining method (Felsenstein 1993) and the Jukes and Cantor (1969) correction to compensate for different evolutionary rates. (b) Maximum-likelihood analysis, using the program DNAML (transition–transversion rate 2.000) (Felsenstein 1993).

compares the topologies of dendrograms of 27 species of *Bacillus* and reference taxa, generated by the NJ dendrogram with the Jukes and Cantor (1969) correction (figure 2.3a) and the ML method (transition–transversion rate 2.000) (figure 2.3b). Both dendrograms are similar in some details but differ significantly in others: most of the closely related species group together in both dendrograms, e.g. *Paenibacillus*, *Brevibacillus*, *Aneurinibacillus*, *B. stearothermophilus* and *Saccharococcus thermophilus*, while significant differences occur at deeper levels of relationship. This is not only demonstrated by the intergeneric relationships of *Paenibacillus*, *Brevibacillus*, *Aneurinibacillus* and most of the non-*Bacillus* reference taxa, but also by members of RNA group 1, which do not appear to form a phylogenetically coherent cluster. Bootstrap values calculated for branching points of the NJ dendrogram are low in most cases, indicating the low statistical significance of the order at which they separate. It should be noted that high bootstrap values are no proof *per se* of exclusive phylogenetic relatedness; they demonstrate that the same branching order is recovered in most of the sub-trees recovered in the analysis. High bootstrap values (>90%) are likely to occur in those cases where lineages are separated from each other by long internodes from neighbouring lineages. The addition of new sequences would most likely change this apparent proof of phylogenetic evidence. For this reason bootstrap values are not indicated in figures 2.5 and 2.6.

Another method of displaying the statistical significance of phylogenies makes use of multiple datasets, generated by bootstrap resembling, which themselves serve as input files for a program that estimates phylogenies by the parsimony method. Figure 2.4 is a consensus tree based upon analyses of 100 bootstrapped datasets included in the DNAPARS program. When compared to the topologies of the NJ and ML dendrograms (figure 2.3), certain topological features are reproduced, such as the separate clustering of *Sulfobacillus*, *Alicyclobacillus*, *B. tusciae* and *B. schlegelii*. Members of *Paenibacillus*, *Brevibacillus* and *Aneurinibacillus* also form recognizable entities, while the branching order of the taxa included in the boxed area differ in all three dendrograms. Considering that only a single representative has been selected from each of the species-rich genera *Lactobacillus*, *Streptococcus* and *Staphylococcus*, additional changes might be expected to occur following their inclusion in the analyses.

As a consequence, phylogenetic patterns derived by any of the several methods available today give no 'proof' that the topologies closely reflect the course of evolution. The closer the matches in the topologies of dendrograms generated by different algorithms, such as distance-matrix analyses, maximum-likelihood and parsimony, the higher the chance that the branching patterns do indeed express phylogenetic evidence. This is clearly the case for the emergence of individual genera described during the past years, as well as for several individual lineages that will probably be described as novel genera in the future, e.g. *B. schlegelii* and *B. tusciae*. The new generation of personal computers will handle even the time-costly maximum-parsimony algorithms better, and future conclusions about taxonomic relatedness among species should be based on more than just a single tree-inferring approach.

Phylogenetic grouping and phenotypic circumscription

The phylogeny-based dissection of *Bacillus* RNA groups is in many cases not supported by clear-cut phenotypic properties. This finding may be of concern to some taxonomists, as affiliation of novel strains to the respective taxa is dependent mainly upon phylogenetic analysis of 16S rDNA (similarity values and signature nucleotides) and the occurrence of PCR fragments in gel electrophoresis following PCR amplification using genus-specific primers (Shida *et al.* 1996). In only a few cases are salient characteristics such as distinct chemotaxonomic properties available for genus affiliation, examples being the amino-acid composition of peptidoglycan in members of *Halobacillus* (Spring *et al.* 1996), or ω-alicyclic fatty acids in *Alicyclobacillus* species (Wisotzkey *et al.* 1992). In the case of *Thermobacillus* (Touzel *et al.* 2000), a sister group of the *Paenibacillus* lineage, the genus description is so poor in descriptive features that its members cannot be affiliated to the genus without the help of 16S rDNA data. Members of yet other genera, such as *Amphibacillus* (Niimura *et al.* 1990), *Paenibacillus* (Shida *et al.*

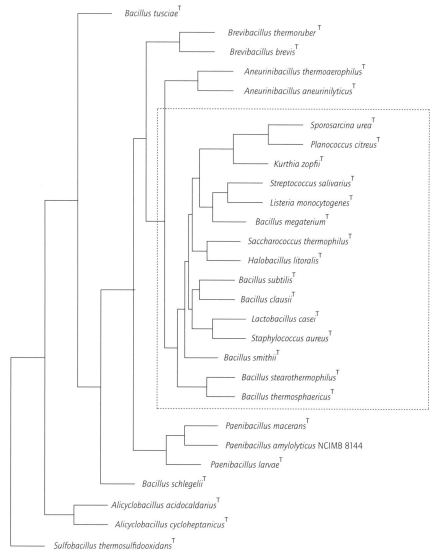

Figure 2.4 Consensus parsimony (DNAPARS; Felsenstein 1993) dendrogram of sequences included in figure 2.3, based upon 100 bootstrapped trees. The topology should be compared to those presented in figure 2.2a,b. The boxed area indicates those organisms whose branching orders are affected most obviously by the selection of treeing algorithms. [T], type strain.

1997a), *Brevibacillus* (Shida *et al.* 1996), *Salibacillus* (Wainö *et al.* 1999), *Gracilibacillus* (Wainö *et al.* 1999), *Virgibacillus* (Heyndrickx *et al.* 1998) and *Aneurinibacillus* differ from each other in some phenotypic characteristics, but these are not exclusive in most cases. Moreover, a comparative listing of genus-specific properties (Heyndrickx *et al.* 1998) indicates that several of these pro-perties have not yet even been elucidated for members of all genera. In fact, the phenotypic properties listed as genus-specific are usually those used to describe species in other genera of Gram-positive bacteria.

Bacillus RNA group 1

This group constitutes the core of *Bacillus*, containing the name-bearing type species *B. subtilis* (figure 2.5). As already indicated by Ash *et al.* (1991), the phylogenetic diversity of this group is huge, encompassing several well-separated species clusters and single-species lineages. When selected single species are included in a larger database of reference organisms, they do not necessarily cluster together (see figures 2.2a, 2.2b and 2.3), which may indicate that this group does not form a coherent phylogenetic entity. Were discriminating phenotypic properties to be available, RNA group 1 could be divided into several genera, as done with similarly remotely related species which were transferred to *Virgibacillus*, *Salibacillus*, *Halobacillus* and *Gracilibacillus* (figure 2.6). One of these subgroups comprises *B. vallismortis*, *B. mojavensis*, *B. subtilis*, *B. amyloliquefaciens*, *B. atrophaeus* and *B. licheniformis*; a second one contains *B. cohnii*, *B. horikoshii* and *B. halmapalus*; a third one harbours *B. cereus*, *B. pseudomycoides*, *B. anthracis* *B. thuringiensis*, *B. weihenstephanensis* and *B. mycoides*; while a fourth one embraces *B. simplex*, *B. psychrosaccharolyticus*, the invalid species '*B. maroccanus*' and two misclassified *Brevibacterium* and *Arthrobacter* species. Most of the other species of this group form more deeply rooting lineages which fan out without allowing the determination of their branching order.

Bacillus RNA group 2

This group constitutes an evolutionary enigma. *Bacillus*-type organisms are intermixed with spherical spore-formers (*Sporosarcina*) and nonspore-forming rods [*Filibacter* (Clausen *et al.* 1985; not shown in figure 2.5), *Kurthia*, *Caryophanon*] and cocci (*Planococcus*). The hallmarks of this group are the presence of either L-lysine or ornithine at position 3 of the peptide subunit and a dicarboxylic amino acid in the interpeptide bridge. The majority of the other taxa covered in this chapter (the exception being *Halobacillus*) contain a directly cross-linked peptidoglycan with *meso*-diaminopimelic acid at position 3 of the subunit. While the non-*Bacillus*-type genera within this RNA group form phylogenetically coherent entities, members of *Bacillus* cluster around *Sporosarcina* and *Caryophanon*. Considering that the ancestors of this group contained *Bacillus* RNA group 1-type characters, this finding may indicate that the *Bacillus* species of RNA group 2 have been prone to significant genomic rearrangements or other genomic changes which lead to the loss of rod-shaped morphology and spore-formation. The intermixing of phenotypically different genera with *Bacillus* species constitutes an interesting taxonomic problem. In order to make classification consistent with phylogeny, the four different lineages of *Bacillus* species (*B. globisporus* and relatives, *B. insolitus*, *B. fusiformis* and *B. silvestris*) would have to

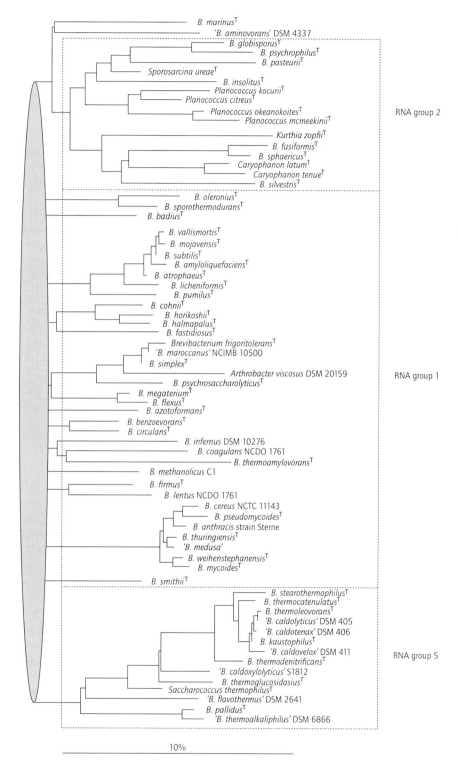

Figure 2.5 Detailed neighbour-joining tree of species of RNA groups 1, 2 and 5. The dotted area indicates the uncertainty of the order at which the lineages diverge from each other. The area was chosen somewhat arbitrarily and may just as well cover more recent branching points. The bar indicates 10% nucleotide substitutions. *B*, *Bacillus*; [T], type strain.

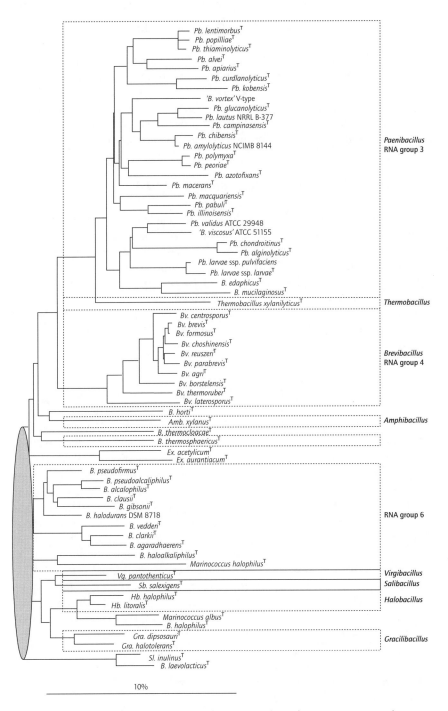

Figure 2.6 Detailed neighbour-joining tree of species members of RNA groups 3, 4 and 6, several other *Bacillus* species, and lineages containing reclassified *Bacillus* species. The dotted areas indicate the uncertainty of the order at which the lineages diverge from each other. The areas are chosen somewhat arbitrarily and may just as well cover more recent branching points. The bar indicates 10% nucleotide substitutions. *Gra, Gracilibacillus; Amb, Amphibacillus; B, Bacillus; Bv, Brevibacillus; Ex, Exiguobacterium; Hb, Halobacillus; Pb, Paenibacillus; Sb, Salibacillus;* [T], type strain; *Vg, Virgibacillus.*

be described as novel genera. The rationale for this conclusion is the lack of monophyly of the lineages and the presence of morphological differences in the non-*Bacillus* taxa. Alternatively, one would abandon ecological and morphological differences (including lack of spore-formation) to include the *Bacillus* species into the neighbouring genera or, in the most progressive (and unlikely) scenario, to describe all members of RNA group 2 as members of a single genus.

Bacillus RNA groups 3 and 4

Members of these two main groups have been reclassified as *Paenibacillus* and *Brevibacillus*, respectively (figure 2.6). These two genera are phylogenetically well separated, although their specific relatedness, indicated in figure 2.6, is doubtful considering the results of maximum-likelihood and parsimony analyses (figures 2.2 and 2.3). With the branching of *Thermobacillus xylanilyticus* at the root of *Paenibacillus*, the phylogenetic distinctness of the latter genus seems to disappear. It should be noted that the type strain of *P. amylolyticus* (NRRL NR-290) clusters with the type strains of *P. illinoisensis* and *P. pabuli* (Shida *et al.* 1977b) (not shown in figure 2.6).

Bacillus RNA group 5

Group 5 constitutes a phylogenetically coherent cluster following distance-matrix analyses (Ash *et al.* 1991; Rainey & Stackebrandt 1993; Studholme *et al.*, 1999; figure 2.5), while according to the consensus tree (figure 2.4), *B. stearo-thermophilus* does not cluster with another representative of the group, *Saccharococcus thermophilus*, but with a further thermophilic organism, *Bacillus thermosphaericus*. The G+C content of 16S rDNA of certain thermophilic bacteria of group 5, including *S. thermophilus*, is, at 59 mol%, about 4–8% higher than those of other members of this group [*B. thermoglucosidasius* (51 mol%), '*B. thermoalkaliphilus*' (56 mol%)], of mesophilic species such as *B. subtilis* (55 mol%), *B. cereus* (52 mol%) and *Halobacillus litoralis* (55 mol%), and also of that of the thermophile *B. thermosphaericus* (55 mol%). It therefore appears unlikely that nucleotide composition-induced artefacts, such as differences in the base ratios of 16S rDNA, are responsible for the branch attraction of certain thermophilic members of group 5 with *B. thermosphaericus*. Likewise, we have no explanation for the grouping of *S. thermophilus* with *Halobacillus litoralis* in the parsimony consensus tree (figure 2.4).

Bacilli of RNA group 5 have recently been reclassified as species of the new genus *Geobacillus* (Nazina *et al.* 2001).

Bacillus RNA group 6

Members of this group, defined by Nielsen *et al.* (1994) are mainly alkaliphilic or alkalitolerant. Three distantly related subclusters emerge from distance-matrix analyses, raising the question of whether this group should be reclassified to constitute one or more genera. As halophilic and halotolerant species are also present in RNA group 1, this phenotypic trait cannot be used as the salient property for the description of genera. More phenotypic data and a more thorough phylogenetic analysis are needed before reclassification can be considered.

New lineages

In addition to these RNA groups, several of which may embrace species clusters worthy of genus status, some other lineages form separate taxonomic entities. The status of *B. schlegelii* and of *B. tusciae* have been mentioned above. Others are *B. thermosphaericus*, *B. thermocloacae*, *B. horti* and *B. laevolacticus* (figure 2.1), *B. halophilus* (figure 2.6), and the pair *B. marinus* and '*B. aminovorans*', which are diverging at the bases of RNA group 2 (figure 2.5). *Bacillus thermosphaericus* has recently been reclassified as *Ureibacillus thermosphaericus* (Fortina *et al.* 2001). This organism exhibits a Lys-D-Asp peptidoglycan variation while its distant phylogenetic neighbour, *B. thermocloacae*, possesses diaminopimelic acid.

Molecular analysis at the species level

Although 16S rDNA methods are appropriate to determine approximate phylogenetic positions at the levels of genera and higher taxa, they are not suitable to classify strains at the species level (Stackebrandt & Goebel 1994). The molecular 'gold standard' for the delineation of species in bacteriology is the determination of reassociation similarity of single-stranded DNA. While this method does not take into account differences in evolutionary rate – i.e. whether strains evolve faster or slower than their nearest phylogenetic neighbours – it is the accepted method either for the allocation of a strain to a described species (recommended value above 70% similarity), or to give evidence that a strain constitutes the nucleus of a new species (recommended value below 70% similarity to existing species). In a polyphasic approach to classification, a new species should be described only if accompanied by the presence of a number of salient phenotypic characteristics that would allow one phenotypically to distinguish the potential new species from those already described. There are no recommendations for the number of species-specific characters that should be determined, but taxonomists are guided by the spectrum of properties available for other members of the genus.

Because of the availability of a cumulative database, determination of 16S rDNA similarities is easier than DNA reassociation values. Comparison of 16S rDNA similarities with DNA reassociation similarities revealed that in no case are the values of the latter method higher than 70% when 16S rDNA similarities are lower than 97%. This means that a strain exhibiting 97% or lower 16S rDNA similarities with other members of the genus, need not be included in DNA–DNA reassociation experiments. It will most certainly represent a new genospecies that, given the existence of diagnostic phenetic markers, can be described as a novel species. Moreover, 16S rDNA analyses will facilitate other taxonomic work, as similarity values above 97% will point towards those species whose type strains should be included in DNA–DNA reassociation experiments.

Molecular analysis at the strain level

Determinations of genomic relationships at the strain level are made mainly by methods that generate a pattern of fragments which can be analysed using standard algorithms, such as the unweighted pair-group method with averages (UPGMA). Numerous studies have been published on medically, biotechnologically and ecologically important aerobic endospore-formers, and only a small selection of the recent literature can be included here. The reader is referred to Istock *et al.* (1996), which covers the importance of typing methods for theoretical and practical aspects of the evolution of bacterial species.

Patterns are generated by a number of different approaches, originating from either whole cell proteins (Heyndrickx *et al.* 1998) and multilocus enzyme electrophoresis of selected alloenzymes (see Istock *et al.* 1996; Helgason *et al.* 1998), or from restriction fragments and PCR products using total DNA (Istock *et al.* 1996), genes (Heyndrickx *et al.* 1998) and/or nontranscribed DNA regions, such as rrn operons (Stroman *et al.* 1998) as a template. Multilocus sequence typing (MLST; Maiden *et al.* 1998) and sequence analysis of genes less conserved than 16S rDNA (Price *et al.* 1999) are of greater importance in epidemiological studies. For taxonomic purposes, the patterns of unclassified and classified strains are compared, dendrograms are produced and the positions of the unclassified strains are determined. These techniques are the methods of choice in population genetics, as a large number of strains can be screened simultaneously. The disadvantage of these approaches is the lack of reproducibility if patterns are not generated under strictly similar conditions, such as those provided by the RiboPrint® system (Qualicon™, Wilmington, DE). Data obtained from MLST and ribotyping systems are currently used to build a cumulative database that will not only allow strain characterization to be used in authentification and certification processes, but also, in the long run, facilitate strain identification. It should be stressed that these patterns are, *per se,* unsuitable to delineate the boundaries of species, but they do provide valuable information on the genomic diversity of the strains presently included within a taxon.

References

Achouak, W., Normand, P. & Heulin, T. (1999) Comparative phylogeny of rrs and nifH genes in the *Bacillaceae*. *International Journal of Systematic Bacteriology* 49, 961–967.

Ash, C., Farrow, J.A.E., Wallbanks, S. & Collins, M.D. (1991) Phylogenetic heterogeneity of the genus *Bacillus* revealed by comparative analysis of small-subunit-ribosomal RNA sequences. *Letters in Applied Microbiology* 13, 202–206.

Clausen, V., Jones, J.G. & Stackebrandt, E. (1985) 16S ribosomal RNA analysis of *Filibacter limicola* indicates a close relationship to the genus *Bacillus*. *Journal of General Microbiology* 131, 2659–2663.

Cohn, F. (1872) Untersuchungen über Bacterien. *Beiträge zur Biologie der Pflanzen* 1, 127–224.

Dale, C.J.H., Moses, E.K., Ong, C.-C. *et al.* (1998) Identification and sequencing of the *groE* operon and flanking genes of *Lawsonia intracellularis*: use in phylogeny. *Microbiology* 144, 2073–2084.

DeSoete, G. (1983) A least squares algorithm for fitting additive trees to proximity data. *Psychometrika* 48, 621–626.

Felsenstein, J. (1993) *Phylip (Phylogeny Inference Package), version 3.5c*. Distributed by the author. Department of Genetics, University of Washington, Seattle, WA.

Fortina, M.G., Pukall, R., Schumann, P. *et al.* (2001) *Ureibacillus* gen. nov., a new genus to accomodate *Bacillus thermosphaericus* (Andersson *et al.* 1995), emendation of *Ureibacillus thermosphaericus* and description of *Ureibacillus terrenus* sp. nov. *International Journal of Systematic and Evolutionary Microbiology* 51, 447–455.

Fox, G.E., Pechman, K.R. & Woese, C.R. (1977) Comparative cataloging of 16 S ribosomal ribonucleic acid: molecular approach to prokaryotic systematics. *International Journal of Systematic Bacteriology* 27, 44–57.

Gruber, T.M. & Bryant, D.A. (1997) Molecular systematic studies of eubacteria, using σ^{70}-type sigma factors of group 1 and group 2. *Journal of Bacteriology* 179, 1734–1747.

Gupta, R.S., Bustard, K., Falah, M. & Singh, D. (1997) Sequencing of heat shock protein 70 (DnaK) homologs from *Deinococcus proteolyticus* and *Thermomicrobium roseum* and their integration in a protein-based phylogeny of prokaryotes. *Journal of Bacteriology* 179, 345–357.

Helgason, E., Caugnant, D.A., Lecadet, M.-M. *et al.* (1998) Genetic diversity of *Bacillus cereus*/*B. thuringiensis* isolates from natural sources. *Current Microbiology* 37, 80–87.

Heyndrickx, M., Lebbe, L., Kersters, K., De Vos, P., Forsyth, G. & Logan, N.A. (1998) *Virgibacillus*: a new genus to accommodate *Bacillus pantothenticus* (Proom and Knight 1950). Emended description of *Virgibacillus pantothenticus*. *International Journal of Systematic Bacteriology* 48, 99–106.

Huang, Y.-P. & Ito, J. (1998) The hyperthermophilic bacterium *Thermotoga maritima* has two different classes of family C DNA polymerases: evolutionary implications. *Nucleic Acids Research* 26, 5300–5309.

Istock, C.A., Bell, J.A., Ferguson, N. & Istock, N.L. (1996) Bacterial species and evolution: theoretical and practical perspectives. *Journal of Industrial Microbiology* 17, 137–150.

Jukes, T.H. & Cantor, C.R. (1969) Evolution of protein molecules. In: *Mammalian Protein Metabolism* (ed. H.N. Munro), pp. 21–132. Academic Press, New York.

Ludwig, W. & Schleifer, K.-H. (1994) Bacterial phylogeny based on 16S and 23S rRNA sequence analysis. *FEMS Microbiology Reviews* 15, 155–173.

Maiden, M.C., Bygraves, J.A., Feil, E. *et al.* (1998) Multilocus sequence typing: a portable approach to the identification of clones within populations of pathogenic microorganisms. *Proceedings of the National Academy of Sciences USA* 95, 3140–3145.

Nazina, T.N., Tourova, T.P., Poltaraus A.B. *et al.* (2001) Taxonomic study of aerobic thermophilic bacilli: descriptions of *Geobacillus subterraneus* gen. nov., sp. nov. and *Geobacillus uzenensis* sp. nov. from petroleum reservoirs and transfer of *Bacillus stearothermophilus*, *Bacillus thermocatenulatus*, *Bacillus thermoleovorans*, *Bacillus kaustophilus*, *Bacillus thermoglucosidasius* and *Bacillus thermodenitrificans* to *Geobacillus* as the new combinations *G. stearothermophilus*, *G. thermocatenulatus*, *G. thermoleovorans*, *G. kaustophilus*, *G. thermoglucosidasius* and *G. thermodenitrificans*. *International Journal of Systematic and Evolutionary Microbiology* 51, 433–446.

Nielsen, P., Rainey, F., Outtrup, H., Priest, F.G. & Fritze, D. (1994) Comparative 16S rDNA sequence analysis of some alkaliphilic bacilli and the establishment of a sixth rRNA group within the genus *Bacillus*. *FEMS Microbiology Letters* 117, 61–66.

Niimura, Y., Koh, E., Yanagida, F., Suzuki, K.-I., Komogata, K. & Kozaki, M. (1990) *Amphibacillus xylanus* gen. nov., sp. nov, a facultatively anaerobic spore-forming xylan-digesting bacterium which lacks cytochrome, quinone, and catalase. *International Journal of Systematic Bacteriology* 40, 297–301.

Ohtani, N., Haruki, M., Morikawa, M., Crouch, R.J., Itaya, M. & Kanaya, S. (1999) Identification of the genes encoding Mn^{2+}-dependent RNase HII and Mg^{2+}-dependent RNase HIII from *Bacillus subtilis*: classification of RNases H into three families. *Biochemistry* 38, 605–618.

Olsen, G.J., Woese, C.R. & Overbeek, R. (1994) The winds of (evolutionary) change: breathing new life into microbiology. *Journal of Bacteriology* 176, 1–6.

Price, L.B., Hugh-Jones, M., Jackson, P.J. & Keim, P. (1999) Genetic diversity in the protective antigen gene of *Bacillus anthracis*. *Journal of Bacteriology* 181, 2358–2362.

Rainey, F.A. & Stackebrandt, E. (1993) Phylogenetic evidence for the relationship of *Saccharococcus thermophilus* to *Bacillus thermoglucosidasius*, *Bacillus kaustophilus* and *Bacillus stearother-*

mophilus. Systematic and Applied Microbiology **16**, 224–226.

Shida, O., Takagi, H., Kadowaki, K. & Komagata, K. (1996) Proposal for two new genera, *Brevibacillus* gen. nov. and *Aneurinibacillus* gen. nov. *International Journal of Systematic Bacteriology* **46**, 939–946.

Shida, O., Takagi, H., Kadowaki, K., Nakamura, L.K. & Komagata, K. (1997a) Transfer of *Bacillus alginolyticus, Bacillus chondroitinus, Bacillus curdlanolyticus, Bacillus glucanolyticus, Bacillus kobensis,* and *Bacillus thiaminolyticus* to the genus *Paenibacillus* and emended description of the genus *Paenibacillus. International Journal of Systematic Bacteriology* **47**, 289–298.

Shida, O., Takagi, H., Kadowaki, K., Nakamura, L.K. & Komagata, K. (1997b) Emended description of *Paenibacillus amylolyticus* and description of *Paenibacillus illinoisensis* sp. nov. and *Paenibacillus chibensis* sp. nov. *International Journal of Systematic Bacteriology* **47**, 299–306.

Spring, S., Ludwig, W., Marquez, M.C., Ventosa, A. & Schleifer, K.-H. (1996) *Halobacillus* gen. nov., with description of *Halobacillus litoralis* sp. nov. and *Halobacillus trueperi* sp. nov., and transfer of *Sporosarcina halophila* to *Halobacillus halophilus* comb. nov. *International Journal of Systematic Bacteriology* **46**, 492–496.

Stackebrandt, E. & Goebel, B.M. (1994) A place for DNA-DNA reassociation and 16S rRNA sequence analysis in the present species definition in bacteriology. *International Journal of Systematic Bacteriology* **44**, 846–849.

Stackebrandt, E., Ludwig, W., Weizenegger, M. *et al.* (1987) Comparative 16S rDNA oligonucleotide analyses and murein types of round-spore-forming bacilli and non-spore-forming relatives. *Journal of General Microbiology* **133**, 2553–2529.

Stroman, D.W., Mclean, C. & Rogers, J. (1998) Discrimination between *B. anthracis, B. cereus, B. mycoides* and *B. thuringiensis* strains by automated rRNA operon ribotyping. British Anthrax Symposium. Source: http://www.qualicon.com/abstracts/01001.htm.

Sudholme, D.J., Jackson, R.A. & Leak, D.J. (1999) Phylogenetic analysis of transformable strains of thermophilic *Bacillus* species. *FEMS Microbiology Letters* **172**, 85–90.

Touzel, J.P., O'Donohue, M., Debeire, P., Samain, E. & Breton, C. (2000) *Thermobacillus xylanilyticus* gen. nov., sp. nov., a new aerobic thermophilic xylan-degrading bacterium isolated from farm soil. *International Journal of Systematic and Evolutionary Microbiology* **50**, 315–320.

Van de Peer, Y., Neefs, J.-M., de Rijk, P., De Vos, P. & de Wachter, R. (1994) About the order of divergence of the major bacterial taxa during evolution. *Systematic and Applied Microbiology* **17**, 32–38.

Viale, A.M., Arakaki, A.K., Soncini, F.C. & Ferreyra, R.G. (1994) Evolutionary relationships among eubacterial groups as inferred from GroEL (chaperonin) sequence comparison. *International Journal of Systematic Bacteriology* **44**, 527–533.

Wainö, M., Tindall, B.J., Schumann, P. & Ingvorsen, K. (1999) *Gracilibacillus* gen. nov., with description of *Gracilibacillus halotolerans* gen. nov., sp. nov.; transfer of *Bacillus dipsosauri* to *Gracilibacillus dipsosauri* comb. nov., and *Bacillus salexigens* to the genus *Salicibacillus* gen. nov., as *Salibacillus salexigens* comb. nov. *International Journal of Systematic Bacteriology* **49**, 821–831.

Wisotzkey, J.D., Jurtshuk, P., Jr, Fox, G.E. Deinhard, G. & Poralla, K. (1992) Comparative sequence analyses on the 16S rRNA (rDNA) of *Bacillus acidocaldarius, Bacillus acidoterrestris,* and *Bacillus cycloheptanicus* and proposal for creation of a new genus, *Alicyclobacillus* gen. nov. *International Journal of Systematic Bacteriology* **42**, 263–269.

Chapter 3

Longstanding Taxonomic Enigmas within the *'Bacillus cereus* group' are on the Verge of being Resolved by Far-reaching Molecular Developments: forecasts on the possible outcome by an *ad hoc* team

Peter C.B. Turnbull, Paul J. Jackson, Karen K. Hill,
Paul Keim, Anne-Brit Kolstø and Douglas J. Beecher

Introduction

Although the term '*Bacillus cereus* group' is unofficial, it is a convenient way to refer collectively to that most important set of aerobic spore-formers – *B. anthracis, B. cereus, B. thuringiensis* and *B. mycoides* – whose precise taxonomic interrelationships have remained elusive since the outset of attempts to put order into the genus *Bacillus*.

This chapter is being composed just as major gene sequencing efforts are in progress and as the results of the application of powerful new molecular tools are just beginning to emerge. Inevitably, therefore, it is a chapter that has to be more predictive than conclusive.

The chapter takes the form of independently written sections representing the outlooks from the separate vantage points of several teams. Each is currently at the forefront of research aimed at finally deciphering the taxonomic ambiguities that have become increasingly urgent to resolve.

Insufficient information on the newly described *B. weihenstephanensis* (Lechner *et al.* 1998) and *B. pseudomycoides* (Nakamura 1998) is available to do any more than acknowledge their membership within this group at this time; therefore, no attempt is made to include them in discussions on the taxonomic ambiguities of the group.

Overview of the issues (P.C.B.T.)

Dependence on extrachromosomal elements

The ease with which *B. anthracis, B. cereus, B. mycoides* and *B. thuringiensis* can generally be differentiated from each other under most routine laboratory circumstances has been covered previously (Turnbull 1999). However, the recognition from early in the history of bacteriology, that these species were closely related to each other and that occasional difficulties may be encountered, particularly in distinguishing *avirulent* forms of *B. anthracis* from *B. cereus*, was also reviewed in that paper. The first authorities on the taxonomy of *Bacillus* regarded these four species as varieties of *B. cereus* (Smith *et al.* 1952; Gordon *et al.* 1973).

With increased understanding of microbial genetics came the realization that the phenotypic differentiation of these species was heavily dependent on virulence factors and that the genes for many of the important determinants – particularly those encoding the toxin and capsule of *B. anthracis* and the Cry toxins of *B. thuringiensis* – lay on extrachromosomal mobile genetic elements (plasmids, insertion sequences and transposons; Mahillon 1997). At the same time, developing DNA- and RNA-based techniques continued to confirm the close associations (Priest 1981; Ash *et al.* 1991; Carlson *et al.* 1994, 1996). This has resulted in the concept that they are subspecies of a single generic species evolved from a common ancestor, having diverged as a result of the exploitation of different niches which led to certain genes common to all being differentially inactivated in one or other of the species.

New pressures for unambiguous identification

For the greater part of the history of bacteriology, the interrelationships of these four species and the occasional problems in differentiating *B. anthracis*, *B. cereus* and *B. thuringiensis* have largely mattered only to taxonomists and academics. However, again as reviewed previously (Turnbull 1999), present-day concerns that isolates of ambiguous identity might actually derive, naturally or artificially, from virulent *B. anthracis* and, therefore, may represent the presence or existence of dangerous parent strains, has placed a new complexion on the importance of categorical identification.

Similarly, the pathogenic potential of *B. cereus* has long been recognized, usually in the context of food poisoning but periodically in relation to severe wound and ocular infections, endocarditis, meningitis, pulmonary infections and so on. Whether or not pathogenic and nonpathogenic forms of *B. cereus* can be distinguished is a well-debated subject and new light is thrown on this below. Because few laboratories will distinguish between *B. cereus* and *B. thuringiensis*, the issue of whether *B. thuringiensis* is the cause of human infections more frequently than is generally recognized has also been long debated, particularly in connection with its use in commercial insecticidal crop sprays. This has been the subject of focus again recently (Damgaard *et al.* 1997; Bernstein *et al.* 1999; Hernandez *et al.* 1999; Helgason *et al.* 2000a). Clear, unambiguous identification and characterization of *B. thuringiensis* as a whole and at the strain level is, therefore, of considerable importance to agricultural industries and public health.

Another concern is the potential ease of genetic exchange between the *B. cereus* group species by means of such entities as conjugative transposons, self-transmissible plasmids, transducing phages and natural competence for transformation. Laboratory demonstration of this potential has taken the form of *B. thuringiensis* containing pXO1 or pXO2, capsule-positive *B. cereus*, and functional *B. thuringiensis* fertility plasmids in *B. anthracis* and *B. cereus* (Thorne 1993). The potential for confusion of identity, if not change of identity, appears to exist here. More sinister possibilities may lie in the ready transfer of virulence factors (Pomerantsev *et al.* 1997).

Coincident with the increasingly urgent need to fully define the relationships among the *B. cereus* group species has been the generation of many of the tools with which it will become possible in the foreseeable future – that is, the next two or three years – to establish unequivocally the criteria for correct speciation of *B. anthracis*, *B. cereus*, *B. thuringiensis* and, if included in the studies performed, *B. mycoides*. [Early evidence indicates that *B. mycoides* may cluster slightly apart from the other three (Keim *et al.* 1997).] Some extremely powerful molecular tools have entered the picture in the last three to five years. In particular, multi-locus enzyme electrophoresis (MLEE), multilocus sequence typing (MLST) and amplified fragment length polymorphism (AFLP) analysis and its derivative multiple-locus variable number tandem repeat (VNTR) analysis (MLVA) have made possible the nucleotide sequencing of alleles and analysis of sequence variation at multiple independent loci of interest. The impact of these novel systems on the developing understanding of the relationships among the *B. cereus* group species is illustrated in the sections that follow.

On the verge of a breakthrough?

This chapter and the '*Bacillus* 2000' meeting to which it is attached, come at a critical point when the sequences of the plasmids of *B. anthracis* have just been revealed (Okinaka *et al.* 1999) and that of the *B. anthracis* chromosome is approaching completion (Read & Peterson 1999). The sequence of the genome of a single *B. cereus* strain is also now underway. The exciting part, in which housekeeping and virulence factor genes in the different species are looked for and compared, is just beginning. Some examples of this are given in the sections that follow.

Bacillus cereus, *B. thuringiensis* and *B. anthracis*: molecular relationships and differentiation (P.J.J. & K.K.H.)

Analysis of the 16S ribosomal DNA sequence does little to resolve *B. cereus*, *B. thuringiensis* and *B. anthracis*, because the sequences within and adjacent to this gene are almost identical among these three species. Moreover, detailed analysis of the variable sites within these sequences shows that multiple ribosomal RNA cistrons within a single isolate can differ at the variable sites. Some 16S rDNA cistrons contain the sequence normally assigned to *B. cereus*/*B. anthracis*, while others contain those minor sequence differences that are characteristic of *B. thuringiensis*.

Carlson *et al.* (1994) examined 24 *B. cereus* and 12 *B. thuringiensis* isolates using pulsed-field gel electrophoresis (PFGE) and multienzyme electrophoresis (MEE) and showed that, contrary to earlier views, there was a high degree of genetic variability within and between the two species. However, neither PFGE nor MEE could effectively differentiate between the two species, so it was suggested that *B. cereus* and *B. thuringiensis* be considered one species. By contrast, 78 *B. anthracis* isolates collected worldwide were analysed by amplified fragment length polymorphism (AFLP) (Keim *et al.* 1997). This study showed distinct differences between *B. cereus*, *B. thuringiensis* and *B. anthracis*. It also demonstrated that, in contrast to *B. cereus* and *B. thuringiensis*, *B. anthracis* is genetically monomorphic with little discernible variation among the many different isolates.

Our analysis of a large collection of *B. cereus*, *B. thuringiensis* and *B. anthracis* isolates using AFLP and 16S rDNA analysis reveals the reason for the controversy regarding the phylogeny of these species. DNA sequence analysis of the 16S rDNA gene confirms that all isolates studied are *B. cereus*, *B. thuringiensis* or *B. anthracis*, yet AFLP analysis shows extensive phylogenetic diversity among the different samples. Analysis of 155 *B. cereus* isolates and 230 *B. thuringiensis* isolates and comparisons of these to representative *B. anthracis* isolates reveals a complex phylogeny for these species. Unlike *B. anthracis*, these results demonstrate extensive genetic diversity among *B. cereus* and *B. thuringiensis* environmental isolates. They also demonstrate that the small number of reference strains normally used to study these two species are not representative of this polymorphism.

Phylogenetic dendrograms based on AFLP analyses and confirmed by MEE and Multilocus Sequence Typing (MLST) analysis show that these three *Bacillus* species map to five major branches of an extensive phylogenetic tree (figure 3.1). *B. cereus* and *B. thuringiensis* isolates are highly polymorphic and map to all five branches of this tree. By contrast, *B. anthracis* maps to a single location. All but one of the *B. cereus* reference strains map to a single branch of the tree, suggesting that the reference strains are not representative of the diversity within the species. All *B. thuringiensis* reference strains map to the same branch of the tree. The reference strain *B. thuringiensis* ATCC 10792 is more closely related to several *B. cereus* isolates – including the *B. cereus* reference strains ATCC 53522 and 6464 – than to most other *B. thuringiensis* isolates. A subset of *B. cereus* and *B. thuringiensis* isolates maps closely to *B. anthracis*. Further analysis of this subset demonstrates that the majority of those tested (96%) are at least opportunistic

---

Figure 3.1 (*Opposite*) Phylogeny of *Bacillus anthracis*, *B. cereus* and *B. thuringiensis* based on AFLP analysis. The dendrogram shows five major branches in a diverse phylogenetic tree. Branches 1 and 5 contain environmental isolates of *B. cereus* and *B. thuringiensis* from diverse locations. Branch 2 contains all but one of the *B. cereus* and *B. thuringiensis* reference strains, suggesting that these strains are not representative of the diversity found in the two species. Branch 3 contains one *B. cereus* reference strain. All others are *B. cereus* and *B. thuringiensis* environmental isolates. Branch 4 contains all the *B. anthracis* isolates. These map to a single locus at the resolution level of this tree. The majority of *B. cereus* and *B. thuringiensis* isolates that map to branch 4 are at least opportunistic pathogens in humans or animals. Several of these pathogens are very closely related to *B. anthracis*.

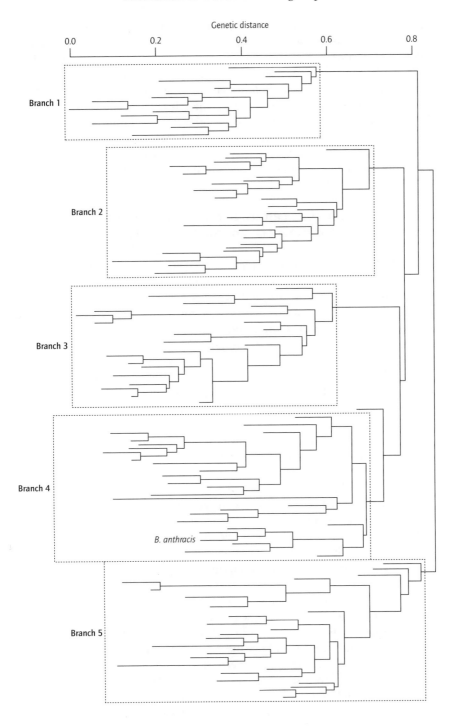

pathogens in humans or animals. This suggests that certain genetic traits common to members of this group contribute to the ability to function successfully within a human or animal host. Based on this analysis, the phylogeny of these three species must be reconsidered. *Bacillus anthracis* is a unique species. The diversity within *B. cereus* and *B. thuringiensis* suggests that perhaps the current classification of these species is in error. There is extensive interspersion of *B. cereus* and *B. thuringiensis* isolates throughout the tree with some isolates very closely related and others distinctly different.

Studies of these *Bacillus* species that rely on comparisons among the different species must take this extensive phylogenetic diversity into account.

Bacillus cereus, B. thuringiensis and *B. anthracis*: one species? (A.-B.K.)

Not only do the genes coding for typical differential properties like the anthrax toxins and the insecticidal toxins reside on plasmids, but also it is well documented that *B. thuringiensis* plasmids with insecticidal toxin genes can be transferred from *B. thuringiensis* to *B. cereus*, and the recipient bacteria then become indistinguishable from *B. thuringiensis*. The converse is also true. When *B. thuringiensis* loses such plasmids, the resulting bacteria become indistinguishable from *B. cereus*.

It has been necessary, therefore, to address the following question: what is the genetic relationship between *B. cereus*, *B. thuringiensis* and *B. anthracis*? To answer this question, multilocus enzyme electrophoresis (MEE) – a well-established and widely used method for the analysis of genetic relationship between bacteria species as well as strains within a bacterial species (Selander *et al.* 1986) – was applied. By analysing the allozyme pattern of 13 housekeeping genes, a phylogenetic investigation of *B. cereus*, *B. thuringiensis* and *B. anthracis* isolates was possible. More recently we have also used PCR and DNA sequencing of nine chromosomal genes to analyse the genetic relationships among selected isolates. The results obtained are summarized in the following four points.

A comparison of *B. thuringiensis* and *B. cereus* strains from different collections showed that the *B. cereus* strains were often more closely related to *B. thuringiensis* insecticidal toxic strains than to other *B. cereus* strains (Carlson *et al.* 1994).

When the genetic relationships of 154 *B. cereus/B. thuringiensis* strains isolated from soil samples were analysed by MEE, they clustered into two major groups. Analysed by serotyping, using *B. thuringiensis* flagellar antisera, 28 serotypes were identified with H34 the most common. However, the serotype patterns did not parallel the genetic relationships among the isolates and only three of those which were serotype-positive produced insecticidal crystals during sporulation. Should the serotype-positive strains that lacked the crystal toxin genes be called *B. thuringiensis* or *B. cereus* (Helgason *et al.* 1998)?

Bacillus cereus isolates from patients (35 isolates) and from dairies (30 *B. cereus/B. thuringiensis* isolates) were analysed. Of 20 isolates from periodontal infections, 12 were identical by MEE and three more had an allele at one locus.

Of the 12 strains, 11 contained a ~300 kb plasmid. We do not know if this plasmid contributes to the virulence of the strains (Helgason *et al.* 2000a).

Thirteen *B. anthracis* isolates and 10 *B. cereus* isolates with the marker Ba813 were analysed by MEE, and found to be very closely related to some *B. thuringiensis* and other *B. cereus* strains (Helgason *et al.* 2000b). One *B. anthracis* isolate was more closely related to some *B. cereus* strains than to the *B. anthracis* cluster. Among the closest relatives of *B. anthracis* were the *B. cereus* cluster isolates from periodontal infections. Sequencing of selected genes showed the same relationship patterns.

In general, then, genetic analysis by MEE clearly showed that each one of the three species contained some members that were more closely related to one of the other species than to the majority of isolates within its own species. On the basis of these results, our conclusions are that *B. cereus*, *B. thuringiensis* and *B. anthracis* should be regarded as one species, with subspecies. How the subspecies should be defined is a matter for debate. One possibility is that the subspecies could be defined according to phenotypic properties, based on important properties coded by plasmids (such as pXO1 and pXO2 in *B. anthracis*). Another possibility is to use their genetic relationships to define the subspecies.

The taxonomic problem and the use of subspecies may have to wait until the *B. anthracis* genome and perhaps a few *B. cereus*/*B. thuringiensis* genomes have been fully sequenced and analysed. In the meantime, it is possible that the repeat structure identified in *B. cereus*, and also found to be present in all those *B. cereus*, *B. thuringiensis* and *B. anthracis* strains analysed thus far (Økstad *et al.* 1999b), may be used as a tool for the identification of the three species.

Systematic relationships within *Bacillus anthracis* (P.K.)

Owing to the great molecular homogeneity of *B. anthracis* isolates, it seems likely that this species is derived from a single recent common ancestor (monophyletic). This is in contrast to the mixed (polyphyletic) evolutionary paths found in *B. cereus* and *B. thuringiensis*. The lack of molecular differences has been documented by a series of reports using molecular markers and nucleotide sequence evaluation (Harrell *et al.* 1995; Henderson *et al.* 1995; Keim *et al.* 1997; Jackson *et al.* 1999; Price *et al.* 1999). It is reasonable to assume that the single ancestral isolate was a *B. cereus* or *B. thuringiensis* that fortuitously obtained both anthrax virulence plasmids. This may have been a virulent strain of *B. cereus* that was already adapted to a pathogenic existence and not a saprophytic soil organism. In all likelihood, acquisition of the vir plasmids was sequential and increased the pathology to the level observed in anthrax. Regardless of the derivation of fully virulent *B. anthracis*, all of the currently known *B. anthracis* isolates are very similar and derived from a common ancestor.

Diversity within *B. anthracis* is rare, but can be found occasionally as molecular marker differences. These were initially discovered as AP-PCR (Henderson *et al.* 1995) and AFLP polymorphisms (Keim *et al.* 1997). Further molecular characterization revealed variable number tandem repeats (VNTR) as the basis for this variation (Andersen *et al.* 1996; Keim *et al.* 2000). VNTRs are short,

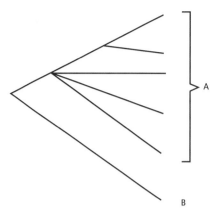

Figure 3.2 Evolution of *Bacillus anthracis*. Isolates of *B. anthracis* can be broken into two major clonal lineages (A and B) with further separation of A lineage into subgroups.

tandemly repeated sequences, which undergo very rapid mutational change (van Belkum *et al.* 1998). The multiple alleles resulting from this rapid change provide great discriminatory power even among the very closely related strains of *B. anthracis*. With the exception of a few rare single nucleotide differences (Price *et al.* 1999), VNTR variation may represent the only molecular variation within this species.

It would seem that few generations separate all examples of *B. anthracis* from a common ancestor. This could be the result of slow evolutionary rates or a recent origin of this pathogen. Slow rates of evolution may be possible in spore-forming pathogens as they may lie dormant for years between bursts of growth in a suitable host. It seems unlikely that *B. anthracis* has significant propagation outside of its victims and this would greatly reduce the number of generations per year or decade, relative to most other bacteria. It is also possible that *B. anthracis* is a relatively new pathogen that emerged with the domestication of livestock. These evolutionary scenarios are not mutually exclusive and a combined model would certainly fit the existing data.

Molecular variation in the VNTRs can be used to identify similarity among *B. anthracis* isolates (Keim *et al.* 2000). This eight-locus VNTR analysis identified only 89 unique genotypes from over 400 isolates. There appear to be two major phylogenetic groups or major clonal lineage in this worldwide collection (figure 3.2). The 'B' lineage accounts for only ~11% of all known strains. The B group is restricted in distribution and rarely found outside southern Africa. Its co-occurrence in this region with members of the genetically distinct 'A' lineage (Smith *et al.* 1999) has led to speculation that Africa is the origin of *B. anthracis* (Keim *et al.* 1997). In contrast, the A lineage is found throughout the world. Minor branching is observed among A types and about five different subgroups can be recognized (Keim *et al.* 2000). Some of these groups have biased geographical distributions, but they are not found exclusively in single sites. The North American endemic group is also found around the world. The Russian STI vaccine strain is closely related to the North American endemic strains and suggests an Asian/European origin for this group. A large cluster of strains, called the Sterne–Ames group after two well-known strains, is responsible for many anthrax outbreaks and is transglobal in distribution. This group may have been spread by human commerce to achieve its wide distribution. Whether its success is the consequence of different biological properties or merely of stochastic processes is unknown.

Bacillus cereus virulence factor genes in the B. anthracis chromosome (D.J.B.)

Aggressins of B. cereus

Although, as discussed earlier, *Bacillus cereus* and *B. anthracis* are generally considered nearly indistinguishable by their physiological and biochemical characteristics, they are glaringly different to those who concentrate primarily on the virulence mechanisms of *B. cereus*. Specifically, *B. anthracis* does not express many of the factors that are considered virulence factors in *B. cereus*.

Bacillus cereus is well armed with weapons capable of damaging or exacerbating damage to human and animal tissue. It is motile and secretes a potent necrotizing tripartite enterotoxin called haemolysin BL (Hbl, comprising B, L_1 and L_2), and at least two homologous tripartite toxins, Hbl_a and nonhaemolytic enterotoxin (Nhe, comprising NheA, NheB and NheC) (Beecher & Wong 1997, 2000; Granum *et al.* 1999; Lund & Granum 1999). It produces at least four additional haemolysins; cereolysin O (CLO), haemolysin-II (Hly-II), Hly-III and Hly-IV. It also secretes three phospholipases C, PC-PLC, PI-PLC and SMase, which hydrolyse membrane phosphatidylcholine, phosphatidylinositol and sphingomyelin, respectively. In combination, PC-PLC and SMase lyse certain erythrocytes. In addition, it secretes a neutral metalloprotease, at least one collagenase and three beta-lactamases (Turnbull 1986; Kramer & Gilbert 1989; Drobniewski 1993).

Preventing *B. cereus* from being an aggressive human pathogen is its lack of a protective mechanism against intact immune systems. Apart from food poisoning, it is most often encountered as a cause of endophthalmitis (intraocular infection). The inner eye is an immune privileged site that does not experience a normal inflammatory response and *B. cereus* causes devastating necrosis, usually resulting in blindness within 24–48 h (Cowan *et al.* 1987; Le Saux & Harding 1987; Schemmer & Driebe 1987; Beecher *et al.* 1995; Callegan *et al.* 1999).

Genes of B. cereus aggressins in the B. anthracis chromosome

Why, it is necessary to ask, does the closely related *B. anthracis* not produce this battery of factors that would seem to enhance its ability to survive in animal hosts? It appears that the acquisition of the two *B. anthracis* virulence plasmids coincided with decreased production of chromosomally encoded virulence factors. The great proportion of *B. anthracis* research has concentrated on these plasmids, and little is known about the chromosomal genes. This situation is changing dramatically with the sequencing of the *B. anthracis* genome. Although not yet completed, BLAST searches of the raw data are now possible through The Institute for Genomic Research (TIGR) at *http://www.tigr.org*. This service was used to make the following observations regarding chromosomal genes.

Two recent developments in the understanding of *B. cereus* virulence will contribute to this discussion. First, as with other bacteria, many *B. cereus* virulence factors are pleiotropically regulated by a protein called PlcR. Regulated proteins include, but are not limited to, PlcR itself, Hbl, Nhe, PC-PLC, SMase, PI-PLC, CLO, Hly-IV (but not Hly-II) and neutral protease (Agaisse *et al.* 1999; Økstad *et al.* 1999a; D. Beecher, unpubl. obs.). Secondly, it appears that *B. cereus/B. thuringiensis* possess at least two distinct secretion pathways for virulence factors and that one of these pathways involves the flagellar transport protein, FlhA (Ghelardi *et al.* 1999; D. Beecher, unpubl. obs.). Nonmotile *flhA* mutants translate but do not secrete a number of proteins (>6) including flagellin, Hbl and PC-PLC (CLO, Hly-II and Hly-IV are unaffected). A similar system in *Yersinia enterocolitica* was reported recently (Young *et al.* 1999). Interestingly, a *B. cereus* mutant with a disrupted *plcR* gene was still motile (mutations generated in the laboratory of D. Lereclus, mutants provided by D. Lereclus and H. Agaisse, analysis of mutants by D. Beecher and E. Ghelardi).

In the present context, a key difference between *B. cereus/B. thuringiensis* and *B. anthracis* is that *B. anthracis* does not produce active *PlcR* because its gene is truncated (Agaisse *et al.* 1999). This should have the immediate effect of reducing, but not eliminating, the expression of all of the genes under its control. Presumably, the large reduction in the expression of *plcR*-regulated proteins would diminish their influence on the life cycle of the organism and deleterious mutations might begin to accumulate.

Consistent with this idea, several disrupted *plcR*-controlled genes appear to be present in *B. anthracis*. The NheA and NheB genes appear to be intact, but NheC is truncated by about 54 amino acids. Two distinct matches to the SMase sequence arose in *B. anthracis*. In one, the gene appeared truncated with a stop codon about 123 amino acids after the start codon. In the other, there was no stop codon at that position, but the gene is not complete on that contig (section of sequenced DNA). There appear, therefore, to be two SMase genes in the genome, although TIGR cautions us that the sequence data are preliminary.

The entire CLO gene is present in *B. anthracis* but its expression is apparently prevented by a mutation in the start codon (ATG to AAG). This raises the question as to whether there are any *B. anthracis* strains lacking that mutation masquerading as haemolytic *B. cereus*. Hbl and Hly-IV appear to be completely absent from the *B. anthracis* genome.

PC-PLC is also a *plcR*-regulated protein, but it is still secreted from *B. anthracis*, which has a positive lecithinase phenotype. As expected, the entire PC-PLC gene is intact. However, *B. anthracis* is considered only weakly positive for PLC production (Parry *et al.* 1983; Claus & Berkeley 1986), apparently as a result of the lack of plcR activity. In addition, the 16-base palindromic PlcR binding site upstream of the PC-PLC gene contains three nucleotide substitutions and PlcR would probably not regulate the gene even if it were present.

As described above, PC-PLC appears to be secreted in *B. thuringiensis* by a mechanism involving the flagellar transporter FlhA. This entire FlhA gene can be found in the *B. anthracis* genome (99% identical to *B. cereus/B. thuringiensis* sequences, unpubl. sequence courtesy of E. Ghelardi). While it is not known whether FlhA is expressed in *B. anthracis*, it can be inferred from the secretion of PC-PLC that it is functional (unless the activity associated with *B. anthracis* is released as a result of cell lysis). It is also not known whether SMase is expressed in *B. anthracis*. However, as PC-PLC is expressed, and the combination of PC-PLC and SMase is haemolytic (Gilmore *et al.* 1989), it is likely that SMase is not expressed by the normally nonhaemolytic *B. anthracis*.

At least one haemolysin that is not affected by either FlhA or PlcR is not secreted by *B. anthracis*. The entire Hly-II gene is present in the *B. anthracis* genome (95–100% identical) but it is disrupted about one quarter of the way from the N-terminus.

Bacillus anthracis is known to secrete a neutral metalloprotease and, as expected, the bacillolysin sequence is present in the *B. anthracis* genome at 94% identity to the published protease sequence. Gelatin hydrolysis among *B. anthracis* strains varies (75–84% positive; Logan & Turnbull 1999) and a collagenase has not been described. An N-terminal collagenase sequence determined by Lund and Granum (1996) was 84, 82 and 53% identical to three distinct sections of the *B. anthracis* genome data. Complete genes could not be found for any of these matches. However, a complete gene was present for the partial putative collagenase sequence, ColA, described by Økstad *et al.* (1999b) (NCBI gi:2462122; Accession, CAA72028.1).

All three beta-lactamases secreted by *B. cereus* have strongly positive BLAST hits against the unfinished genome of *B. anthracis*, which is sensitive to penicillin. Beta-lactamase I, II and III are 88, 90 and 61% identical (respectively) to sequences of 209, 168 and 208 amino acids. The N-termini of these genes could not be located.

Ironically, the genome data for nonmotile *B. anthracis* can be used to gain insight into the motility and chemotaxis system of *B. cereus*. Only about five motility and chemotaxis genes have been identified in *B. cereus* thus far. A cursory comparison of 55 known motility genes from *B. subtilis* with the available *B. anthracis* genome produced very high BLAST scores against 48 of them (D. Beecher, unpubl. obs.). Because *B. anthracis* is nonmotile these probably do not all represent functional genes, but it does suggest that the *B. cereus* group possesses a complex motility and chemotaxis system similar to *B. subtilis* and *Escherichia coli*.

It remains to be seen whether the FlhA is actually functional in *B. anthracis*. If so, it would be interesting to determine its function and identify the proteins whose secretion it influences. Relative to *B. cereus*, *B. anthracis* represents a reservoir of natural mutations in a chromosomal virulence system.

Conclusions

Whatever name is given to the agent that causes anthrax, and whatever molecular identities it shares with other *Bacillus* species, it will continue to be distinguished emotively from those other organisms. Similarly, a multimillion dollar industry is centred around those *B. thuringiensis* strains which are the effective ingredients of large-scale agricultural insecticides, and strong resistance will develop to any changes in taxonomy that might lead to a confusion of identity.

It is still too early to predict the outcome of molecular taxonomic studies on this group of aerobic endospore-formers. New groupings are likely to emerge, but perhaps more than with any other microbial species, any proposed changes in taxonomic identity and nomenclature could become entangled in a mesh of far-reaching political and/or economic consequences.

References

Agaisse, H., Gominet, M., Økstad, O.A., Kolstø, A.B. & Lereclus, D. (1999) PlcR is a pleiotropic regulator of extracellular virulence factor gene expression in *Bacillus thuringiensis*. *Molecular Microbiology* **32**, 1043–1053.

Andersen, G.L., Simchock, J.M. & Wilson, K.H. (1996) Identification of a region of genetic variability among *Bacillus anthracis* strains and related species. *Journal of Bacteriology* **178**, 377–384.

Ash, C. & Collins, M.D. (1992) Comparative analysis of 23S ribosomal RNA gene sequences of *Bacillus anthracis* and emetic *Bacillus cereus* determined by PCR-direct sequencing. *FEMS Microbiology Letters* **94**, 75–80.

Ash, C., Farrow J.A.E., Dorsch, M., Stackebrandt, E. & Collins, M.D. (1991) Comparative analysis of *Bacillus anthracis*, *Bacillus cereus*, and related species on the basis of reverse transcriptase sequencing of 16S rRNA. *International Journal of Systematic Bacteriology* **41**, 343–346.

Beecher, D.J. & Wong, A.C.L. (1997) Tripartite hemolysin BL from *Bacillus cereus*: hemolytic analysis of component interactions and a model for its characteristic paradoxical zone phenomenon. *Journal of Biological Chemistry* **272**, 233–239.

Beecher, D.J. & Wong, A.C.L. (2000) Tripartite haemolysin BL: isolation and characterization of two distinct homologous sets of components from a single *Bacillus cereus* isolate. *Microbiology* **146**, 1371–1380.

Beecher, D.J., Pulido, J.S., Barney, N.P. & Wong, A.C.L. (1995) Extracellular virulence factors in *Bacillus cereus* endophthalmitis: methods and implication of involvement of hemolysin BL. *Infection and Immunity* **63**, 632–639.

Bernstein, I.L., Bernstein, J.A., Miller, M. *et al.* (1999) Immune responses in farm workers after exposure to *Bacillus thuringiensis* pesticides. *Environmental Health Perspectives* **107**, 575–582.

Callegan, M.C., Booth, M.C., Jett, B.D. & Gilmore, M.S. (1999) Pathogenesis of gram-positive bac-

terial endophthalmitis. *Infection and Immunity* **67**, 3348–3356.

Carlson, C.R., Caugant, D. & Kolstø, A.B. (1994) Genotypic diversity among *Bacillus cereus* and *Bacillus thuringiensis* strains. *Applied and Environmental Microbiology* **60**, 1719–1725.

Carlson, C.R., Johansen, T. & Kolstø A.B. (1996) The chromosome map of *Bacillus thuringiensis* subsp. *canadensis* HD224 is highly similar to that of the *Bacillus cereus* type strain ATCC 14579. *FEMS Microbiology Letters* **141**, 163–167.

Claus, D. & Berkeley, R.C.W. (1986) Genus *Bacillus*. In: *Bergey's Manual of Systematic Bacteriology* (eds P.H. Sneath *et al.*), Vol. 2, pp. 1105–1139. Williams & Wilkins, Baltimore, MD.

Cowan, C.L., Madden, W.M., Hatem, G.F. & Merritt, J.C. (1987) Endogenous *Bacillus cereus* endophthalmitis. *Annals of Ophthalmology* **9**, 65–68.

Damgaard, P.H., Granum, P.E., Bresciani, J., Torregrossa, M.V., Eilenberg, J. & Valentino, L. (1997) Characterization of *Bacillus thuringiensis* isolated from infections in burn wounds. *FEMS Immunology and Medical Microbiology* **18**, 47–53.

Drobniewski, F.A. (1993) *Bacillus cereus* and related species. *Clinical Microbiology Reviews* **6**, 324–338.

Ghelardi, E., Beecher, D.J., Celandroni, F. *et al.* (1999) Requirement of flhA for flagella assembly, swarming differentiation, phospholipases and hemolysin BL secretion in *Bacillus thuringiensis*. In: *The 2nd International Workshop on the Molecular Biology of* Bacillus cereus, Bacillus anthracis, *and* Bacillus thuringiensis, *11–13 August 1999, Taos, NM, USA*. Los Alamos National Laboratory, Los Alamos, NM.

Gilmore, M.S., Cruz-Rodz, A.L., Leimeister-Wächter, M., Kreft, J. & Goebel, W. (1989) A *Bacillus cereus* cytolytic determinant, cereolysin AB, which comprises the phospholipase C and sphingomyelinase genes: nucleotide sequence and genetic linkage. *Journal of Bacteriology* 171, 744–753.

Gordon, R.E., Haynes, W.C. & Pang, C.H-N. (1973) *The Genus* Bacillus. Agriculture Handbook 427. Agricultural Research Service, USDA, Washington, DC.

Granum, P.E., O'Sullivan, K. & Lund, T. (1999) The sequence of the non-haemolytic enterotoxin operon from *Bacillus cereus*. *FEMS Microbiology Letters* 177, 225–229.

Harrell, L.J., Andersen, G.L. & Wilson, K.H. (1995) Genetic variability of *Bacillus anthracis* and related species. *Journal of Clinical Microbiology* 33, 1847–1850.

Helgason, E., Caugant, D.A., Lecadet, M.M. *et al.* (1998) Genetic diversity of *Bacillus cereus*/*B. thuringiensis* isolates from natural sources. *Current Microbiology* 37, 80–87.

Helgason, E., Caugant, D.A., Olsen, I. & Kolstø, A-B. (2000a) Genetic structure of population of *Bacillus cereus* and *B. thuringiensis* isolates associated with periodontitis and other human infections. *Journal of Clinical Microbiology* 38, 1615–1622.

Helgason, E., Økstad, O.A., Caugant, D.A. *et al.* (2000b) *Bacillus anthracis*, *Bacillus cereus* and *Bacillus thuringiensis* – one species on the basis of genetic evidence. *Applied and Environmental Microbiology* 66, 2627–2630.

Henderson, I., Yu, D. & Turnbull, P.C. (1995) Differentiation of *Bacillus anthracis* and other '*Bacillus cereus* group' bacteria using IS231-derived sequences. *FEMS Microbiology Letters* 128, 113–118.

Hernandez, E., Ramisse, F., Cruel, T., le Vagueresse, R. & Cavallo, J.D. (1999) *Bacillus thuringiensis* serotype H34 isolated from human and insecticidal strains serotypes 3a3b and H14 can lead to death of imunocompetent mice after pulmonary infection. *FEMS Immunological and Medical Microbiology* 24, 43–47.

Jackson, P.J., Hill, K.K., Laker, M.T., Ticknor, L.O. & Keim, P. (1999). Genetic comparison of *B. anthracis* and its close relatives using AFLP and PCR analysis. *Journal of Applied Microbiology* 87, 263–269.

Keim, P., Kalif, A., Schupp, J.M. *et al.* (1997) Molecular evolution and diversity in *Bacillus anthracis* as detected by amplified fragment length polymorphism markers. *Journal of Bacteriology* 179, 818–824.

Keim, P., Price, L.B., Klevytska, A.M. *et al.* (2000) Multiple-locus VNTR analysis (MLVA) reveals genetic relationships within *Bacillus anthracis*. *Journal of Bacteriology* 182, 2928–2936.

Kramer, J.M. & Gilbert, R.J. (1989) *Bacillus cereus* and other *Bacillus* species. In: *Foodborne Bacterial Pathogens* (ed. M.P. Doyle), pp. 21–70. Marcel Dekker, New York.

Le Saux, N. & Harding, G.K. (1987) *Bacillus cereus* endophthalmitis. *Canadian Journal of Surgery* 30, 28–29.

Lechner, S., Mayr, R., Francis, K.P. *et al.* (1998) *Bacillus weihenstephanensis* sp. nov. is a new psychrotolerant species of the *Bacillus cereus* group. *International Journal of Systematic Bacteriology* 48, 1373–1382.

Logan, N.A. & Turnbull, P.C.B. (1999) *Bacillus* and recently derived genera. In: *Manual of Clinical Microbiology* (eds P.R. Murray *et al.*), 6th edn, pp. 357–369. ASM Press, Washington, DC.

Lund, T. & Granum, P.E. (1996). Characterisation of a non-haemolytic enterotoxin complex from *Bacillus cereus* isolated after a foodborne outbreak. *FEMS Microbiology Letters* 141, 151–156.

Lund, T. & Granum, P.E. (1999) The 105-kDa protein component of *Bacillus cereus* non-haemolytic enterotoxin (Nhe) is a metalloprotease with gelatinolytic and collagenolytic activity. *FEMS Microbiological Letters* 178, 355–361.

Mahillon, J. (1997) Mobile DNA within the *Bacillus cereus* group. In: *Abstracts of the First International Workshop on the Molecular Biology of* B. cereus, B. anthracis *and* B. thuringiensis, 23–25 May 1997, Oslo, Norway, p. 13. BiO-ami, Oslo.

Nakamura, L.K. (1998) *Bacillus pseudomycoides* sp. nov. *International Journal of Systematic Bacteriology* 48, 1031–1035.

Okinaka, R., Cloud, K., Hampton, O. *et al.* (1999) Sequence and organization of pXO1, the large *Bacillus anthracis* plasmid harboring the anthrax toxin genes. *Journal of Bacteriology* 181, 6509–6515.

Økstad, O.A., Gominet, M., Purnelle, B., Rose, M., Lereclus, D. & Kolstø, A-B. (1999a) Sequence analysis of three *Bacillus cereus* loci carrying PlcR-regulated genes encoding degradative enzymes and enterotoxin. *Microbiology* 145, 3129–3138.

Økstad, O.A., Hegna, I., Lindback, T., Rishovd, A-L. & Kolstø, A-B. (1999b) Genome organization is not conserved between *Bacillus cereus* and *Bacillus subtilis*. *Microbiology* 145, 621–631.

Parry, J.M., Turnbull, P.C.B. & Gibson, J.R. (1983) *A Colour Atlas of* Bacillus *Species*. Atlas No. 19. Wolfe Medical Publications, London.

Pomerantsev, A.P., Staritsin, N.A., Mockov, Y.V. & Marinin, L.I. (1997) Expression of cereolysine AB genes in *Bacillus anthracis* vaccine strain ensures protection against experimental hemolytic anthrax infection. *Vaccine* 15, 1846–1850.

Price, L.B., Hugh-Jones, M., Jackson, P.J. & Keim, P. (1999) Natural genetic diversity in the protective antigen gene of *Bacillus anthracis*. *Journal of Bacteriology* 181, 2358–2362.

Priest, F.G. (1981) DNA homology in the genus *Bacillus*. In: *The Aerobic Endospore-forming Bacteria* (eds R.C.W. Berkeley & M. Goodfellow), pp. 33–57. Academic Press, London.

Read, T. & Peterson, S. (1999) Whole-genome sequencing of *Bacillus anthracis*. In: *The 2nd International Workshop on the Molecular Biology of* Bacillus cereus, Bacillus anthracis, *and* Bacillus thuringiensis, *11–13 August 1999, Taos, NM, USA*, p. 20. Los Alamos National Laboratory, Los Alamos, NM.

Schemmer, G.B. & Driebe, W.T. (1987) Posttraumatic *Bacillus cereus* endophthalmitis. *Archives of Ophthalmology* **105**, 342–344.

Selander, R.K., Caugant, D.A., Ochman, H., Musser, J.M., Gilmour & Whittam, T.S. (1986) Methods of multilocus enzyme electrophoresis for bacterial population genetics and systematics. *Applied Environmental Microbiology* **51**, 873–884.

Smith, K.L., deVos, V., Bryden, H. *et al.* (1999) Meso-scale ecology of anthrax in southern Africa: a pilot study of diversity and clustering. *Journal of Applied Microbiology* **87**, 204–207.

Smith, N.R., Gordon, R.E. & Clark, F.E. (1952) *Aerobic Sporeforming Bacteria*. Agricultural Monograph 16. USDA, Washington, DC.

Thorne, C.B. (1993) *Bacillus anthracis*. In: Bacillus subtilis *and other Gram-positive Bacteria* (eds A.L. Sonenshein *et al.*), pp. 113–124. American Society for Microbiology, Washington, DC.

Turnbull, P.C.B. (1986) *Bacillus cereus* toxins. In: *Pharmacology of Bacterial Toxins* (eds F. Dorner & J. Drews), pp. 397–448. Pergamon Press, Oxford.

Turnbull, P.C.B. (1999) Definitive identification of *Bacillus anthracis* – a review. *Journal of Applied Microbiology* **87**, 237–240.

van Belkum, A., Scherer, S., van Alphen, L. & Verbrugh, H. (1998) Short-sequence DNA repeats in prokaryotic genomes. *Microbiology and Molecular Biology Review* **62**, 275–293.

Young, G.M., Schmiel, D.H. & Miller, V.L. (1999) A new pathway for the secretion of virulence factors by bacteria: the flagellar export apparatus functions as a protein-secretion system. *Proceedings of the National Academy of Science of the USA* **96**, 6456–6461.

Chapter 4

Bacillus cereus and Food Poisoning

Per Einar Granum

Introduction

Bacillus cereus is a food poisoning bacterium of growing concern, although it does not cause a type of illness that might make newspaper headlines. However, outbreaks of both the emetic and the diarrhoeal type have caused deaths during the last few years (Mahler *et al.* 1997; Lund *et al.* 2000). In Norway, where there have been relatively few outbreaks of campylobacteriosis and salmonellosis, we have seen a steady increase in the number of *B. cereus* food poisoning cases. As the Norwegian reference laboratory for spore-forming food poisoning bacteria we can follow this trend closely. In Europe and the USA, the same increase in *B. cereus* outbreaks is probably hidden behind the usually more serious outbreaks caused by other bacteria. Indeed, it has been reported to be the most important cause of foodborne disease in the Netherlands, together with *Salmonella* spp., between 1985 and 1991 (Schmidt 1995).

Bacillus cereus is a Gram-positive, spore-forming, motile, aerobic rod that also grows well anaerobically. It is a common soil saprophyte and is spread easily to many types of foods, especially those of plant origin, but is also frequently isolated from meat, eggs and dairy products (Granum & Baird-Parker 2000). *Bacillus cereus* and other members of the *B. cereus* group can cause two different types of food poisoning: the diarrhoeal type and the emetic type. The diarrhoeal type of food poisoning is caused by complex enterotoxins (Granum 1997; Granum & Brynestad 1999) during vegetative growth of *B. cereus* in the small intestine (Granum *et al.* 1993), while the emetic toxin is preformed during the growth of cells in the foods (Kramer & Gilbert 1989; Granum 1997). For both types of food poisoning the food involved has usually been heat-treated, and surviving spores are the source of the food poisoning. *Bacillus cereus* is not a competitive microorganism, but grows well after cooking and cooling (<46–50°C). The heat treatment will cause spore germination, and in the absence of competing flora *B. cereus* grows well. Invasion by psychrotolerant strains from the *B. cereus* group in the dairy industry has led to increasing surveillance of *B. cereus* in recent years (Granum 1997).

Bacillus cereus food poisoning is a nonreportable disease across Europe. As a consequence of this, and the fact it is usually a relatively mild disease of short duration (<24 h), it is highly underreported in official statistics (Granum & Baird-Parker 2000). However, occasional reports have described more severe forms of the diarrhoeal type of *B. cereus* food poisoning (Andersson *et al.* 1998), including a necrotic enteritis causing three deaths (Lund *et al.* 2000).

The closely related *B. thuringiensis* is reported to produce enterotoxins (Damgaard *et al.* 1996; Rivera *et al.* 2000) and has been shown to cause food poisoning symptoms when given to human volunteers (Ray 1991). It has also been reported to cause food poisoning in regular outbreaks (Jackson *et al.* 1995). The extensive use of this organism as a protective agent against insect attacks on crops may soon lead to serious problems for the food industry. Normal procedures for confirmation of *B. cereus* would not differentiate between the two species – if indeed it is possible from heat-treated food products – because *B. thuringiensis* tends to throw out the insecticidal plasmids when grown above 30–37°C. This makes it difficult to investigate the real numbers of food poisoning cases caused by commercially used *B. thuringiensis*. In order to assure safe spraying with *B. thuringiensis*, the organism in use should be unable to produce food-poisoning toxins. The Health and Consumer Protection Directorate-General (European Commission) has already accepted that only nontoxin-producing *Bacillus* spp. should be allowed to be used in animal nutrition (see http://www.europa.eu.int/comm/dg24, expressed on 12.02.00).

Taxonomy of the *Bacillus cereus* group

The aerobic endospore-forming bacteria have traditionally been placed in the genus *Bacillus*. Over the past three decades, this genus has expanded to accommodate more than 100 species. Analysis of 16S rRNA sequences from numerous *Bacillus* species indicated that the genus *Bacillus* should be divided into at least five genera or rRNA groups (Ash *et al.* 1991).

Bacillus anthracis, *B. cereus*, *B. mycoides*, *B. thuringiensis* and more recently *B. pseudomycoides* (Nakamura 1998) and *B. weihenstephanensis* (Lechner *et al.* 1998) comprise the *B. cereus* group. These bacteria have highly similar 16S and 23S rRNA sequences, indicating that they have diverged from a common evolutionary line relatively recently. Extensive genomic studies of *B. cereus* and *B. thuringiensis* have shown that there is no taxonomic basis for separate species status (Carlson *et al.* 1996). Nevertheless, the name *B. thuringiensis* is retained for those strains that synthesize a crystalline inclusion (Cry protein) or δ-endotoxin that may be highly toxic to insects. The *cry* genes are usually located on plasmids and loss of the relevant plasmid(s) makes the bacterium indistinguishable from *B. cereus*. It is now clear that most strains in the *B. cereus* group, including *B. thuringiensis*, carry enterotoxin genes (Rivera *et al.* 2000).

Food-borne outbreaks

Bacillus cereus is widely recognized as a food poisoning organism. Outbreaks can be divided into two types according to their symptoms. The diarrhoeal type is far more frequent in Europe and USA, while the emetic type appears more prevalent in Japan (Granum 1997). Typical foods implicated are stews, puddings, sauces, and flour and rice dishes (Granum & Baird-Parker 2000). When expressed as a proportion of all reported food poisonings, outbreaks ascribed to *B. cereus* seem to be concentrated in Scandinavia and Canada (10–47% of total) and are less frequent in Central Europe, UK, USA and the Far East (1–5% of total) (Schmidt 1995; Granum 1997). Although these differences might partly be a consequence of different consumer habits, they are also not comparable owing to dissimilar reporting practices. Thus, in the Netherlands in 1991, *B. cereus* was responsible in 27% of outbreaks in which the causative agent was identified. However, the incidence was only 2.8% of the total, because the majority of cases of food poisoning were of unknown aetiology (Schmidt 1995). In addition, when the number of food poisoning cases ascribed to *B. cereus* are expressed on a per-head-of-population basis, many of the large regional differences in incidence disappear.

Characteristics of disease

The emetic toxin causes vomiting, while the other type of illness, caused by enterotoxins, is characterized by diarrhoea (Granum & Lund 1997). In a small number of cases both types of symptoms are recorded (Kramer & Gilbert 1989), probably owing to production of both types of toxins. There has been some debate about whether or not the enterotoxin(s) can be preformed in foods, and cause an intoxication. By reviewing the literature it is obvious that the incubation time is a little too long for that (>6 h; average 12 h) (Granum 1997), and in model experiments it has been shown that the enterotoxin(s) is degraded on its way to the ileum (Granum *et al.* 1993). Although the enterotoxin(s) can be preformed, the number of *B. cereus* cells in the food would need to be at least two orders of magnitude higher than that necessary for causing food poisoning (Granum 1997), and such products would no longer be acceptable to the consumer. The characteristics of *B. cereus* food poisoning are given in table 4.1.

Table 4.1 Characteristics of the two types of disease caused by *Bacillus cereus* (from Granum & Lund 1997).

	Diarrhoeal syndrome	Emetic syndrome
Infective dose	10^5–10^7 (total)	10^5–10^8 (cells per gram food to produce enough emetic toxin)
Toxin produced	In the small intestine of the host	Preformed in foods
Type of toxin	Protein(s)	Cyclic peptide
Incubation period	8–16 h (occasionally >24 h)	0.5–5 h
Duration of illness	12–24 h (occasionally several days)	6–24 h
Symptoms	Abdominal pain, watery diarrhoea (occasionally bloody diarrhoea), sometimes with nausea	Nausea, vomiting and malaise (sometimes followed by diarrhoea, owing to additional enterotoxin production?)
Foods most frequently implicated	Meat products, soups, vegetables, puddings/sauces and milk/milk products	Fried and cooked rice, pasta, pastry and noodles

Infectious dose

Counts ranging from 10^4 to 10^9/g (or mL) *B. cereus* (Granum & Baird-Parker 2000) have been reported in the incriminated foods after food poisoning, giving total infective doses ranging from about 5×10^4–10^{11}. The variation in the infective dose is partly because of differences in the amount of enterotoxin produced by different strains (at least two orders of magnitude), and partly because of the difference in infective dose between vegetative cells and spores, as all the spores will survive the stomach acid barrier. Thus, any food containing more than 10^3 *B. cereus*/g cannot be considered completely safe for consumption.

Virulence factors/mechanisms of pathogenicity

Very different types of toxins cause the two types of *B. cereus* food poisoning. The emetic toxin, causing vomiting, is a ring-formed small peptide while the diarrhoeal disease is caused by several different enterotoxins (table 4.2).

The emetic toxin

The emetic toxin causes emesis (vomiting) only and its structure was for many years a mystery, as the only detection system involved living primates (Kramer & Gilbert 1989). The discovery that the toxin could be detected (vacuolation activity) by the use of HEp-2 cells (Hughes *et al.* 1988) led to its isolation and structure determination (Agata *et al.* 1994). The emetic toxin has been named cereulide, and consists of a ring structure of three repeats of four amino- and/or oxy-acids: [D-O-Leu-D-Ala-L-O-Val-L-Val]₃. This ring structure (dode-cadepsi-peptide) has a molecular mass of 1.2 kDa, and is chemically closely related to the potassium ionophore valinomycin (Agata *et al.* 1994). The emetic toxin is resistant to heat,

Table 4.2 Toxins known to be produced by *Bacillus cereus*.

Toxin	Type/size	Food poisoning	References
Haemolysin BL (Hbl)	Protein, 3 components	Probably	Beecher & Wong (1997); Heinrichs *et al.* (1993); Ryan *et al.* (1997); Beecher *et al.* (1995)
Nonhaemolytic enterotoxin (Nhe)	Protein, 3 components	Yes	Lund & Granum (1996, 1997); Granum *et al.* (1999)
Enterotoxin T (BceT)	Protein, 1 component, 41 kDa	?	Agata *et al.* (1995b)
Enterotoxin FM (EntFM)	Protein, 1 component, 45 kDa	??	Asano *et al.* (1997)
Cytotoxin K (CytK)	Protein, 1 component, 34 kDa	Yes, 3 deaths	Lund *et al.* (submitted)
Emetic toxin (cereulide)	Cyclic peptide, 1.2 kDa	Yes, 1 death	Agata *et al.* (1994, 1995a); Yokoyama *et al.* (1999); Mahler *et al.* (1997)

pH and proteolysis but is not antigenic (Kramer & Gilbert 1989). The biosynthetic pathway and mechanism of action of the emetic toxin remain uncertain, although it has recently been shown that it stimulates the vagus afferent through binding to the 5-HT$_3$ receptor (Agata *et al.* 1995a). It is not clear how this peptide is synthesized, but with such a structure it is most likely that cereulide is an enzymatically synthesized peptide and not a direct genetic product.

Cereulide was responsible for the death of a 17-year-old Swiss boy a few years ago, following fulminant liver failure (Mahler *et al.* 1997). A large amount of *B. cereus* emetic toxin was found in the residue from the pan used to reheat the food (pasta) and in the boy's liver and bile. Recently, mice were injected i.p. with synthetic cereulide and the development of histopathological changes was examined (Yokoyama *et al.* 1999). At high cereulide doses, massive degeneration of hepatocytes occurred. The serum values of hepatic enzymes were highest on days 2–3 post-inoculation, and decreased rapidly thereafter. General recovery from the pathological changes and regeneration of hepatocytes was observed after 4 weeks.

Enterotoxins

Five different enterotoxins have been characterized to date (see table 4.2). The three-component haemolysin (Hbl; consisting of three proteins: B, L$_1$ and L$_2$) with enterotoxin activity was the first to be fully characterized (Beecher & Wong 1994, 1997). This toxin also has dermonecrotic and vascular permeability activities, and causes fluid accumulation in ligated rabbit ileal loops. Hbl has been suggested to be a primary virulence factor in *B. cereus* diarrhoea (Beecher *et al.* 1995). Convincing evidence has shown that all three components are necessary for maximal enterotoxin activity (Beecher *et al.* 1995). It has been suggested, from studies of interactions with erythrocytes, that the B protein is the component that binds Hbl to the target cells, and that L$_1$ and L$_2$ have lytic functions (Beecher & Macmillan 1991). More recently another model for the action of Hbl has been proposed, suggesting that the components of Hbl bind to target cells independently and then constitute a membrane-attacking complex resulting in a colloid osmotic lysis mechanism (Beecher & Wong 1997). Substantial heterogeneity has been observed in the components of Hbl, and individual strains produced various combinations of single or multiple bands of each component (Schoeni & Wong 1999).

A nonhaemolytic three-component enterotoxin (Nhe) was more recently characterized by Lund and Granum (1996). The three components of this toxin were different from the components of Hbl, although there are similarities. The three components of Nhe enterotoxin were first purified from a *B. cereus* strain isolated after a large food-poisoning outbreak in Norway in 1995. The strain used to characterize Nhe did not produce the L$_2$-component (Granum *et al.* 1996). We expected to show that the L$_2$-component was unnecessary for biological activity of the Hbl enterotoxin, but instead we came across another three-component enterotoxin. Binary combinations of the components of this enterotoxin possess some biological activity, but not nearly as high as when all the components are present (Lund & Granum 1997).

Almost all tested *B. cereus*/*B. thuringiensis* strains produce Nhe and about 50% produce Hbl (P.E. Granum, unpubl. results). Presently we do not know how important each of them is in relation to food poisoning. The enterotoxin T gene was shown to be absent in 57 of 95 strains of *B. cereus*, and in 5 out of 7 strains involved in food poisoning (Granum *et al.* 1996). Nothing is known about the role of enterotoxin FM in food poisoning; this 'toxin' has only been cloned without any biological characterization (table 4.2) .

The newly discovered cytotoxin K (CytK) is similar to the β-toxin of *Clostridium perfringens* (and other related toxins) and was the cause of the symptoms in a severe outbreak of *B. cereus* food poisoning in France in 1998 (Lund *et al.* 2000). In this outbreak several people developed bloody diarrhoea and three died. It would be fair to call this an outbreak of *B. cereus* necrotic enteritis, although it is not nearly as severe as the *C. perfringens* type C food poisoning (Granum 1990).

There is significant sequence identity between the three proteins of Nhe and between the Nhe and Hbl proteins. The identity is highest in the N-terminal third of the proteins. The most pronounced gene sequence similarities are found between *nheA* and *hblC*, *nheB* and *hblD*, and *nheC* and *hblA*. This is not only in direct comparison of the sequences, but also in predicted transmembrane helices for the six proteins. NheA and HblC have no predicted transmembrane helices, while NheB and HblD have two each. Finally NheC and HblA each have one predicted transmembrane helix, in the same position in the two proteins (Granum *et al.* 1999).

Other possible virulence factors

The *B. cereus* spore is more hydrophobic than the spores of any other *Bacillus* species, which makes it adhesive to several types of surfaces (Andersson *et al.* 1995). This makes it difficult to remove during cleaning, and a difficult target for disinfection. The *B. cereus* spores also contain appendages and/or philli (Andersson *et al.* 1995; Granum 1997) that are, at least partly, involved in adhesion (Husmark & Rönner 1992). Not only can these properties of the *B. cereus* spore make them survive sanitation – and thus remain available for contamination of different foods – but also aid adherence to epithelial cells. Experiments have shown that spores, at least from one strain isolated after one outbreak (Andersson *et al.* 1998), can indeed adhere to Caco-2 cells in culture, and that these properties are linked to hydrophobicity and possibly to the appendages. A longer incubation period than normal was observed in this case (Granum 1997), as would be expected, as the spore would first have to germinate.

Gene organization of the enterotoxins (figure 4.1)

All three proteins of the Hbl are transcribed from one operon (*hbl*) (Ryan *et al.* 1997), and Northern blot analysis has shown RNA transcripts of 5.5 kb *hblC* (transcribing L$_2$) and *hblD* (transcribing L$_1$) are only separated by 37 bp and

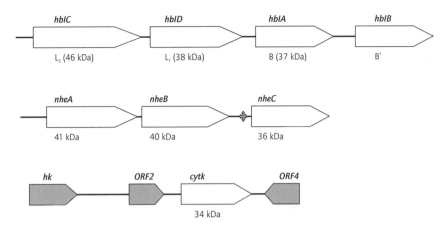

Figure 4.1 Genetic organization of the three enterotoxins: *hbl*, *nhe* and *cytK*. The two three-component enterotoxins are both regulated by PlcR that binds upstream of *hblC* and *nheA*. There is an inverted repeat of 13 bp between *nheB* and *nheC*. *cytK* has an open reading frame (ORF2) upstream, encoding a protein of unknown function, and a histidine kinase (*hk*) further upstream. ORF4 encodes for a long-chain-fatty-acid-CoA ligase (orientated in opposite direction to that of *cytK*). The arrowheads indicate the orientation of the genes. The gaps between the different genes are indicated.

encode proteins of 447 aa and 384 aa, respectively. L_2 has a signal peptide of 32 aa and L_1 a signal peptide of 30 aa. The B-protein, transcribed from *hblA*, consists of 375 aa, with a signal peptide of 31 aa (Heinrichs *et al.* 1993). The exact spacing between *hblD* and *hblA* is at least 100 bp (overlapping sequence not published), but is claimed to be approximately 115 bp (Ryan *et al.* 1997). The spacing between *hblA* and *hblB* is 381 bp, and the length of *hblB* is not known (Heinrichs *et al.* 1993). However, based on the size judged from a Northern blot, a size similar to *hblA* could be suggested for hblB. The B- and the putative B'-protein (transcribed by *hblA* and *hblB*, respectively) are very similar in the first 158 aa (the known sequence of B'-protein based on the DNA sequence). The function of this putative protein is not yet known, but it is possible that it may substitute for the B-protein. The *hbl* operon is mapped to the unstable part of the *B. cereus* chromosome (Carlson *et al.* 1996).

The *nhe* operon contains three open reading frames (figure 4.1): *nheA*, *nheB* and *nheC* (Granum *et al.* 1999). The first two gene products have earlier been addressed as the 45- and 39-kDa proteins respectively. The last possible transcript from the *nhe* operon (*nheC*) has not been purified and its function is not known, although we know that a third component is involved in *nhe* activity. The three proteins transcribed from the *nhe* operon have properties as shown in table 4.3. The correct size of the 45-kDa protein and 39-kDa proteins are 41 kDa and 39.8 kDa, respectively. The size of NheC is 36.5 kDa. There is a gap of 40 bp between *nheA* and *nheB*, and a gap of 109 bp between *nheB* and *nheC*. In the 109 bp between *nheB* and *nheC* there is an inverted repeat of 13 bp (Granum *et al.* 1999). This structure may result in little production of NheC, compared to that of NheA and NheB. This might again be the reason why we have not yet managed to isolate the NheC protein.

Table 4.3 Properties of the Nhe proteins (from Granum *et al.* 1999).

Proteins	Signal peptide	Active protein	Molecular weight (active protein)	pI
NheA	26 aa	360 aa	41.019	5.13
NheB	30 aa	372 aa	39.820	5.61
NheC	30 aa	329 aa	36.481	5.28

The *B. cereus* strain producing CytK did not contain genes for other known *B. cereus* enterotoxins. CytK is a protein of 34 kDa and is also haemolytic. The *cytK* is organized genetically as shown in figure 4.1.

Regulation of enterotoxin production

For both Hbl and Nhe, maximal enterotoxin activity is found during the late exponential or early stationary phase, and indeed it has been shown that both these enterotoxin operons are under regulation of *plcR* (Agaisse *et al.* 1999), a gene first described to regulate the *plcA* gene encoding for phospholipase C (Lereclus *et al.* 1996). The binding sequence for this regulatory protein is TATG8NCATG (Agaisse *et al.* 1999).

Commercial methods for detection of the *Bacillus cereus* toxins

Neither of the two commercially available immunoassays will quantify the toxicity of the enterotoxins from *B. cereus*. The assay from Oxoid measures the presents of the HblC (L_2) component while the Tecra kit mainly detects the NheA (45-kDa) component (Granum & Lund 1997). However, if one or both of the commercial kits reacts positively with proteins from *B. cereus* supernatants it is likely that the strain is enterotoxin-positive, specifically if the strains are cytotoxic on epithelial cells. If the supernatants are also shown to be cytotoxic the strains can be regarded as enterotoxin-positive. At present, there is no commercial method available for detecting CytK, and, unfortunately, there is no commercial method for detection of the emetic toxin either. However, a specific, sensitive, semiautomated, and quantitative Hep-2 cell culture-based 3-(4,5-dimethylthiazol-2-yl)-2, 5-diphenyltetrazolium bromide assay for *B. cereus* emetic toxin has been developed (Finlay *et al.* 1999).

References

Agaisse, H., Gominet, M., Okstad, O.A., Kolstø, A-B. & Lereclus, D. (1999) PlcR is a pleiotropic regulator of extracellular virulence factor gene expression in *Bacillus thuringiensis*. *Molecular Microbiology* 32, 1043–1053.

Agata, N., Mori, M., Ohta, M., Suwan, S., Ohtani, I. & Isobe, M. (1994) A novel dodecadepsipeptide, cereulide, isolated from *Bacillus cereus* causes vacuole formation in Hep-2 cells. *FEMS Microbiology Letters* 121, 31–34.

Agata, N., Ohta, M., Mori, M. & Isobe, M. (1995a) A novel dodecadepsipeptide, cereulide, is an emetic toxin of *Bacillus cereus*. *FEMS Microbiology Letters* **129**, 17–20.

Agata, N., Ohta, M., Arakawa, Y. & Mori, M. (1995b) The *bceT* gene of *Bacillus cereus* encodes an enterotoxic protein. *Microbiology* **141**, 983–988.

Andersson, A., Rönner, U. & Granum, P.E. (1995) What problems does the food industry have with the sporeforming pathogens *Bacillus cereus* and *Clostridium perfringens*? *International Journal of Food Microbiology* **28**, 145–156.

Andersson, A., Granum, P.E. & Rönner, U. (1998) The adhesion of *Bacillus cereus* spores to epithelial cells might be an additional virulence mechanism. *International Journal of Food Microbiology* **39**, 93–99.

Asano, S.I., Nukumizu, Y., Bando, H., Iizuka, T. & Yamamoto, T. (1997) Cloning of novel enterotoxin genes from *Bacillus cereus* and *Bacillus thuringiensis*. *Applied and Environmental Microbiology* **63**, 1054–1057.

Ash, C., Farrow, J.A., Wallbanks, S. & Collins, M.D. (1991) Phylogenetic heterogeneity of the genus *Bacillus* revealed by comparative analysis of small subunit ribosomal RNA sequences. *Letters in Applied Microbiology* **13**, 202–206.

Beecher, D.J. & Macmillan, J.D. (1991) Characterization of the components of hemolysin BL from *Bacillus cereus*. *Infection and Immunity* **59**, 1778–1784.

Beecher, D.J. & Wong, A.C.L. (1994) Improved purification and characterization of hemolysin BL, a hemolytic dermonecrotic vascular permeability factor from *Bacillus cereus*. *Infection and Immunity* **62**, 980–986.

Beecher, D.J. & Wong, A.C.L. (1997) Tripartite hemolysin BL from *Bacillus cereus*. Hemolytic analysis of component interaction and model for its characteristic paradoxical zone phenomenon. *Journal of Biological Chemistry* **272**, 233–239.

Beecher, D.J., Schoeni, J.L. & Wong, A.C.L. (1995) Enterotoxin activity of hemolysin BL from *Bacillus cereus*. *Infection and Immunity* **63**, 4423–4428.

Carlson, C.R., Johansen, T. & Kolstø, A.B. (1996) The chromosome map of *Bacillus thuringiensis* subsp. *canadensis* HD224 is highly similar to that of *Bacillus cereus* type strain ATCC 14579. *FEMS Microbiology Letters* **141**, 163–167.

Damgaard, P.H., Larsen, H.D., Hansen, B.M., Bresciani, J. & Jorgensen, K. (1996) Enterotoxin-producing strains of *Bacillus thuringiensis* isolated from food. *Letters in Applied Microbiology* **23**, 146–150.

Finlay, W.J., Logan, N.A. & Sutherland, A.D. (1999) Semiautomated metabolic staining assay for *Bacillus cereus* emetic toxin. *Applied and Environmental Microbiology* **65**, 1811–1812.

Granum, P.E. (1990) *Clostridium perfringens* toxins involved in food poisoning. *International Journal of Food Microbiology* **10**, 101–112.

Granum, P.E. (1997) *Bacillus cereus*. In: *Food Microbiology. Fundamentals and Frontiers* (eds M. Doyle *et al.*), pp. 327–336. ASM Press, Washington, DC.

Granum, P.E. & Baird-Parker, T.C. (2000) *Bacillus* spp. In: *The Microbiological Safety and Quality of Food* (eds B. Lund *et al.*), pp. 1029–1039. Aspen Publishers, Gaithersburg, MD.

Granum, P.E. & Brynestad, S. (1999) Bacterial toxins as food poisons. In: *The Comprehensive Sourcebook of Bacterial Protein Toxins* (eds J.E. Alouf & J.H. Freer), pp. 669–681. Academic Press, London.

Granum, P.E. & Lund, T. (1997) *Bacillus cereus* enterotoxins. *FEMS Microbiology Letters* **157**, 223–228.

Granum, P.E., Brynestad, S., O'Sullivan, K. & Nissen, H. (1993) The enterotoxin from *Bacillus cereus*: production and biochemical characterization. *Netherlands Milk and Dairy Journal* **47**, 63–70.

Granum, P.E., Andersson, A., Gayther, C. *et al.* (1996) Evidence for a further enterotoxin complex produced by *Bacillus cereus*. *FEMS Microbiology Letters* **141**, 145–149.

Granum, P.E., O'Sullivan, K. & Lund, T. (1999) The sequence of the non-haemolytic enterotoxin operon from *Bacillus cereus*. *FEMS Microbiology Letters* **177**, 225–229.

Heinrichs, J.H., Beecher, D.J., MacMillan, J.M. & Zilinskas, B.A. (1993) Molecular cloning and characterization of the *hblA* gene encoding the B component of hemolysin BL from *Bacillus cereus*. *Journal of Bacteriology* **175**, 6760–6766.

Hughes, S., Bartholomew, B., Hardy, J.C. & Kramer, J.M. (1988) Potential application of a Hep-2 cell assay in the investigation of *Bacillus cereus* emetic-syndrome food poisoning. *FEMS Microbiology Letters* **52**, 7–12.

Husmark, U. & Rönner, U. (1992) The influence of hydrophobic, electrostatic and morphologic properties on the adhesion of *Bacillus* spores. *Biofouling* **5**, 335–344.

Jackson, S.J., Goodbrand R.B., Ahmed, R. & Kasatiya, S. (1995) *Bacillus cereus* and *Bacillus thuringiensis* isolated in a gastroenteritis outbreak investigation. *Letters in Applied Microbiology* **21**, 103–105.

Kramer, J.M. & Gilbert, R.J. (1989) *Bacillus cereus* and other *Bacillus* species. In: *Foodborne Bacterial Pathogens* (ed. M.P. Doyle), pp. 21–70. Marcel Dekker, New York.

Lechner, S., Mayr, R., Francis, K.P. *et al.* (1998) *Bacillus weihenstephanensis* sp. nov. is a new psychrotolerant species of the *Bacillus cereus* group. *International Journal of Systematic Bacteriology* **48**, 1373–1382.

Lereclus, D., Agaisse, H., Gominet, M., Salamitou, S. & Sanchis, V. (1996) Identification of a *Bacillus thuringiensis* gene that positively regulates transcription of the phosphatidyl-inositol-specific phospholipase C gene at the onset of the stationary phase. *Journal of Bacteriology* **178**, 2749–2756.

Lund, T. & Granum, P.E. (1996) Characterisation of a non-haemolytic enterotoxin complex from *Bacillus cereus* isolated after a foodborne outbreak. *FEMS Microbiology Letters* **141**, 151–156.

Lund, T. & Granum, P.E. (1997) Comparison of biological effect of the two different enterotoxin complexes isolated from three different strains of *Bacillus cereus*. *Microbiology* **143**, 3329–3336.

Lund, T., De Buyser, M.L. & Granum, P.E. (2000) A new enterotoxin from *Bacillus cereus* that may cause necrotic enteritis. *Molecular Microbiology* **38**, 254–261.

Mahler, H., Pasi, A., Kramer, J.M. *et al.* (1997) Fulminant liver failure in association with the emetic toxin of *Bacillus cereus*. *New England Journal of Medicine* **336**, 1142–1148.

Nakamura, L.K. (1998) *Bacillus pseudomycoides* sp. nov. *International Journal of Systematic Bacteriology* **48**, 1031–1035.

Ray, D.E. (1991) Pesticides derived from plants and other organisms. In: *Handbook of Pesticide Toxicology* (eds W.J. Hayes & E.R. Laws, Jr), pp. 585–636. Accademic Press, New York.

Rivera, A.M.G., Granum, P.E. & Priest, F.G. (2000) Common occurrence of enterotoxin genes and enterotoxicity in *Bacillus thuringiensis*. *FEMS Microbiology Letters* **190**, 151–155.

Ryan, P.A., Macmillan, J.M. & Zilinskas, B.A. (1997) Molecular cloning and characterization of the genes encoding the L_1 and L_2 components of hemolysin BL from *Bacillus cereus*. *Journal of Bacteriology* **179**, 2551–2556.

Schmidt, K. (ed.) (1995) *WHO Surveillance Programme for Control of Foodborne Infections and Intoxications in Europe* (Sixth Report), pp. 162–168. FAO/WHO Collaborating Centre for Research and Training in Food Hygiene and Zoonoses, Berlin.

Schoeni, J.L. & Wong, A.C.L. (1999) Heterogeneity observed in the components of hemolysin BL, an enterotoxin produced by *Bacillus cereus*. *International Journal of Food Microbiology* **53**, 159–167.

Yokoyama, K., Ito, M., Agata, N. *et al.* (1999) Pathological effect of synthetic cereulide, an emetic toxin of *Bacillus cereus*, is reversible in mice. *FEMS Immunology and Medical Microbiology* **24**, 115–120.

Chapter 5

Thermophilic *Bacillus* Isolates from Antarctic Environments

Barbara Nicolaus, Licia Lama and Agata Gambacorta

Introduction

Temperature is one of the most important variables in terrestrial environments. The classification of living organisms based on their relationship with temperature has therefore always been considered one of the most basic elements of biological systematics. Microorganisms have thus been simply divided into three main groups: psychrophiles, mesophiles and thermophiles. The thermophilic phenotype still needs to divided further, and a relatively simple division might be to define thermophilic, extremely thermophilic prokaryotes and hyperthermophiles, depending upon cardinal temperatures of growth. Generally speaking, the thermophilic *Bacteria*, with few exceptions, lie within the thermophilic range, while the thermophilic *Archaea* are in the hyperthermophilic range (Kristjansson 1991).

The main interest in thermophiles in recent years has been on two fronts. On the scientific front, studies on thermophiles have helped redraw the evolutionary tree of life, with the discovery of hyperthermophilic archaea able to grow at temperatures higher than the boiling point of water (Danson *et al.* 1992). On the biotechnological front, there has been the realization that thermophilic microorganisms – including the thermophilic endospore-forming bacilli – can serve as excellent sources for new molecules and/or more thermostable biocatalysts than are presently available (Kristjansson 1991; Mora *et al.* 1998). Even if the proteins of some thermophiles do not prove to be useful, chemists hope to learn from them how to redesign conventional enzymes to perform in harsh conditions. Recent developments have shown clearly that thermozymes are good sources of novel catalysts of industrial interest. Some of these catalysts have been isolated and their genes successfully cloned and expressed in mesophilic hosts (Vieille *et al.* 1996).

The majority of the thermophilic bacteria investigated belong to the genus *Bacillus*, with strains that have been isolated from mesophilic and thermophilic environments (Sunna *et al.* 1997).

Although the genus *Bacillus* is the largest and best-known member of the family *Bacillaceae*, its taxonomy is very confusing, and new species often have been described on the basis of only a few physiological or ecological features. In 1981, Logan and Berkeley concluded from their extensive studies on *Bacillus* that further information was needed before the genus could be subdivided into three or more different genera. The polyphasic approach, including phenotypic, genotypic and chemical characterization, has allowed us to establish a new taxonomy of thermophilic aerobic endospore-formers. Nevertheless, there are still several difficulties in the identification and characterization of new thermophilic isolates, and this results mainly from the great genotypic and phenotypic variability that characterizes strains belonging to some species, such as *B. stearothermophilus* and *Alicyclobacillus* (formerly *Bacillus*) *acidocaldarius*, which have traditionally been placed in the thermophilic aerobic spore-forming bacilli. Recently, new genera have been proposed and several new species have been validly proposed, while some others have lost their standing in bacterial nomenclature (Mora *et al*. 1998).

Thermozymes are not only of interest to industry, but also the membranes of thermophiles may contain surfactants bearing unique stabilities that can be valuable in pharmaceutical formulations, e.g. for preparation of liposomes with useful mechanical and thermal properties.

In recent years, increasing attention has been paid to microbial exopolysaccharides (EPS), that are of widespread occurrence and which can be prepared readily by fermentation. Such molecules are of interest because of their bio-active roles and their wide ranges of commercial applications; these include uses as food additives and in biopharmaceutical industries. Bio-polymers from prokaryotes in both the *Bacteria* and *Archaea* offer a number of novel material properties and commercial opportunities. In fact, these molecules are an important resource in several biotechnological applications (Nicolaus *et al*. 1999a).

However, despite the widespread occurrence of polysaccharide-producing microorganisms, the number of bio-polymers with actual or potential industrial value is limited. Some extremophiles, including *Methanosarcina* species, produce extracellular polysaccharides as do some species of *Haloferax*, *Haloarcula*, *Sulfolobus* and *Bacillus* (Manca *et al*. 1996; Nicolaus *et al*. 1999a,b).

In this chapter we report the identification and distribution of species of thermophilic bacteria isolated from several geographical regions in Antarctica. Lipids, exopolysaccharides and hydrolytic enzymes from these species are also described.

Sampling sites

Thermophilic microorganisms were isolated from several different geographical areas of the Antarctic Continent. Regions characterized by geothermal activity are common in Antarctica. They have been found in the Antarctica Peninsula and in proximity to the Ross Sea coast. Fumaroles exist on Mount Erebus, an active volcano on Ross Island, on Cryptogam Ridge near the crater of Mount Melbourne in northern Victoria Land, at the end of the eastern slope of Mount Melbourne near the seashore (named north of Edmonson Point), and near the crater of Mount Rittmann (figure 5.1).

During the austral summers of 1986–7, 1987–8, 1993–4 and 1998–9 samples of soil, sediments and water were aseptically collected from four sites in the vicinity of Terra Nova Bay Station (74°42′S, 164°07′E) on the Ross Sea coast, northern Victoria Land, Antarctica (Nicolaus *et al*. 1996, 1998):

Figure 5.1 Caldera near the crater of Mt Rittmann from where soil samples were collected by Enrico Esposito during the Italian Antarctic Expedition, in the austral summer of 1993–4.

1 Cryptogam Ridge is located near the crater of Mount Melbourne (74°22′S, 164°40′E). Isolates M1 (*B. thermoantarcticus*), M2, M3 and M5 were from samples of soil collected from a depth of 5–10 cm, where the temperature reaches 50–60°C.

2 North of Edmonson Point (74°20′S, 165°07′E), which is located near Mount Melbourne. Isolates M4 and M6 were collected in close proximity to the seashore.

3 Mount Erebus, near Tranway Ridge (77°32′S, 167°10′E). Isolate M7 was from a sample collected near the crater of the volcano.

4 Mount Rittmann (73°28′S, 165°36′E) is about 100 km from Terra Nova Bay Station. The isolates MR1–MR5 were obtained from geothermal soil collected near the crater.

The sites and the temperature conditions of isolations are shown in table 5.1.

Bacterial strains

A wide range of media at different pH values was used for the isolation and characterization of thermophilic and thermoacidophilic Antarctic strains.

'*Bacillus thermoantarcticus*' (M1) and the other strains M2–M7 were grown optimally at 60°C, at pH 6.0 in 0.8% yeast extract and 0.3% NaCl (Nicolaus *et al.* 1996). For exopolysaccharide production the culture medium contained 0.3% NaCl, 0.6% glucose and 0.1% yeast extract. The glucose medium supplemented with 2% agar was used for agar plate preparations. The colonies on

Table 5.1 Sites and microorganisms isolated in Antarctica.

| Site | Temperature (°C) | | pH | Microorganisms isolated |
	External	Soil		
Mount Melbourne, Cryptogam Ridge	−30	+60	4–6	Thermophiles: 4 strains: M1– M3, M5 '*B. thermoantarcticus*'
Mount Melbourne	−20	+50	6–7	*Bacillus* species as above
Mount Rittmann	−20	+62	4.5–6.0	Thermoacidophiles: 5 strains: '*Alicyclobacillus acidocaldarius* ssp. *rittmannii*'
Mount Rittmann downhill	−10	+60	5–6	*Alicyclobacillus* species
North of Edmonson Point	−5	+40	5	*Bacillus* species: 2 strains: M4 and M6
Mount Erebus, Tranway Ridge	−20	+40	5–6	*Bacillus* species: 1 strain: M7

plates, observed using a Leica Wild M 8 stereomicroscope, showed the presence of a mucous layer (Manca *et al.* 1996; Nicolaus *et al.* 1996). Production of the enzymes was investigated on media containing 0.1% yeast extract, 0.3% NaCl and 0.6% of a different carbon source: glucose, xylose, xylan, galactose or arabinose.

'*Alicyclobacillus acidocaldarius* ssp. *rittmannii*' (MR1) and the other related strains were grown optimally at 63°C, pH 4.0 using 0.4% yeast extract, 0.2% $(NH_4)_2SO_4$, 0.3% KH_2PO_4, 4 mL/L sol. A, 4 mL sol. B (Nicolaus *et al.* 1998).

Strain characterization

Spore production and strain characterization were performed using the methods reported in Nicolaus *et al.* (1996). Electron microscopy was performed as reported in Nicolaus *et al.* (1998).

Evaluation of 16S rDNA sequence, DNA–DNA hybridization and % G+C content were carried out as described previously (Nicolaus *et al.* 1996, 1998).

Chemical analysis

Freshly harvested cells were lyophilized and extracted continuously by Soxhlet with $CHCl_3$/MeOH (1 : 1 v/v) for 5 h at 70°C.

Lipid analysis was performed according to Nicolaus *et al.* (1998). Lipid hydrolysis was performed by acid methanolysis. Fatty acid methyl esters were analysed by separation on TLC and on GC-MS as reported in Nicolaus *et al.* (1995). Compounds were identified by interpretation of mass spectra using standards for comparison (Goodfellow & O'Donnell 1994).

For the production and characterization of exopolysaccharides (EPS), cells were harvested in the stationary phase of growth by centrifugation (1 L, 9800 g, 20 min). The supernatant was treated with 1 volume of cold absolute ethanol added drop-wise while stirring. The alcoholic solution was kept at −18°C overnight and then centrifuged at 15 000 *g* for 30 min. The pellet was dissolved in hot distilled water. The procedure up to this point was then repeated. The final water solution was dialysed against tap water (48 h) and distilled water (20 h), then freeze-dried and weighed. The sample was tested for carbohydrate, protein and nucleic acid contents. Polysaccharide fractions were purified by gel chromatography as reported in Manca *et al.* (1996).

Sugar analysis was performed by hydrolysis of EPS with 2 M trifluoroacetic acid (TFA) at 120°C for 2 h. The sugar mixture was identified by TLC and HPAE-PAD using standards for identification and calibration curves (for details see Manca *et al.* 1996).

The infrared spectrum of polymer (KBr tablet, 10 mg) was recorded at room temperature using a FT-IR Bio-Rad spectrometer. Ultraviolet spectra of EPS were obtained reading the absorbance of aqueous solutions (3 mg/mL) from 350 to 210 nm on a Varian DMS-90 instrument. The optical rotation value was obtained on a Perkin-Elmer 243 B polarimeter at 25°C. NMR spectra were obtained on a Bruker AMX-500 (500,13 MHz for ^{1}H and 125,75 MHz for ^{13}C) at 70°C.

Enzyme assays

Xylanase. Xylanase activity was assayed by using 2% birch-wood xylan, suspended in 50 mM sodium acetate buffer pH 5.6, as substrate (Lama *et al.* 1996).

D-*Xylose isomerase.* The formation of D-xylulose from D-xylose was measured using a colorimetric assay in which 20 μL of the enzyme solution (0.2 mg/mL) was added to 20 μL solution of 0.2 M D-xylose and 0.4 mM $MnSO_4$ in 20 mM Tris-HCl pH 7.0, and incubated for 10 min at 80°C. After this period the sample was put on ice to stop the enzyme reaction and 40 μL 1.5% (w/v) cysteine hydrochloride in distilled water, 40 μL 0.12% (w/v) carbazole in absolute ethanol, and 1.25 mL 70% (v/v) sulphuric acid were added. After standing for 10 min, the absorbance at 560 nm was determined. The enzyme activity was determined using a standard curve of D-xylulose.

D-*Glucose isomerase.* The test for D-glucose isomerase activity was the same as that for D-xylose isomerase, except that D-xylose and $MnSO_4$ were replaced by 0.4 M D-glucose, 10 mM $MgSO_4$ and 1 mM $CoCl_2$ in 20 mM Tris-HCl pH 7.0. The enzyme activity was calculated on the basis of the standard curve obtained with D-fructose. Reaction products were identified by HPAE-PAD Dionex, equipped with Carbo-Pac PA–1 column, using 100 mM NaOH as the solvent system (Lama *et al.* 1996).

α-Glucosidase. The activity was determined at 70°C for 10 min in the reaction system (1 mL) consisting of 2 mM *p*-nitrophenyl-α-D-glucopyranoside 0.8 mL of sodium acetate buffer pH 5.6 and 0.1 mL of the enzyme. The reaction was stopped by adding 1 mL of 1M Na_2CO_3 and 3 mL H_2O. The liberated *p*-nitrophenol was measured at 420 nm.

Table 5.2 Biochemical and physiological characteristics of the Antarctic isolates.

Characteristics	*'Bacillus thermoantarcticus'* (7 strains)	*'Alicyclobacillus acidocaldarius* ssp. *rittmannii'* (5 strains)
Gram staining	+	+
Cellular morphology	Rods	Rods
Spore morphology	Oval, terminal	Oval, central to terminal
CG%	53.7%	64.9%
Temperature range (°C)	37–65	45–70
Optimum temperature (°C)	60	63
pH range	5.5–9.0	2.5–5.0
Optimum pH	6.0	4.0
Motility	+	–
EPS production	+	–
Gelatin hydrolysis	+	–
Oxidase test	+	–
Catalase test	–	–
Casein hydrolysis	–	–
α-Amylase	–	–

Taxonomic features

Thermophilic bacilli, defined as aerobic or facultatively anaerobic, endospore-forming organisms with growth temperature optima in the range of 45–70°C were isolated from the sites indicated in table 5.1 and classified in the two genera *Bacillus* and *Alicyclobacillus*.

Most of the features of Antarctic isolates from Mount Melbourne and Mount Erebus, such as their optimal growth temperatures, mol % G+C contents, some physiological properties and membrane lipids, indicate that they should be assigned to the genus *Bacillus* (table 5.2).

In particular, strain M1, '*B. thermoantarcticus*', isolated from Mount Melbourne, was a novel, thermophilic, Gram-positive rod (figure 5.2a) (Nicolaus *et al.* 1996). The organism grew at an optimal temperature of 60°C at pH 6.0, and was oxidase-positive and catalase-negative. Glucose, trehalose, sucrose and xylose supported growth as sole carbon sources. The DNA base ratio was 53.7 mol % G+C (table 5.2).

A detailed study, comprising genetic characterization with reference strains of *B. thermoglucosidasius* and of mesophilic and facultatively thermophilic bacilli, was performed to define the taxonomic position of the isolate. A complete 16S rDNA sequence was determined and compared with all sequences currently available for members of *Bacillus* rDNA group 5. The phylogenetic dendrogram indicating the position of strain M1 is shown in figure 5.3a (Nicolaus *et al.* 1996).

During the Italian expedition in the Antarctic summer of 1993–4, it was possible to reach the caldera of Mount Rittmann for the first time (figure 5.1) and to collect samples in the deglaciated, geothermal soil. In the areas analysed only sporulating species belonging to the genus *Alicyclobacillus* were found.

(a)

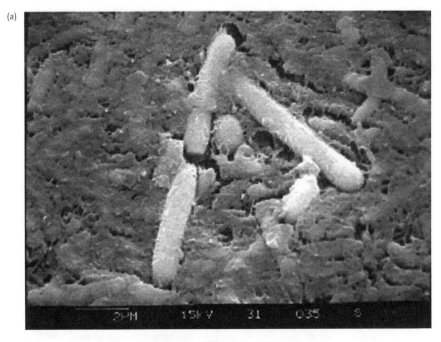

(b)

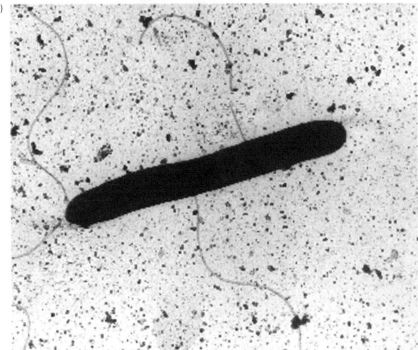

Figure 5.2 Electron micrographs of (a) '*Bacillus thermoantarcticus*' and (b) '*Alicyclobacillus acidocaldarius* ssp. *rittmannii*'.

(a)

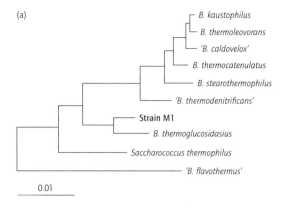

(b)

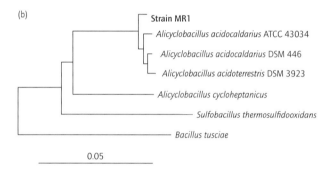

Figure 5.3 Phylogenetic dendrograms, based on comparison of complete 16S rDNA sequences of the Antarctic strains and all sequences currently available of the related taxa, indicating the position of strains (a) '*Bacillus thermoantarcticus*' and (b) '*Alicyclobacillus acidocaldarius* ssp. *rittmannii*'. Scale bar indicates nucleotide substitutions per 100 nucleotides. *B*, *Bacillus*.

Strain MR1, '*Alicyclobacillus acidocaldarius* ssp *rittmannii*', was a nonmotile, Gram-positive, spore-former (figure 5.2b). The isolate grew aerobically between 45 and 70°C with an optimum at 63°C and in the pH range of 2.5–5.0 with an optimum at pH 4.0, and was oxidase-, catalase-, urease-, starch-, gelatin- and hippurate-hydrolysis-negative. It was negative for extra- and endocellular α-amylase and showed a G+C content of 64.9 mol % (table 5.2). A complete 16S rDNA sequence was determined and compared with all sequences available for members of the genus *Alicyclobacillus* and related taxa. The phylogenetic dendrogram is shown in figure 5.3b.

Enzymes as industrial tools

The ability of many bacteria to grow at high temperatures has for a long time held a particular fascination for microbiologists and biochemists interested in the production of thermozymes. These thermozymes are not only more thermostable, but they are also more resistant than their mesophilic homologues to chemical agents, and this makes them very interesting for industrial processes. Despite this, most of the enzymes used at present in industrial processes have been isolated from mesophiles owing to the limited knowledge of thermophiles and the difficulties of growing them on a large scale.

Today, the most promising thermophilic biocatalysts are thermophilic enzymes from *Bacillus* species. In fact, the most famous and best-investigated thermostable enzyme is probably thermolysin, the protease from '*B. thermoproteolyticus*'. Therefore, a large series of thermophilic *Bacillus* strains was investigated as sources of industrial enzymes for biotechnological processes, especially those hydrolases able to degrade polysaccharides as renewable resources. Recently, xylose and xylo-oligosaccharides have become important food additives. In order to satisfy the industrial need for them, it is necessary to explore more effective methods for the production of xylose and xylo-oligosaccharides from xylan – a major constituent of the hemicellulose complex of the plant cell wall – by the actions of, for example, xylanolytic enzymes that are active at high temperatures. Many microorganisms produce xylan-degrading enzymes. The enzymes involved, xylanase (endo-β-1,4-xylanase, EC 3.2.1.8) and β-xylosidase (EC 3.2.1.37), have been purified and characterized from fungi and bacteria. However, the provenances of these enzymes were limited to those microorganisms that grow at moderate temperatures, and the enzymes did not show stability at higher temperatures. Few examples of thermostable xylan-degrading enzymes have been described. *Bacillus stearothermophilus* was the first microorganism in which the secreted thermostable xylanase and β-xylosidase were purified and characterized (Nanmori *et al.* 1990). Also '*B. thermoantarcticus*' was able to produce thermostable xylanase and β-xylosidase simultaneously. The concerted action of the two enzymes seemed to enhance the efficiency of degradation of xylan into xylose by this bacterium. The main physicochemical properties of these enzymes are summarized in table 5.3.

The extracellular enzymes from '*B. thermoantarcticus*' were separated by gel filtration with Sephacryl S-200 and further purified. The optimum temperature was 80°C for xylanase and 70°C for β-xylosidase. The molecular masses were 40 kDa for xylanase and 150 kDa for β-xylosidase. The xylanase was stable for 24 h at 60°C, whereas at 70°C about half of the initial activity remained after 24 h; at 80°C the half-life of the xylanase activity was 50 min. Heat treatment at 60°C for 1 h did not cause inhibition of the activity of β-xylosidase. The action of two enzymes on xylan gave only xylose.

'*Bacillus thermoantarcticus*' also possessed a D-xylose isomerase (D-xylose ketol isomerase, EC 5.3.1.5). The exoenzymes (xylanase and β-xylosidase) degraded the polymer to D-xylose, which was transported into the cell, isomerized to D-xylulose and phosphorylated to xylulose 5-phosphate, which then entered either the pentose phosphate pathway or the phosphoketolase pathway. Owing to the industrial significance of the enzyme, xylose isomerases from various microorganisms have been studied, and their catalytic and physicochemical properties have been reviewed. In fact, this enzyme has been used commercially because of its capacity to produce a fructose-enriched syrup (HFCS) by

Table 5.3 Physicochemical properties of biotechnological enzymes in thermophilic *Bacillus*.

Enzyme	'B. thermoantarcticus' DSM 9572	B. thermoglucosidasius KP 1006 DSM 2542	B. stearothermophilus DSM 21
Xylanase	+	n.d.	+
Optimum T (°C)	80		75
Optimum pH	5.6		6.5–7.0
M.W.	40 kDa		40 kDa
β-Xylosidase	+	n.d.	+
Optimum T (°C)	70		70
Optimum pH	6.0		6.0
M.W.	150 kDa		150 kDa
Xylose isomerase	+	n.d.	+
Optimum T (°C)	90		80
Optimum pH	7.0		7.5–8.0
M.W.	200 kDa		130 kDa
α-Glucosidase	+	+	+
Optimum T (°C)	70	75	70
Optimum pH	5.6	5.0–6.0	6.3
M.W.	67 kDa	60 kDa	47 kDa
Linkage-bond	α-1,6	α-1,6	α-1,4

M.W., molecular weight; n.d., not determined; T, temperature.

converting D-glucose into fructose. D-Xylose isomerase is also of industrial interest for the fermentation of hemicelluloses to ethanol. The production of ethanol
from D-xylose can be accomplished by using yeast and commercially available
xylose isomerase in a simultaneous isomerization and fermentation process
(Bhosale *et al.* 1996).

The xylose isomerase from '*B. thermoantarcticus*' was purified 73-fold to
homogeneity and its biochemical properties were determined (table 5.3). It was a
homotetramer with a native molecular mass of 200 kDa and a subunit molecular
mass of 47 kDa, with an isoelectric point at 4.8. The enzyme had a Km of 33 mM
for xylose and also accepted D-glucose as substrate (Km 167 mM). An Arrhenius
plot of the enzyme activity of xylose isomerase was linear up to a temperature of
85°C. Its optimum pH was around 7.0, showing 80% of its maximum activity at
pH 6.0. Like other xylose isomerases, this enzyme required divalent cations for its
activity and thermal stability. Mn^{2+}, Co^{2+} or Mg^{2+} were of comparable efficiency
for enzyme action, while Mg^{2+} was necessary for glucose isomerase reaction (Lama
et al. 1999). However, no significant differences in pH and temperature behaviour
could be observed when D-xylose was compared with D-glucose as substrate.

'*Bacillus thermoantarcticus*' can synthesize an extracellular, thermostable
α-glucosidase capable of causing the rapid hydrolysis of *p*-nitrophenyl-α-D-
glucopyranoside. This enzyme, such as that of *B. thermoglucosidasius,* can split
the nonreducing terminal α-1,6-glucosidic linkages involved in panose and
isomalto-saccharides (Suzuki *et al.* 1979) (table 5.3).

Exopolysaccharide production

Little information about polysaccharide production by thermophiles has been reported in the literature and until recently it was not clear whether or not these organisms were likely to prove useful sources for polymer production. Therefore, a wide search for bacterial strains able to produce good yields of new polysaccharides with potentially useful properties has been undertaken, which also involved thermophiles because of their well-known capability for biotechnological applications.

'*Bacillus thermoantarcticus*' produced two different sulphated polysaccharides named EPS1 and EPS2. Their production was carried out in 3-L fermenters with an aeration flux of 20 mL/min. The yield reached 400 mg/L of polymer in the presence of mannose as the carbon source. The production of EPSs increased with increasing cell density, reached a maximum at the beginning of the stationary phase and the EPS content was proportional to total biomass. On a weight basis, EPS1 and EPS2 represented about 27 and 71% (respectively) of the total carbohydrate fraction. Analysis of hydrolysis products revealed the presence of a terminal glucose in EPS1; the chain sugars were 1,2-linked mannose, 1,4-linked glucose, 1,3-linked mannose and 1,6-linked mannose, while 1,3,4-linked glucose and 1,2,6-linked mannose represented branch points in the molecule; in the relative proportion of 0,9 : 0,5 : 0,1 : 0,2 : 0,5 : 0,1 : 1,0 (respectively). The same analysis revealed for EPS2, the presence of a terminal mannose; the chain sugars were 1,2-linked mannose, 1,4-linked mannose and 1,6-linked mannose, and the branch point was 1,2,6-linked mannose. Their relative proportions were 1,0 : 0,5 : 0,2 : 0,1 : 0,8, respectively. EPS2 showed a molecular weight of 3.0×10^5 Da and an optical rotation $[\alpha]_D^{20} = -90°$. Sulphate group and pyruvate presence were also detected.

The absolute configurations of biopolymer hexoses were shown to be D when analysed as their respective acetylated (+)-2-butyl glycosides from methylation analysis and NMR spectra.

^{1}H and ^{13}C NMR spectroscopy were performed on both EPSs and showed that EPS1 was a heteropolysaccharide whose repeating unit consisted of four different α-D-mannoses and three different β-D-glucoses. This structure was closely related to xanthan polymers.

EPS2 was a mannan with four different α-D-mannoses and a trace of pyruvic acid as the repeating unit.

Studies of the chemical structure of these molecules, substituent identification and physical properties are essential for understanding their possible application. Increasing attention is being paid to bio-polymers because of their bio-active role and their wide range of commercial applications, which include uses as food additives (xanthan, alginate, dextran, glucomannan) and in nonfood applications such as viscosity control, gelation and flocculation (agar, glucan, mannan) (Nicolaus *et al.* 1996).

It has been hypothesized that the synthesis of exocellular polysaccharides in microorganisms plays a major role in protecting cells from stress in extreme habitats (De Philippis & Vincenzini 1998). Therefore, the production of exopolysaccharides by '*B. thermoantarcticus*' might serve as a boundary between the bacterial cell and its immediate environment.

Lipids

The lipid analysis of microorganisms is an important chemotaxonomic approach for classification and identification. In fact, the structure of the membrane lipids permitted us to discriminate rapidly between the domains *Bacteria* and *Archaea*, and very often allowed the classification of a new isolate at the genus level.

The Antarctic thermophiles and thermoacidophiles considered here had membrane lipids based on ester derivatives of 1,2-*sn*-glycerol, and therefore they belonged to the *Bacteria* domain (Nicolaus *et al.* 1996).

More attention was usually given to the analysis of fatty acid composition, rather than the polar lipid pattern, both with respect to the organisms' chemotaxonomies and their responses to stressful environmental conditions. These microorganisms synthesize their lipids rapidly and, therefore, must respond to a changing environment with equal rapidity, altering their membrane lipid compositions in order to ensure the appropriate membrane fluidity and their survival in any growth conditions.

The pattern of polar lipids of the seven strains of thermophiles and of four strains of thermoacidophiles (table 5.2), as well as their fatty acid compositions, were analysed. The seven strains of thermophiles showed similar polar lipid patterns to those of *Bacillus* species, with the sole exception that in five strains glycolipids were not detected. This last lipid type was present in isolates M2 and M4, as minor components. The absence of glycolipids, reported only for few species of the genus *Bacillus* (Nicolaus *et al.* 1995), could be considered as a chemotaxonomic marker for Antarctic isolates. One major phosphoaminolipid, three major closely migrating phospholipids and one phosphoglycolipid were the main components of the Antarctic thermophiles.

Fatty acid methyl ester (FAME) composition was analysed by GLC and mass spectrometry after methanolysis of polar lipids. Branched FAMEs were the predominant acyl chains of lipids of all of the Antarctic isolates (figure 5.4a). The *iso*-family ranged between 56% in isolate M1 and 82% in isolate M7; in contrast, the straight-chain FAMEs were from 4% up to 10% in M6. The isolate M1 and M2 were the richest in *anteiso*-family components (*c.* 38%), while these FAMEs were only 12% in M7 strain (figure 5.4a, b). The FAMEs were a combination of *iso*-C15, *iso*-C16, *iso*-C17 and *anteiso*-C17 as major components, while *anteiso*-C15 and *anteiso*-C16 were present in smaller amounts. The fatty acid profile found in the Antarctic thermophiles was consistent with that reported for the genus *Bacillus*, but it should be noted that in Antarctic strains the branched C17 fatty acids were present in very high amounts with respect to the content usually found in other *Bacillus* representatives (Kämpfer 1994). Detailed analysis of fatty acids in the seven strains of Antarctic thermophiles showed that all strains possessed high amounts of *iso*-C17, suggesting a possible relationship of Antarctic strains with *B. thermoglucosidasius*, while the high content of *anteiso*-C17 in some Antarctic isolates was unusual (Kämpfer 1994).

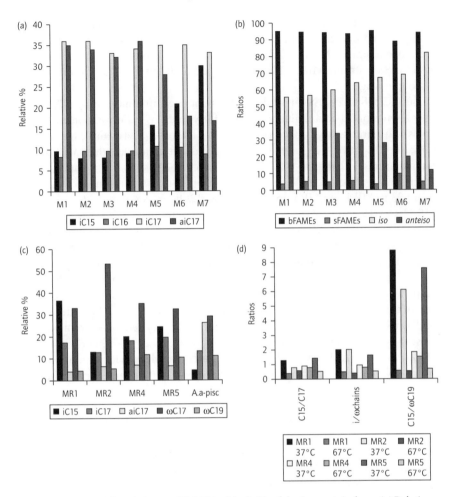

Figure 5.4 Fatty acid methyl esters (FAMEs) of the lipids of the Antarctic isolates. (a) Relative percentage of major FAMEs of the thermophilic strains (M1–M7). (b) Relative percentage of the of *iso*, *anteiso*, straight and branched FAMEs of thermophilic Antarctic isolates (bFAMEs, branched fatty acids; sFAMEs, straight fatty acids). (c) Relative percentage of the major FAMEs of alicyclobacilli from Antarctica and *Alicyclobacillus acidocaldarius* Pisciarelli strain (A.a-pisc). (d) Ratios of C15 and C17 FAMEs, *iso* and ω FAMEs, C15 and ω-C19 FAMEs in Antarctic alicyclobacilli (MR1, MR2, MR4, MR5) grown at different temperatures. ai, *anteiso*; i, *iso*.

The analysis of polar lipid pattern and fatty acid composition was useful in a rapid assignment of the thermoacidophilic isolates from Mount Rittmann to the genus *Alicyclobacillus*. In fact, all isolates had sulfonoglycosyldiacylglycerol as one of the major polar lipids, quinones of MK-7 type, polar lipids based on hopanoids, and fatty acids with a terminal cyclohexane; the simultaneous presence of all these compounds is a specific chemotaxonomic marker for the genus *Alicyclobacillus* (Wisotzkey *et al.* 1992).

The pattern of polar lipids appeared complex and similar to that reported for *Alicyclobacillus* species (Nicolaus *et al.* 1998). Two major glycolipids with higher chromatographic mobility, a sulfonoglycosyldiacylglycerol, two phospholipids and one glycolipid with slower Rf, were the main components of all thermoacidophiles (MR1, MR2, MR4, MR5) from Mount Rittmann (Nicolaus *et al.* 1998). Strains MR2 and MR4 showed additional spots positive for the glycolipid reagent.

The GC/MS and the NMR of the FAMEs obtained after methanolysis, showed the presence of *iso*-acyl chains and ω-acyl fatty acids, as main components of the lipids of Antarctic alicyclobacilli (figure 5.4c). The other acyl chains were *anteiso*-C15, *iso*-C16 and *n*-C16, and in all strains these were each less than 3.5% of the total mixtures. In MR1 *iso*-C15 and ω-C17 occurred in similar amounts, while in other strains, although in a different proportion, ω-C17 was the main component (from 33 in MR4 and MR5 up to 53% in MR2, figure 5.4c). In MR2, MR4 and MR5, *iso*-C15 and *iso*-C17 occurred in similar amounts; ω-C19 was at a higher level, above 10% in MR4 and MR5, than in MR1 and MR2 (figure 5.4c). It is noteworthy that in Antarctic isolates the *iso*-family FAMEs occurred in quite similar amounts to ω-cyclohexyl FAMEs, in contrast to that reported for the related strain, *A. acidocaldarius* Pisciarelli, in which the more abundant branched FAMEs were those of the *anteiso*-series instead of the *iso*-series (figure 4c). In this last microorganism *n*-C16 was up to 6% of total FAMEs and stearic acid was also present. In Antarctic isolates, these acyl chains were absent or present only as a minor component (Nicolaus *et al.* 1998). Finally, in isolate MR2, two new FAMEs were identified, as 2-methylcyclohexylundecanoic and ω-cyclohexylnonanoic acids, not previously reported for the genus *Alicyclobacillus*.

Lipid modulation by growth temperature

Significant differences in the lipid pattern of the thermophiles (M1–M7) and the thermoacidophiles (MR1, MR2, MR4, MR5) were observed under stress conditions at temperatures lower and higher than the optimum growth temperature. In fact, the relative proportions of different polar lipids in the Antarctic isolates were affected by temperatures. In the isolates belonging to the genus *Bacillus* (M1–M7), the phosphoglycolipid (PG) content increased with increasing growth temperature, at the expense of phosphoaminolipid and phospholipids. In the Antarctic alicyclobacilli, the phospho-positive compounds increased with increasing growth temperature, paralleled by the decrease of a glyco-positive compound.

In both thermophiles and thermoacidophiles, the highest growth temperatures favoured the biosynthesis of higher-melting point acyl chains such as *iso*-C17 for M1–M7 isolates and ω-cyclohexyl fatty acids for alicyclobacilli from Mount Rittmann.

In thermophiles, the *iso*-C15/*iso*-C17 ratio showed that the *iso*-C17 were preferred at the maximum growth temperatures and *iso*-C15 at the minimum one. C18 fatty acids were found only at higher growth temperatures. Finally, the unsaturated FAMEs were found only at the lowest growth temperature.

In isolates MR1 and MR2 the *iso*-chains both as *iso*-C15 and *iso*-C17 decreased with increased growth temperature, this decrease was paralleled by a similar increase of the ω-family, with a major increase of the C19 chain. In MR4 and in MR5 the behaviour was a little different: the *iso*-C17 remained constant with changing growth temperature, whereas *iso*-C15 increased from 37 to 50°C, but both strains decreased at 67°C. At increasing growth temperatures, the C15/C17 ratios decreased (although with different values in different strains), as did the *iso*/ω-chain ratios; the C15/C19 ratios greatly decreased in comparison with the other FAME ratios (figure 5.4d).

For *Alicyclobacillus* (formerly *Bacillus*) *acidocaldarius*, growing around an optimal pH (*c.* 4), the percentage of cyclohexyl fatty acids increases if the growth temperature is raised. Our results confirmed that the ω-family of fatty acids is of special physiological importance for cells exposed to high temperatures. Krischke and Poralla (1990) showed the importance of ω-cyclohexyl fatty acids for the growth and resistance to different chemical and physical agents.

The change in lipid compositions of Antarctic isolates, both polar and fatty acids, with growth temperature has been viewed as a compensatory or homovis- cous mechanism, often referred to as a homoviscous adaptation for maintaining a critical degree of fluidity of membranes at any temperature (Russell & Fukunaga 1990).

Ecological considerations

The biodiversities of ecosystems have become objects of intensive study, and con- sequently a wealth of information has been gathered regarding the distribution of microorganisms in the world. These studies are particularly valuable when attempting to construct a map of world distribution. The extremophiles in gen- eral, and the thermophiles in particular, are widely distributed in the world. They have been isolated in all continents and in different types of ecosystems. The aero- bic spore-forming thermophiles have been isolated frequently from environments where the temperature was not very high, yet they are able to grow at high tem- perature in the laboratory. Hudson *et al.* (1989) reported studies on the distri- bution of thermophilic microorganisms in Antarctica, but detailed taxonomic, genetic, physiological and biochemical properties were not evaluated.

In our studies we have described two new species and several related strains. The thermophiles belong to the genus *Bacillus* and the thermoacidophiles belong to the genus *Alicyclobacillus*. These isolations have extended our knowledge of the biodiversity of thermophiles in continental Antarctica. Although our isolates show several features different from the related organisms isolated in other ecosystems, it is noteworthy that phylogenetically and metabolically related representatives of thermophilic bacilli were found to occur in geograph- ically distant geothermal soils. The Antarctic isolates showed a wide range of growth temperatures, perhaps in order to tolerate the environment in which they lived; the soil temperatures were around 50°C while the external temperature was −30°C. In this respect it is not surprising that sporulating organisms were found in such hostile environments. As yet, there are no reports of the isolation and characterization of thermophilic members of the domain *Archaea* from Antarctica.

Acknowledgements

This research was supported partially by the Italian National Programme for Antarctic Research. We are grateful to Mr Enrico Esposito for the work carried out in Terra Nova Bay Station. We thank also R. Turco for drafting the original artwork.

References

Bhosale, S.H., Rao, M.B. & Deshpande, V. (1996) Molecular and industrial aspects of glucose isomerase. *Microbiological Reviews* 60, 280–300.

Danson, M.J., Hough, D.W. & Lunt, G.G. (eds) (1992) *The Archaebacteria: Biochemistry and Biotechnology.* Portland Press, London.

De Philippis, R. & Vincenzini, M. (1998) Exocellular polysaccharides from cyanobacteria and their possible applications. *FEMS Microbiology Reviews* 22, 151–175.

Goodfellow, M. & O'Donnell, A.G. (eds) (1994) *Chemical Methods in Prokaryotic Systematics.* Wiley, Chichester.

Hudson, J.A., Daniel, R.M. & Morgan, H.W. (1989) Acidophilic and thermophilic *Bacillus* strains from geothermally heated antarctic soil. *FEMS Microbiology Letters* 60, 279–282.

Kämpfer, P. (1994) Limits and possibilities of total fatty acids analysis for classification and identification of *Bacillus* species. *Systematic and Applied Microbiology* 17, 86–98.

Krischke, W. & Poralla, K. (1990) Properties of *Bacillus acidocaldarius* mutants deficient in ω-cyclohexyl fatty acid biosynthesis. *Archives of Microbiology* 153, 463–469.

Kristjansson, J.K. (ed.) (1991) *Thermophilic Bacteria.* CRC Press, Boca Raton, FL.

Lama, L., Nicolaus, B., Calandrelli, V., Esposito, E. & Gambacorta, A. (1996) Xylanase produced by *Bacillus thermoantarcticus,* a new thermophilic *Bacillus. Annals of the New York Academy of Sciences: Enzyme Engineering XIII* 799, 285–289.

Lama, L., Nicolaus, B., Calandrelli, V., Esposito, E. & Gambacorta, A. (1999) *Bacillus thermoantarcticus* producing enzymes for bioconversion of xylan (abstract). In: *FEBS' 99 Nice, 19–24 June* (ed. M. Grunberg-Manago), abstract 17/307, p. 356. Elsevier, Paris.

Logan, N. & Berkeley, R.C. (1981) Classification and identification of the genus *Bacillus* using API tests. In: *The Aerobic Endospore-forming Bacteria: Classification and Identification* (eds R.C.W. Berkeley & M. Goodfellow), pp. 106–140. Academic Press, New York.

Manca, M.C., Lama, L., Improta, R., Esposito, E., Gambacorta, A. & Nicolaus, B. (1996) Chemical composition of two exopolysaccharides from *Bacillus thermoantarcticus. Applied and Environmental Microbiology* 62, 3265–3269.

Mora, D., Fortina, M.G., Nicastro, G., Parini, C. & Manachini, P.L. (1998) Genotypic characterization of thermophilic bacilli: a study on new soil isolates and several reference strains. *Research in Microbiology* 149, 711–722.

Nanmori, T., Watanabe, T., Shinke, R., Kohno, A. & Kawamura, Y. (1990) Purification and properties of thermostable xylanase and β-xylosidase produced by a newly isolated *Bacillus stearothermophilus* strain. *Journal of Bacteriology* 172, 6669–6672.

Nicolaus, B., Manca, M.C., Lama, L., Esposito, E. & Gambacorta, A. (1995) Effects of growth temperature on the polar lipid pattern and fatty acid composition of seven thermophilic isolates from the Antarctic continent. *Systematic and Applied Microbiology* 18, 32–36.

Nicolaus, B., Lama, L., Manca, M.C., Esposito, E., di Prisco, G. & Gambacorta, A. (1996) 'Bacillus thermoantarcticus' sp. nov., from Mount Melbourne, Antarctica: a novel thermophilic species. *Polar Biology* 16, 101–104.

Nicolaus, B., Improta, R., Manca, M.C., Lama, L., Esposito, E. & Gambacorta, A. (1998) *Alicyclobacilli* from an unexplored geothermal soil in Antarctica: Mount Rittmann. *Polar Biology* 19, 133–141.

Nicolaus, B., Lama, L., Manca, M.C. & Gambacorta, A. (1999a) Extremophiles: polysaccharides and enzymes degrading polysaccharides. *Recent Research Develepments in Biotechnology and Bioengineering* 2, 37–64.

Nicolaus, B., Lama, L., Esposito, E. *et al.* (1999b) *Haloarcula* spp able to biosynthesize exo- and endopolymers. *Journal of Industrial Microbiology and Biotechnology* 23, 489–486.

Russell, N.J. & Fukunaga, N. (1990) A comparison of thermal adaptation of membrane lipids in psycrophilic and thermophilic bacteria. *FEMS Microbiology Reviews* 75, 171–182.

Sunna, A., Tokajlan, S., Burghardt, J., Rainey, F., Antranikian, G. & Hashwa, F. (1997) Identification of *Bacillus kaustophilus, Bacillus thermocatenulatus* and *Bacillus* strain HSR as members of *Bacillus thermoleovorans*. *Systematic and Applied Microbiology* 20, 232–237.

Suzuki, Y., Ueda, Y., Nakamura, N. & Abe, S. (1979) Hydrolysis of low molecular weight isomaltosaccharides by a *p*-nitrophenyl-α-D-glucopyranoside-hydrolyzing α-glucosidase from a thermophile, *Bacillus thermoglucosidius* KP 1006. *Biochimica Biophysica Acta* 566, 62–66.

Vieille, C., Burdette, D.S. & Zeikus, J.G. (1996) Thermozymes. In: *Biotechnology Annual Review* (ed. M.R.E. Gewely), Vol. 2, p. 1. Elsevier, New York.

Wisotzkey, J.D., Jurtshuk, P.I.R., Fox, G.E., Deinhard, G. & Poralla, K. (1992) Comparative sequence analyses on the 16S tRNA (rDNA) of *Bacillus acidocaldarius, Bacillus acidoterrestris,* and *Bacillus cycloheptanicus* and proposal for creation of a new genus, *Alicyclobacillus* gen. nov. *International Journal of Systematic Bacteriology* 42, 263–269.

Chapter 6

Bacilli Associated with Spoilage in Dairy Products and Other Food

M. Heyndrickx and P. Scheldeman

Introduction

Aerobic spore-forming bacteria belonging to the genus *Bacillus* and allied genera are important in the food industry, and there are several reasons for this (Andersson *et al.* 1995). First, spores of bacilli are ubiquitous, which makes it practically impossible to prevent their presence in raw food and ingredients. Post-treatment contamination can occur easily when deficiencies occur in the filling apparatus or in the sterilization of the packaging material. Secondly, common heat treatments, such as pasteurization, are adequate to kill most vegetative cells but insufficient to kill spores. Surviving spores in heat-treated food products have no or little competition from vegetative, faster-growing Gram-negative bacteria. Thirdly, several spores have adhesive characteristics, which facilitate their attachment to the surfaces of pipelines and processing equipment, with the formation of bio-films. Finally, and probably of more recent concern, is the (increasing) tolerance or resistance that spores or vegetative cells may show to conditions or treatments generally presumed to stop growth (low temperatures and low pH), or to inactivate all living material, such as ultra heat treatment (UHT) and commercial sterilization. Increasingly, the food industry seems to be confronted with tolerant or resistant bacilli that might be side-effects of the use of new ingredients and the application of new processing and packaging technologies. These result in the selection of adapted or new organisms with intrinsic tolerance or resistance attributes.

Bacilli present two kinds of problems to the food industry. First, there is the pathogen *Bacillus cereus*, which may cause food-borne intoxications (discussed in Chapter 4). Secondly, there is the reduction of shelf-life, or food spoilage during production, storage and distribution (Huis in 't Veld 1996). The most important *Bacillus* species causing food spoilage (or other problems) are listed in table 6.1. Food spoilage can be considered as any change that renders a product unacceptable for human consumption. Although there are no exact figures for total economic losses resulting from food spoilage, it is estimated that it constitutes enormous financial losses despite modern food technology and preservation techniques. Microbial spoilage of food is usually indicated by changes in texture or the development of off-flavours. Food defects, however, can also simply be unwanted microbial growth in commercially sterile products.

The *Bacillus* species responsible for the major problems in the last decade, with a special emphasis on the dairy sector, will be discussed here in detail, including

Table 6.1 Major *Bacillus* species causing spoilage or other problems in food.

Species	Foods implicated	Possible spoilage or problem	Further reading
Bacillus cereus (*B. weihenstephanensis*), *B. mycoides*	Pasteurized milk and cream	Bitty cream, sweet curdling, off-flavours	Meer *et al.* (1991)
Bacillus stearothermophilus, *B. coagulans*, *B. licheniformis*	Evaporated milk, low-acid canned vegetables, tomato juice	Flat-sour, increasing risk of botulinic intoxication by elevating pH	Kalogridou-Vassiliadou (1992), Montville (1982)
Bacillus subtilis	Bread	Ropy bread (unpleasant odour, discoloured, sticky and soft bread crumb)	Rosenkvist & Hansen (1995)
Bacillus sporothermodurans	UHT and sterilized milk	Nonsterility	Hammer *et al.* (1995)
Alicyclobacillus acidoterrestris	Fruit juice and beverages	Off-flavours	Jensen (1999)
Paenibacillus larvae subsp. *larvae*	Honey	Spread of American foulbrood in honeybee larvae	Govan *et al.* (1999)

the current taxonomy, identification and typing methods. These methods are important because of the increasing need in the food industry for the establishment of an early warning system and insight into the contamination routes of food-spoiling *Bacillus* species.

Alicyclobacillus problems in the fruit juice industry

Until recently, it was believed that the high acidity of fruit juices (pH <4), combined with a hot-fill and hold process in the fruit beverage industry, was sufficient to render these products commercially sterile. However, a large-scale spoilage of shelf-stable apple juice by unusual bacilli has been reported in Germany (Cerny *et al.* 1984), and since then the focus has centred on *Alicyclobacillus acidoterrestris*, a species commonly isolated from soil. In the 1990s, the episodic spoilage of fruit juice and similar products by this organism was reported in Europe, the USA and Japan (Jensen 1999). Spoilage is associated with an increase in turbidity, white sediment at the bottom of packages and strong, medicinal flavours or taints caused by the production of guaiacol (2-methoxyphenol) at levels up to 100 ppb (taste threshold 2 ppb) (Pettipher *et al.* 1997) and 2,6 di-bromo-phenol at levels of several ppt (taste threshold 0.5 ppt) (Borlinghaus & Engel 1997). Affected products are pasteurized fruit juices (e.g. apple, orange) and hot-filled fruit juice blends. Good growth was also observed in tomato juice, white grape juice and various blends of juices (Splittstoesser *et al.* 1994). Although spoilage does not occur very often, as it requires a combination of adequate conditions such as available oxygen, high temperature and taint precursors, most fruit juice manufacturers have started quality assurance programmes to monitor and reduce levels of alicyclobacilli in raw materials. Insufficiently washed raw fruit (contaminated with soil) and poor hygiene measures in fruit juice concentrate production are the most likely contamination sources. *Alicyclobacillus acidoterrestris* has also been isolated from water used in a fruit processing plant (McIntyre *et al.* 1995) and alicyclobacilli have been found in liquid sugar (Splittstoesser *et al.* 1998; M. Heyndrickx, unpubl. obs.).

Alicyclobacillus acidoterrestris survives pasteurization, or hot-fill and hold processes (*c.* 2 min at 88–96°C), and grows in the product because of the heat resistance of the spores (*D*-value 65.6 min at 85°C and 11.9 min at 91°C in orange juice; Silva *et al.* 1999; see table 6.3 for explanation of *D*) and its physiological characteristics (growth from pH 2.5 to 6.0 with an optimum around pH 3.5–5.0 and from 20 to 60°C with an optimum of 42–53°C). It is believed that the presence of ω-alicyclic fatty acids as major lipids in the cell membrane, a unique phenotype amongst bacilli (Wisotzkey *et al.* 1992), contributes to the thermal resistance of the vegetative cells, and that this might explain their survival in lab-pasteurization conditions (M. Heyndrickx, unpubl. results). Spore numbers in raw materials and processed products may be very low (<1/100 g), requiring sensitive detection procedures, including heat shock to induce germination and enrichment or filtration to test larger volumes (Jensen 1999). Identification of *A. acidoterrestris* by biochemical tests (including API 50 CHB) seems unsatisfactory (Eiroa *et al.* 1999). With 16S rDNA-based specific primers, *A. acidoterrestris* can be detected by RT-PCR with high sensitivity after enrichment (Yamazaki *et al.* 1996).

Bacillus problems in the dairy sector

Bacillus flora of raw and processed milk

The dairy industry is one of the few where the *Bacillus* flora has been determined in several steps of the production chain. This is not surprising, because the production line is selecting for spore-formers through several heating processes and for psychrophiles during cold storage. The heating processes used in the dairy industry, to increase shelf-life and guarantee microbiological safety, are thermization (57–68°C for 15 s), pasteurization (a minimum of 71.7°C for 15 s), UHT (a minimum of 135°C for 1 s) and sterilization (110–120°C for 10–20 min). In table 6.2, the compositions of the psychrophilic or psychrotolerant, mesophilic and/or thermophilic *Bacillus* floras of milk at several stages of processing are summarized. Despite some regional, seasonal, sampling and methodological differences, general trends can be observed. *Bacillus licheniformis*, *B. pumilus* and *B. subtilis* are the predominant mesophilic species (Phillips & Griffiths 1986; Sutherland & Murdoch 1994; Tatzel *et al.* 1994), with *B. licheniformis* far outnumbering the other species. *Bacillus cereus*, *B. circulans* and *B. mycoides* are the predominant psychrophilic or (better) psychrotolerant species (Phillips &

Table 6.2 Psychrophilic or psychrotolerant, mesophilic and/or thermophilic *Bacillus* composition[a] of milk classified according to decreasing predominance in the samples investigated (lab-pasteurized for 10 min at 80°C, unless otherwise indicated). When monthly samples were taken, the mean value over the whole period investigated was used for the classification.

Phillips & Griffiths (1986) Farm bulk tank & creamery silo raw milk[b,e]	Griffiths & Phillips (1990) Farm bulk tank raw milk[c,e]	Griffiths & Phillips (1990) Retail packs pasteurized milk[c,e]	Crielly et al. (1994) Farm bulk tank raw milk[d,e,g] nonheated	Sutherland & Murdoch (1994) Farm, tanker & dairy raw milk & pasteurized milk[d,e]	Tatzel et al. (1994) Dairy plant milk & cream[b,e]	Cosentino et al. (1997) Pasteurized consumer milk[b,e]	Cosentino et al. (1997) UHT consumer milk[b,e]	Scheldeman (unpubl. obs.) Farm bulk tank raw milk[b,f], heated 30 min at 100°C
B. licheniformis	B. cereus	B. cereus	B. cereus	B. pumilus	B. licheniformis	B. licheniformis	Br. laterosporus	B. licheniformis
B. stearothermophilus	B. circulans	B. mycoides	B. licheniformis	B. licheniformis	B. subtilis	B. coagulans	B. sphaericus	Thermophilic Bacillus sp.
B. cereus	B. mycoides	B. circulans	B. sphaericus	B. subtilis	B. pumilus	B. subtilis	Br. brevis	Paenibacillus spp.
B. circulans	B. thuringiensis	B. lentus	B. firmus	B. lentus	Br. laterosporus	B. cereus	B. coagulans	B. thermosphaericus
B. pumilus	B. amyloliquefaciens	B. thuringiensis	B. circulans	V. pantothenticus	B. circulans	B. sphaericus	B. licheniformis	Brevibacillus spp.
B. subtilis	B. lentus	Br. brevis	B. coagulans	B. megaterium	V. pantothenticus	B. pumilus	B. pumilus	Br. agri
B. lentus	B. pasteurii / sphaericus	'B. carotarum'	B. subtilis	B. amyloliquefaciens	P. polymyxa	P. macerans		B. smithii
B. mycoides	P. polymyxa	B. pumilus		B. mycoides	B. badius	B. circulans		Br. borstelensis
Br. brevis		B. firmus		B. stearothermophilus	Br. brevis	Br. laterosporus		P. macerans
B. thuringiensis		P. polymyxa		B. cereus	P. alvei	B. lentus		B. subtilis / amyloliquefaciens
B. pasteurii		B. stearothermophilus		B. circulans	B. sphaericus	B. megaterium		V. pantothenticus
B. coagulans				B. firmus	B. stearothermophilus	B. mycoides		B. megaterium
B. sphaericus				B. sphaericus	B. cereus	B. stearothermophilus		A. thermoaerophilus
'B. carotarum'				P. polymyxa	B. firmus			V. proomii
B. amyloliquefaciens				P. macerans	P. macerans			B. sporothermodurans
P. polymyxa				Br. laterosporus	B. megaterium			
B. firmus					B. insolitus			
Br. laterosporus								
P. alvei								

[a] Genus abbreviations: A., Aneurinibacillus; B., Bacillus; Br., Brevibacillus; P., Paenibacillus; V., Virgibacillus.

[b] Psychrophiles or psychrotolerant organisms, mesophiles and thermophiles.

[c] Psychrophiles or psychrotolerant organisms.

[d] Mesophiles.

[e] Identification by biochemical tests (mostly API 50CHB).

[f] Identification by gas chromatographic analysis of cellular fatty acids and amplified ribosomal DNA restriction analysis.

[g] After pasteurization, B. licheniformis, B. cereus, B. firmus and B. subtilis were found in decreasing order of predominance.

Griffiths 1986; Griffiths & Phillips 1990), with *B. cereus* being the most common (Sutherland & Murdoch 1994). *Bacillus stearothermophilus* and thermotolerant isolates of *B. licheniformis* are the most common thermophiles or thermotolerant species (Phillips & Griffiths 1986). A lot of strains remained unidentified [e.g. 1.6% (Phillips & Griffiths 1986) up to 48% (Sutherland & Murdoch 1994)], because of the limitations of biochemical identification systems and the presence of as yet unknown *Bacillus* species (see below).

Seasonal variations in total aerobic spore counts and in spore counts of individual species have been observed. The monthly average of the mesophilic aerobic spore count is usually in the range of 10–10^2 colony forming units (cfu)/mL of raw milk (Waes 1976; Phillips & Griffiths 1986; Crielly *et al.* 1994; Slaghuis *et al.* 1997). A higher incidence of mesophilic isolates (Sutherland & Murdoch 1994) and of *B. licheniformis* in particular has been observed in the winter, while *B. cereus* tended to be more common during the summer (Crielly *et al.* 1994). The incidence of psychrotolerant spore-formers (average of 460 cfu/L of raw milk; Griffiths & Phillips 1990) and of *B. cereus* in particular seems to be highest in the summer/autumn (Phillips & Griffiths 1986; Sutherland & Murdoch 1994). The level of thermophilic spore-formers (Phillips & Griffiths 1986) and of thermoresistant spores isolated after a heat treatment of 30 min at 100°C (average of 0.68 cfu/mL; Waes 1976) is higher in the winter. The identity of the thermoresistant spore-formers isolated from raw milk in the winter of 1998–9 has been determined (table 6.2) (P. Scheldeman, unpubl. obs.).

Pasteurized milk and other heat-treated milk products

Spoilage caused by psychrotolerant bacilli

The term psychrotolerant microorganisms is now preferred to psychrotrophs – meaning 'cold eaters' – although the latter term is still used by the International Dairy Federation (IDF), where it is defined as microorganisms that can grow at 7°C or less, irrespective of their optimal growth temperatures (Collins 1981). The importance of the psychrotolerant bacilli for the keeping quality of milk has increased significantly. This is a consequence of extended refrigerated storage of raw milk before processing on the farm and in the dairy, higher pasteurization temperatures, reduction of post-pasteurization contamination by principally Gram-negative organisms such as *Pseudomonas* spp. (excluding competition) and prolonged shelf-life requirements of the consumer product (Meer *et al.* 1991). It should be remembered that pasteurization activates germination of spores and thus enhances vegetative cell growth. A significantly higher mean count of psychrotolerant *B. cereus* was found in cream at the end of its shelf-life, and this was attributed to more efficient spore activation by the higher pasteurization temperature of double cream (Larsen & Jørgensen 1997). It has been estimated that 25% of all shelf-life problems in pasteurized milk and cream in the USA are caused by psychrotolerant spore-formers (Sørhaug & Stepaniak 1997). The shelf-life can be determined as the average time in days for the psychrotolerant organisms to reach 1×10^7 cfu/mL during storage at 6°C.

The defects caused by psychrotolerant bacilli are quite similar to those of other spoilage bacteria, namely off-flavours and structural defects caused by proteolytic, lipolytic and/or phospholipase enzymes (Meer *et al.* 1991). The off-flavours can be bitter, putrid, unclean, stale, rancid, fruity, yeasty or sour (Washam *et al.* 1977). Bitter flavour is caused by protease activity on the milk proteins, while rancid and fruity flavours are caused by lipases. *Brevibacillus laterosporus* and *Bacillus coagulans* are associated with fruity flavours, often in combination with an unclean characteristic, while *Paenibacillus polymyxa* causes a sour, yeasty and gassy defect in milk. The latter organism was found as the main Gram-positive spoilage organism (accounting for 38% of the milk spoilage) of Swedish and Norwegian pasteurized milk stored at 5 or 7°C (Ternström *et al.* 1993). Two other possible problems associated with this and other milk *Bacillus* species for the cheese industry, are fermentative growth with gas production and vigorous denitrification diminishing the anticlostridial effect of added nitrite.

The structural defects are sweet curdling in fluid milk and bitty cream, mainly caused by pyschrotolerant *B. cereus* strains, although sweet curdling may also be caused by *Br. laterosporus* or *B. subtilis* strains. Bitty or broken cream in non-homogenized milk is characterized by fat destabilization (flocs) when cream is added to a hot beverage and is caused by phospholipase C or lecithinase of *B. cereus*. A higher pasteurization temperature increases the germination rate of *B. cereus* and accelerates the formation of flocs (Van Heddeghem & Vlaemynck 1992). Sweet curdling of milk is caused by proteinases. The first sign is the appearance of small buttons or 'pellicles' on the bottom of the container, which may be unnoticed by the consumer. Upon further storage (especially at >7°C), a curd formation may appear over the entire bottom surface of the container. This defect is rather common and affects more containers upon prolonged refrigerated storage.

Yoghurt can be defective (lumpy structure, separation of whey, off-flavours) in exceptional cases where *B. cereus* develops to 10^6 cfu/mL before the pH is reduced to 6.1 by the starter bacteria. In custards, porridges and desserts, which are semi-solid and made with a longer period at higher temperature, more thermoresistant types such as *B. licheniformis*, *B. circulans* and sometimes *B. coagulans* and *P. polymyxa* may develop anaerobically (Stadhouders & Driessen 1992).

Sources of *Bacillus cereus* contamination

Bacillus cereus is the major concern of the dairy industry because it seems impossible completely to avoid its presence in milk, leading to potential threats both to the quality and safety of the pasteurized product. This is largely a problem for countries with a high consumption of pasteurized milk, such as the Netherlands and the Scandinavian countries, whereas in Belgium most consumer milk is UHT or sterilized. In Sweden, there is a legal limit of 1000 *B. cereus*/mL pasteurized milk after storage at 8°C until the 'best before' date (Christiansson *et al.* 1999). Fifty-six percent of samples of Danish pasteurized milk cartons contained *B. cereus* and both prevalence and counts per mL (after storage at 7°C for 8 days) were significantly higher in the summer than in the winter (Larsen & Jørgensen 1997). In Canada, more than 90% of pasteurized final products contained *B. cereus*, with an average count of 5.5×10^6 cfu/mL after storage at 8°C for 14 days

(Lin *et al.* 1998). Comparison of cellular fatty acid compositions of isolates from the whole milk processing line suggested that spores in raw milk were a major source of *B. cereus* contamination in pasteurized milk, but that post-pasteurization contamination also occurred. Recent evidence has been obtained for the raw milk route of contamination using sensitive detection methods and/or molecular typing. A higher incidence of *B. cereus* in raw milk was observed when cows were at pasture (23%) compared to cows housed during summer (4%), despite a higher median spore count in the latter situation (Slaghuis *et al.* 1997). Soil (containing up to 380 000 *B. cereus* spores/g), and to some extent also dung, was found to be the major contamination source of milk (mean of 41 spores/L) via dirty teats, and this was influenced by weather conditions (Christiansson *et al.* 1999). On the other hand, milking equipment, feed and air were of minor importance. te Giffel *et al.* (1995) came to similar conclusions concerning farms in the Netherlands, with higher levels of *B. cereus* spores in faeces (up to 10^4/g). In both studies, the percentage of psychrotolerant strains on the total number of *B. cereus* isolates was comparable (e.g. 30–55% in milk, 60% in soil and faeces, 0% in concentrate).

Besides the raw milk route on the farm, evidence has been obtained of contamination at the dairy processing plant by an increased incidence of *B. cereus* in pasteurized milk (71% of the samples) compared to raw milk in farm and dairy storage tanks (35% of the samples) (te Giffel *et al.* 1996). Evidence for post-pasteurization contamination was the detection of new biotypes following the pasteurization process. It has been shown that *B. cereus* spores can adhere to, germinate and multiply on the stainless steel surfaces of heat exchangers (up to $2600/cm^2$) (te Giffel *et al.* 1997b). The high adhesion capacity of *B. cereus* spores compared to other species on hydrophobic surfaces such as stainless steel is caused by the high hydrophobicity and the extending hair-like structures of the spore surface (Rönner *et al.* 1990).

Recent taxonomy of the *Bacillus cereus* group

The classical *B. cereus* group consists of four valid species: *B. cereus*, *B. mycoides*, *B. thuringiensis* and *B. anthracis*. These organisms share a high degree of DNA–DNA homology (Kaneko *et al.* 1978) and sequence similarity of the 16S (>99% similarity) and 23S rDNA (Ash *et al.* 1991; Ash & Collins 1992) and of the 16S–23S ribosomal spacer region (Bourque *et al.* 1995). Although some authors (e.g. Priest *et al.* 1988) have proposed that these organisms be merged (eventually as subspecies) into a single species, there are still strong economic and medical reasons for maintaining the four species. In Chapter 3, the results of the application of new molecular tools are presented which might resolve the taxonomic enigma of the *B. cereus* group in the near future. The *B. cereus* group can easily be distinguished from the other species of the morphological group 1 (Gordon *et al.* 1973) on the basis of physiological properties (Stadhouders 1992). The selective isolation and confirmation of *B. cereus* group isolates is based on some of these properties (ISO 1993), but atypical *B. cereus* strains reacting negatively in one or more of the confirmatory tests or non-*B. cereus* strains reacting positively in all these tests can and do occur in dairy samples (te Giffel *et al.* 1995, 1996). Alternatively, PCR tests based on primers from the sphingomyelinase gene (Hsieh *et al.* 1999) or a RAPD marker (Daffonchio *et al.* 1999) can be used for rapid identification or detection in food samples.

A few easily determined morphological and physiological characteristics have been used to distinguish the members of the classical *B. cereus* group (Stadhouders 1992; Drobniewski 1993) but some of them are disputable as taxonomic markers because of instability or transferability between the species. A more reliable differentiation is reported to be possible using modern but expensive apparatus requiring techniques such as cellular fatty acid analysis (von Wintzingerode *et al.* 1997), Fourier transform infrared spectroscopy (Beattie *et al.* 1998) and pyrolysis mass spectrometry (Helyer *et al.* 1997). Because DNA hybridization and PCR assays are more amenable to routine laboratories, putative specific DNA probes or primers have been designed based on the 16S rDNA for *B. cereus* and *B. thuringiensis* (te Giffel *et al.* 1997a) and for *B. mycoides* (von Wintzingerode *et al.* 1997), and based on the gyrase B subunit gene for *B. cereus*, *B. thuringiensis* and *B. anthracis* (Yamada *et al.* 1999) (see also legend to figure 6.1).

In 1998, two new species were added to the *B. cereus* group, *B. pseudomycoides* (Nakamura 1998) and *B. weihenstephanensis* (Lechner *et al.* 1998). The former organism can be distinguished from *B. mycoides* and *B. cereus* by differences in specific fatty acid levels. *Bacillus weihenstephanensis* comprises psychrotolerant *B. cereus* strains able to grow at 4°C and is genetically more closely related to (psychrotolerant) *B. mycoides* than to mesophilic strains of the species *B. cereus* and to *B. thuringiensis*. Consequently, some but not all psychrotolerant *B. cereus* strains responsible for pasteurized milk spoilage now have to be regarded as belonging to *B. weihenstephanensis* (Stenfors & Granum 2001). On the basis of signature sequences found in both the 16S rDNA and the major cold shock protein gene *csp*A, discriminatory primers (figure 6.1) were designed to distinguish psychrotolerant (i.e. belonging to *B. cereus*, *B. weihenstephanensis* or *B. mycoides*) from mesophilic *B. cereus* group strains (i.e. belonging to *B. cereus* or *B. thuringiensis*) (Francis *et al.* 1998; von Stetten *et al.* 1998). According to its 16S rDNA signatures, *B. pseudomycoides* also belongs to the latter (mesophilic) group (minimum growth temperature 15°C). In comparison with the slow conventional detection procedure based on monitoring growth at 6–7°C for up to 14 days, these PCR assays allow an early detection of psychrotolerant bacilli in dairy products. In the 16S rDNA-based assay, intermediate strains were also detected which carried both psychrotolerant and mesophilic signatures and a correlation was found between percentage of psychrotolerant signatures and growth at low temperatures (Prüß *et al.* 1999).

Ultra high heat (UHT) treated and sterilized milk

Spoilage caused by (highly) heat-resistant bacilli

UHT or sterilization combined with aseptic filling is intended to eliminate all viable microorganisms and should result in commercially sterile liquid milk products, having long shelf-lives without refrigeration. According to the Milk

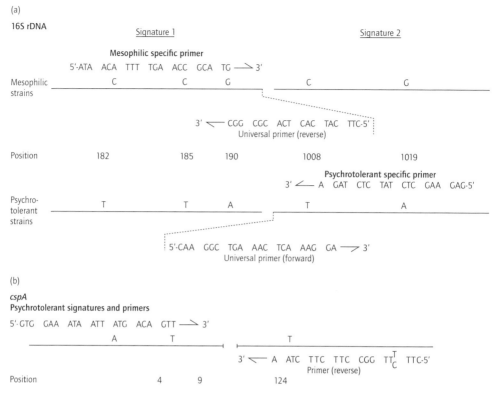

Figure 6.1 Discrimination between psychrotolerant and mesophilic strains of the *Bacillus cereus* group with specific primers based on signature sequences on on (a) the 16S rDNA (von Stetten *et al.* 1998) and (b) the major cold shock protein gene *csp*A (Francis *et al.* 1998). Positions are according to the *E. coli* sequence. On the 16S rDNA, some *B. thuringiensis* strains (a.o. type strain) carry a psychrotolerant signature (position 185) and *B. pseudomycoides* carries a mismatch (C) with the mesophilic primer (position 189). A *B. mycoides*-specific probe (von Wintzingerode *et al.* 1997) is based on the 16S rDNA psychrotolerant signature 1 and cross-reacts with *B. weihenstephanensis*. A *B. cereus*-specific probe (te Giffel *et al.* 1997), based on the V1 region (not shown) of 16S rDNA, cross-reacts with *B. mycoides*, *B. pseudomycoides* and *B. weihenstephanensis*.

Hygiene EC directive 92/46, the colony count at 30°C after incubation of unopened packages for 15 days at 30°C should be <10 cfu/0.1 mL. In 30% of fresh Sardinian UHT milk samples, *Bacillus* spores (species as shown in table 6.1) were found below this level (Cosentino *et al.* 1997). Spoilage of UHT and sterilized milk occurs only occasionally as a result of recontamination during filling and is caused mostly by proteolytic activity of some *Bacillus* species (Westhoff & Dougherty 1981; Foschino *et al.* 1990). However, more massive occurrences of contamination with mesophilic spore-forming bacteria were first reported in 1985 in Italy and Austria and in 1990 in Germany. It was soon obvious that this was a problem of survival of the UHT process by spores of a new mesophilic, highly heat-resistant or heat-resistant spore-former (HHRS or HRS), because the organism could be isolated from a bypass directly after the heating section of an indirect heating device (Hammer *et al.* 1995a). Later on, this problem spread to other European (France, Benelux, Spain) and extra-European countries (Mexico,

USA). Upon colony counting of incubated commercially-sterile milk on (milk) plate count agar at 30°C according to the EC regulation, dairies and food inspection services detected small, glossy, pin-point colonies of the HRS organism up to 10^5 vegetative cells/mL and 10^3 spores/mL. The nonsterility was also detected by measurement of bacterial ATP or the redox-potential, but not by pH control or microcalorimetry because the organism grows only poorly in milk and does not affect the pH of the product. Affected milk products included whole, skimmed, evaporated or reconstituted UHT milk, UHT cream and chocolate milk in polyethylene, cardboard or aluminium packaging (Klijn *et al.* 1997); milk powder was also found to be contaminated (Hammer *et al.* 1995a). The taxonomic position of the HRS organism was clarified with the description of *Bacillus sporothermodurans* as a new species (Pettersson *et al.* 1996). Three working groups found no pathogenic or toxigenic properties for several strains against mice, a cell-culture model and embryonized chicken eggs (Hammer *et al.* 1995b; Hammer & Walte 1996).

Real spoilage defects in consumer milk are only rarely noticed as a slight pink colour change, off-flavours and coagulation, especially in containers with a low oxygen barrier (e.g. plastic bottles) (Lembke 1995). During aerobic incubation for 72 h with a high cell concentration of *B. sporothermodurans*, breakdown of the casein and an occurrence of small peptides were observed, but because of poor growth of the organism in the milk, this proteolytic activity caused only minor changes in the product (Klijn *et al.* 1997). *Bacillus sporothermodurans* is thus less important from a hygienic point of view, but is most important from a technological perspective.

Occurrence and origin of *Bacillus sporothermodurans*

It seems that UHT milk has been occupied as a special ecological niche, for which several contamination sources have been suggested (Hammer *et al.* 1995a). The first likely source is raw farm milk. Hammer *et al.* (1995a) reported for the first time the detection of *B. sporothermodurans* in ex-farm milk. A temporal, local occurrence and a very low contamination level in raw farm milk (only three of 100 samples positive at the 100 mL level) was found using a PCR detection method (see section below) (Herman *et al.* 2000). However, neither the latter author nor others (e.g. Klijn *et al.* 1997) were unsuccessful in isolating the organism from raw milk after heat treatment (30 min at 100°C). In a polyphasic characterization study, including DNA–DNA hybridization and 16S rDNA sequence comparison (figure 6.2), one of 12 isolates from lab-pasteurized raw milk with colony morphologies similar to *B. sporothermodurans* was identified as the first reported raw milk isolate of *B. sporothermodurans* (Scheldeman *et al.*, 2002). Four other isolates were found to belong to *Bacillus oleronius*, a species previously isolated from the hindgut of a termite (Kuhnigk *et al.* 1995) and the closest phylogenetic relative of *B. sporothermodurans*, while six other isolates represented four potentially new *Bacillus* species (phylogenetic positions shown in figure 6.2) and one isolate was found to belong to *Brevibacillus borstelensis*. Contamination of raw milk at the farm is possible via feed or the milking equipment. Hammer *et al.* (1995a) found high numbers of *Bacillus sporothermodurans*

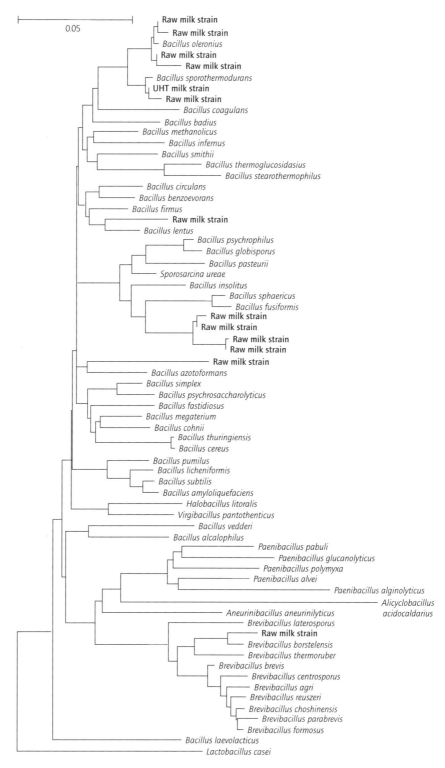

Figure 6.2 Phylogenetic tree based on 16S rDNA sequences (neighbour-joining) showing the position of one UHT milk and 12 raw milk *Bacillus* strains (in bold). *Lactobacillus casei* was used as an outgroup.

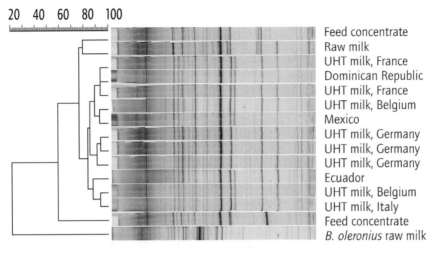

20 40 60 80 100

Feed concentrate
Raw milk
UHT milk, France
Dominican Republic
UHT milk, France
UHT milk, Belgium
Mexico
UHT milk, Germany
UHT milk, Germany
UHT milk, Germany
Ecuador
UHT milk, Belgium
UHT milk, Italy
Feed concentrate
B. oleronius raw milk

Figure 6.3 Clustering of REP-PCR patterns (Pearson coefficient) of *Bacillus sporothermodurans* isolates from different origins and/or countries. The Latin-American isolates are from UHT or sterilized milk. *Bacillus oleronius* is an outgroup.

spores only in only one or two out of 120 samples of corn silage, grass silage and sugar beet. De Silva *et al.* (1998) obtained one *B. sporothermodurans* isolate from silage heat treated for 60 min at 100°C. From one of six concentrate (a dried, pelleted supplementary feed ingredient) samples (heat treated for 30 min at 100°C) from five different farms, two *B. sporothermodurans* isolates were obtained, while from another sample *B. oleronius* was isolated (Vaerewijck *et al.* 2001). Concentrate often contains tropical ingredients (e.g. coconut, citrus pulp, manioc), and it can thus be speculated that *B. sporothermodurans* and *B. oleronius* strains had a (sub)tropical origin, which is for the latter species supported by its original isolation source.

A second route of contamination is possible through reprocessing of contaminated lots of UHT milk. Reprocessing of one contaminated package (containing 10^3 spores/mL after incubation for sterility checking) can contaminate a considerable fraction of the whole day's production at a level of 1 spore/L, which is regarded as the common contamination level. Finally, contamination is also possible via contaminated milk powder.

The relative importance of these different contamination sources was revealed by molecular typing. With REP-PCR based on the presence of intragenically located REP-like elements in *B. sporothermodurans* (Herman & Heyndrickx 2000), individual strains could be distinguished within a large collection of isolates from nonsterile UHT or sterilized milk from different European countries (Herman *et al.* 1998). Nevertheless, the REP patterns were very homogeneous (figure 6.3), suggesting that the contamination had originated from a clonal source, or that the extreme heat resistance is a property, either natural or acquired, of a specific HRS clone of *B. sporothermodurans*. Evidence for the circulation of contamination within or between dairy production units was obtained by observing an identical REP pattern for isolates of different units in

Table 6.3 Heat resistance of spores of *B. sporothermodurans* and other *Bacillus* species. Decimal reduction time *D* is the time required to reduce the number of spores by a factor of 10 at a given temperature; the *z*-value is the temperature change required to cause a 10-fold change in the *D*-value.

Species	Source	D-value (s)						z-value (°C)	Reference
		120°C	125°C	130°C	135°C	140°C			
Bacillus sporothermodurans	UHT milk	135		42.2	14.3	5.0		13.1–	Huemer et al. (1998)
		(79)[a]		(23.9)	(6.7)	(1.6)		14.2	
	Naturally contaminated UHT milk[b]		45.8	28.9	7.5	4.7		14	Scheldeman (unpubl. obs.)
	UHT milk[c]		13						Scheldeman (unpubl. obs.)
	Raw milk[c]		2						Scheldeman (unpubl. obs.)
	Feed concentrate			7.2	1.3				Scheldeman (unpubl. obs.)
Bacillus stearothermophilus	NIZO B469	191				0.9		9.1	Huemer et al. (1998), Klijn et al. (1997)
	ATCC 12980		56	13				7.0	Ocio et al. (1996)
	ATCC 12980			19.8		1.08		9.0	Rodrigo et al. (1997)
Bacillus oleronius	Raw milk[c]	2				<0.8			Klijn et al. (1997)
Brevibacillus spp.	Raw milk[c]	5				<0.8			Klijn et al. (1997)

[a] Values in brackets determined after 10 culture passages of the original stock culture.
[b] Heat resistance of spores naturally occurring in contaminated UHT milk, without prior laboratory isolation.
[c] Phylogenetic position shown in figure 6.2.

the same region over a 1-year period. In figure 6.3, it is shown that isolates from Latin America have the same REP pattern as the European HRS clone, suggesting intercontinental spread possibly by import of contaminated milk powder. On the other hand, the raw milk and concentrate isolates each have a unique pattern, and seem thus to be genetically different from the HRS clone.

Heat resistance of *Bacillus sporothermodurans* spores

Spores of *B. sporothermodurans* display different degrees of heat resistance (table 6.3). The highest heat resistance is observed for spores of freshly isolated UHT-milk isolates, or spores present in naturally contaminated UHT milk without isolation. Compared with *B. stearothermophilus*, these spores seem to be about equally or slightly less heat resistant at temperatures of up to 125°C, but they are considerably more heat resistant at UHT temperatures. This phenomenon is also reflected by the exceptionally high z-value (see table 6.3 for explanation of z-value). On the other hand, UHT milk isolates repeatedly cultured in the laboratory partially lose their natural heat-resistance properties, while farm isolates (from raw milk or concentrate) are not UHT-resistant, which corresponds with their non-HRS-REP pattern. This may indicate the importance of sublethal stress conditions prevailing at the farm, or (more likely) in the dairy, for the induction of UHT resistance.

Detection and identification of *Bacillus sporothermodurans*

Bacillus sporothermodurans is phylogenetically related (around 97% 16S rDNA sequence similarity) to *B. lentus*, *B. firmus*, *B. benzoevorans* and *B. circulans* (Pettersson *et al.* 1996). A sequence heterogeneity in the V1 and V2 region of the 16S rDNA seems to be characteristic for this species, but may hamper identification by direct sequencing (Pettersson *et al.* 1996; Klijn *et al.* 1997). Phenotypically, *B. sporothermodurans* resembles *Aneurinibacillus* spp. and *B. badius* because of its negative reactions for most of the standard and API 50CHB tests, but the phenotypic behaviour of the farm isolates recently isolated from raw milk and concentrate is not yet known. Some of the positive phenotypic characters, such as hydrolysis of casein, gelatin and urea, seem to be variable between strains or variable in the hands of different authors (Pettersson *et al.* 1996; Klijn *et al.* 1997). Unequivocal phenotypic identification of *B. sporothermodurans* thus seems to be difficult, and it is also hampered by the presence of phenotypically similar yet undescribed *Bacillus* species in raw milk.

Identification by molecular typing (RAPD, REP-PCR) has been proposed based on the assumed high genetic homogeneity of the species (Klijn *et al.* 1997; Herman *et al.* 1998) and a PCR identification test using primers derived from subtractive hybridization of *B. sporothermodurans* DNA with DNA of an unknown related strain from raw milk has been developed (see legend to figure 6.5). However, it now seems that this specific PCR is not valid for the farm strains but only for the HRS clone from UHT milk (Scheldeman *et al.*, 2002). De Silva *et al.* (1998) used a probe based on the 16S–23S rDNA spacer region for identification of a *B. sporothermodurans* isolate from silage. We have developed

Position	589	605	1237	1252
B. sporothermodurans	AC GGA TCA ACC GTG GAG		TAG ACC GCG AGG TTA C	
B. oleronius	G. . . . . .CA AGC		A. . . . . .T A	
B. circulans	. . . .		A. A . . . . .CG A	
B. firmus	C. . . . . .G. . . .		A. . . . . A	
B. lentus	. . . . .A		A. . . . . .T A	

Figure 6.4 Specific primers (reverse primer in complement and inverse) for *Bacillus sporothermodurans sensu lato* based on the 16S rDNA. Positions are according to the *B. sporothermodurans* sequence. In the aligned sequences of related species, only differentiating bases are indicated.

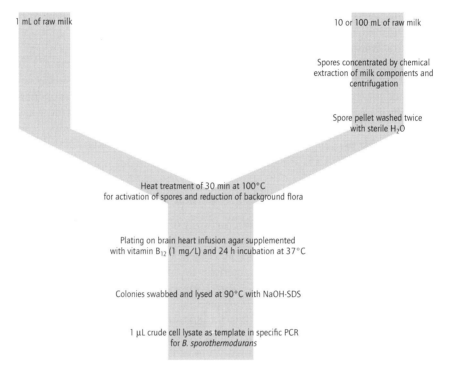

1 mL of raw milk

10 or 100 mL of raw milk

Spores concentrated by chemical extraction of milk components and centrifugation

Spore pellet washed twice with sterile H₂O

Heat treatment of 30 min at 100°C for activation of spores and reduction of background flora

Plating on brain heart infusion agar supplemented with vitamin B₁₂ (1 mg/L) and 24 h incubation at 37°C

Colonies swabbed and lysed at 90°C with NaOH-SDS

1 μL crude cell lysate as template in specific PCR for *B. sporothermodurans*

Figure 6.5 Simplified scheme of the detection procedure of *Bacillus sporothermodurans*. In the specific PCR, either the HRS-specific primers (5′-GAT TCA GGC AGA ATG TAG CA-3′ and 5′-TTT CGC CAC TTG ATG GTA CA-3′) can be used with a detection limit of 0.22–9 cfu/mL (Herman *et al.* 1997), or the general primers for *B. sporothermodurans* given in figure 6.4 can be used.

a new specific PCR based on the 16S rDNA for the identification of *B. sporothermodurans sensu lato* (i.e. including HRS and non-HRS clone strains) (figure 6.4; Scheldeman *et al.*, 2002).

While *B. sporothermodurans* is easily isolated from contaminated UHT products after pre-incubation at 37°C, isolation from raw or pasteurized milks and feeds with high background floras is more problematic. Herman *et al.* (1997) developed a sensitive detection method in raw milk (and feed) based on the specific PCR test without the need of prior purification of the strain (figure 6.5).

It can be concluded that *B. sporothermodurans* seems to be more heterogeneous (at least at the genotypic level) than was initially assumed. This means that an emended description of this industrially important organism is probably necessary. Furthermore, more knowledge is needed about the molecular mechanisms controlling heat-resistance induction of the spores, which is so important for the development of effective preventative measures.

References

Andersson, A., Rönner, U. & Granum, P.E. (1995) What problems does the food industry have with the spore-forming pathogens *Bacillus cereus* and *Clostridium perfringens*? *International Journal of Food Microbiology* **28**, 145–155.

Ash, C. & Collins, M.D. (1992) Comparative analysis of 23S ribosomal RNA gene sequences of *Bacillus anthracis* and emetic *Bacillus cereus* determined by PCR-direct sequencing. *FEMS Microbiology Letters* **94**, 75–80.

Ash, C., Farrow, J.A.E., Dorsch, M., Stackebrandt, E. & Collins, M.D. (1991) Comparative analysis of *Bacillus anthracis, Bacillus cereus*, and related species on the basis of reverse transcriptase sequencing of 16S rDNA. *International Journal of Systematic Bacteriology* **41**, 343–346.

Beattie, S.H., Holt, C., Hirst, D. & Williams, A.G. (1998) Discrimination among *Bacillus cereus, B. mycoides* and *B. thuringiensis* and some other species of the genus *Bacillus* by Fourier transform infrared spectroscopy. *FEMS Microbiology Letters* **164**, 201–206.

Borlinghaus, A. & Engel, R. (1997) *Alicyclobacillus* incidence in commercial apple juice concentrate (AJC) supplies – Method development and validation. *Fruit Processing* **7**, 262–266.

Bourque, S.N., Valero, J.R., Lavoie, M.C. & Levesque, R.C. (1995) Comparative analysis of the 16S to 23S ribosomal intergenic spacer sequences of *Bacillus thuringiensis* strains and subspecies and of closely related species. *Applied and Environmental Microbiology* **61**, 1623–1626.

Cerny, G., Hennlich, W. & Poralla, K. (1984) Fruchtsaftverderb durch Bacillen: Isolierung und Charakterisierung des Verderbserregers. *Zeitschrift Lebensmitteln Untersuchungen und Forschungen* **179**, 224–227.

Christiansson, A., Bertilsson, J. & Svensson, B. (1999) *Bacillus cereus* spores in raw milk: factors affecting the contamination of milk during the grazing period. *Journal of Dairy Science* **82**, 305–314.

Collins, E.B. (1981) Heat resistant psychrotrophic organisms. *Journal of Dairy Science* **64**, 157–160.

Cosentino, S., Mulargia, A.F., Pisano, B., Tuveri, P. & Palmas, F. (1997) Incidence and biochemical characteristics of *Bacillus* flora in Sardinian dairy products. *International Journal of Food Microbiology* **38**, 235–238.

Crielly, E.M., Logan, N.A. & Anderton, A. (1994) Studies on the *Bacillus* flora of milk and milk products. *Journal of Applied Bacteriology* **77**, 256–263.

Daffonchio, D., Borin, S., Frova, G. *et al.* (1999) A randomly amplified polymorphic DNA marker specific for the *Bacillus cereus* group is diagnostic for *Bacillus anthracis*. *Applied and Environmental Microbiology* **65**, 1298–1303.

De Silva, S., Petterson, B., De Muro, M.A. & Priest, F.G. (1998) A DNA probe for the detection and identification of *Bacillus sporothermodurans* using the 16S–23S rDNA spacer region and phylogenetic analysis of some field isolates of *Bacillus* which form highly heat resistant spores. *Systematic and Applied Microbiology* **21**, 398–407.

Drobniewski, F.A. (1993) *Bacillus cereus* and related species. *Clinical Microbiology Reviews* **6**, 324–338.

Eiroa, M.N.U., Junqueira, V.C.A. & Schmidt, F.L. (1999) *Alicyclobacillus* in orange juice: occurrence and heat resistance of spores. *Journal of Food Protection* **62**, 883–886.

Foschino, R., Galli, A. & Ottogali, G. (1990) Research on the microflora of UHT milk. *Annales de Microbiologie* **40**, 47–59.

Francis, K.P., Mayr, R., Von Stetten, F., Stewart, G.S.A.B. & Scherer, S. (1998) Discrimination of psychrotrophic and mesophilic strains of the *Bacillus cereus* group by PCR targeting of major cold shock protein genes. *Applied and Environmental Microbiology* **64**, 3525–3529.

Gordon, R.E., Haynes, W.C. & Pang, C.H.-N. (1973) *The Genus* Bacillus. US Department of Agriculture, Washington, DC.

Govan, V.A., Allsopp, M.H. & Davison, S. (1999) A PCR detection method for rapid identification of *Paenibacillus larvae*. *Applied and Environmental Microbiology* **65**, 2243–2245.

Griffiths, M.W. & Phillips, J.D. (1990) Incidence, source and some properties of psychrotrophic

Bacillus spp found in raw and pasteurized milk. *Journal of the Society of Dairy Technology* **43**, 62–66.

Hammer, P. & Walte, H.G. (1996) Zur Pathogenität hitzeresistenter mesophiler Sporenbildner aus UHT-Milch. *Kieler Milchwirtschaftliche Forschungsberichte* **48**, 151–161.

Hammer, P., Lembke, F., Suhren, G. & Heeschen, W. (1995a) Characterization of a heat resistant mesophilic *Bacillus* species affecting the quality of UHT-milk – a preliminary report. *Kieler Milchwirtschaftliche Forschungsberichte* **47**, 303–311.

Hammer, P., Suhren G. & Heeschen, W. (1995b) Pathogenicity testing of unknown mesophilic heat resistant bacilli from UHT-milk. *Bulletin of the International Dairy Federation* **302**, 56–57.

Helyer, R.J., Kelley, T. & Berkeley, R.C.W. (1997) Pyrolysis mass spectrometry studies on *Bacillus anthracis, Bacillus cereus* and their close relatives. *International Journal of Medical Microbiology, Virology, Parasitology and Infectious Diseases* **285**, 319–328.

Herman, L. & Heyndrickx, M. (2000) The presence of intragenically located REP-like elements in *Bacillus sporothermodurans* is sufficient for REP-PCR typing. *Research in Microbiology* **151**, 255–261.

Herman, L.M.F., Vaerewijck, M.J.M., Moermans, R.J.B. & Waes, G.W.V.J. (1997) Identification and detection of *Bacillus sporothermodurans* spores in 1, 10, and 100 milliliters of raw milk by PCR. *Applied and Environmental Microbiology* **63**, 139–143.

Herman, L., Heyndrickx, M. & Waes, G. (1998) Typing of *Bacillus sporothermodurans* and other *Bacillus* species isolated from milk by repetitive element sequence based PCR. *Letters in Applied Microbiology* **26**, 183–188.

Herman, L., Heyndrickx, M., Vaerewijck, M. & Klijn, N. (2000) *Bacillus sporothermodurans* – a *Bacillus* forming highly heat-resistant spores. 3. Isolation and methods of detection. *Bulletin of the International Dairy Federation* **357**, 9–14.

Hsieh, Y.M., Sheu, S.J., Chen, Y.L. & Tsen, H.Y. (1999) Enterotoxigenic profiles and polymerase chain reaction detection of *Bacillus cereus* group cells and *B. cereus* strains from foods and foodborne outbreaks. *Journal of Applied Microbiology* **87**, 481–490.

Huemer, I.A., Klijn, N., Vogelsang H.W.J. & Langeveld, L.P.M. (1998) Thermal death kinetics of spores of *Bacillus sporothermodurans* isolated from UHT milk. *International Dairy Journal* **8**, 851–855.

Huis 't Veld, J.H.J. (1996) Microbial and biochemical spoilage of foods: an overview. *International Journal of Food Microbiology* **33**, 1–18.

ISO (1993) *ISO/DIS 7932. Microbiology – General Guidance for the Enumeration of* Bacillus cereus – *Colony-count Technique at 30 degrees C.* International Organization for Standardization, Switzerland.

Jensen, N. (1999) *Alicyclobacillus* – a new challenge for the food industry. *Food Australia* **51**, 33–36.

Kalogridou-Vassiliadou, D. (1992) Biochemical activities of *Bacillus* species isolated from flat sour evaporated milk. *Journal of Dairy Science* **75**, 2681–2686.

Kaneko, T., Nozaki, R. & Aizawa, K. (1978) Deoxyribonucleic acid relatedness between *Bacillus anthracis, Bacillus cereus*, and *Bacillus thuringiensis*. *Microbiology and Immunology* **22**, 639–641.

Klijn, N., Herman, L., Langeveld, L. *et al.* (1997) Genotypical and phenotypical characterization of *Bacillus sporothermodurans* strains, surviving UHT sterilization. *International Dairy Journal* **7**, 421–428.

Kuhnigk, T., Borst, E.-M., Breunig, A. *et al.* (1995) *Bacillus oleronius* sp. nov., a member of the hindgut flora of the termite *Reticulitermes santonensis* (Feytaud). *Canadian Journal of Microbiology* **41**, 699–706.

Larsen, H.D. & Jørgensen, K. (1997) The occurrence of *Bacillus cereus* in Danish pasteurized milk. *International Journal of Food Microbiology* **34**, 179–186.

Lechner, S., Mayr, R., Francis, K.P. *et al.* (1998) *Bacillus weihenstephanensis* sp. nov. is a new psychrotolerant species of the *Bacillus cereus* group. *International Journal of Systematic Bacteriology* **48**, 1373–1382.

Lembke, F. (1995) Highly heat-resistant spores in UHT-milk. *Bulletin of the International Dairy Federation* **302**, 60–61.

Lin, S., Schraft, H., Odumeru, J.A. & Griffiths, M.W. (1998) Identification of contamination sources of *Bacillus cereus* in pasteurized milk. *International Journal of Food Microbiology* **43**, 159–171.

McIntyre, S., Ikawa, J.Y., Parkinson, N., Haglund, J. & Lee, J. (1995) Characteristics of an acidophilic *Bacillus* strain isolated from shelf-stable juices. *Journal of Food Protection* **58**, 319–321.

Meer, R.R., Bakker, J., Bodyfelt, F.W. & Griffiths, M.W. (1991) Psychrotrophic *Bacillus* spp. in fluid milk products: a review. *Journal of Food Protection* **54**, 969–979.

Montville, T.J. (1982) Metabiotic effect of *Bacillus licheniformis* on *Clostridium botulinum*: implications for home-canned tomatoes. *Applied and Environmental Microbiology* **44**, 334–338.

Nakamura, L.K. (1998) *Bacillus pseudomycoides* sp. nov. *International Journal of Systematic Bacteriology* **48**, 1031–1035.

Ocio, M.J., Fernàndez, P., Rodrigo, F. & Martinez, A. (1996) Heat resistance of *Bacillus stearothermophilus* spores in alginate-mushroom puree mixture. *International Journal of Food Microbiology* 29, 391–395.

Pettersson, B., Lembke, F., Hammer, F., Stackebrandt, E. & Priest, F.G. (1996) *Bacillus sporothermodurans*, a new species producing highly heat-resistant endospores. *International Journal of Systematic Bacteriology* 46, 759–764.

Pettipher, G.L., Osmundson, M.E. & Murphy, J.M. (1997) Methods for the detection and enumeration of *Alicyclobacillus acidoterrestris* and investigation of growth and production of taint in fruit juice and fruit juice-containing drinks. *Letters in Applied Microbiology* 24, 185–189.

Phillips, J.D. & Griffiths, M.W. (1986) Factors contributing to the seasonal variation of *Bacillus* spp. in pasteurized dairy products. *Journal of Applied Bacteriology* 61, 275–285.

Priest, F.G., Goodfellow, M. & Todd, C. (1988) A numerical classification of the genus *Bacillus*. *Journal of General Microbiology* 134, 1847–1882.

Prüß, B.M., Francis, K.V., Von Stetten, F. & Scherer, S. (1999) Correlation of 16S ribosomal DNA signature sequences with temperature-dependent growth rates of mesophilic and psychrotolerant strains of the *Bacillus cereus* group. *Journal of Bacteriology* 181, 2624–2630.

Rodrigo, E., Fernandez, P.S., Rodrigo, M., Ocio, M.J. & Martinez, A. (1997) Thermal resistance of *Bacillus stearothermophilus* heated at high temperatures in different substrates. *Journal of Food Protection* 60, 144–147.

Rönner, U., Husmark, U. & Henriksson, A. (1990) Adhesion of *Bacillus* spores in relation to hydrophobicity. *Journal of Applied Bacteriology* 69, 550–556.

Rosenkvist, H. & Hansen, A. (1995) Contamination profiles and characterisation of *Bacillus* species in wheat bread and raw materials for bread production. *International Journal of Food Microbiology* 26, 353–363.

Scheldeman, P., Herman, L., Goris, J., De Vos, P. & Heyndrickx, M. (2002) Polymerase chain reaction (PCR) identification of *Bacillus sporothermodurans* from dairy sources. *Journal of Applied Microbiology*, in press.

Silva, F.M., Gibbs, P., Vieira, C. & Silva, C.L.M. (1999) Thermal inactivation of *Alicyclobacillus acidoterrestris* spores under different temperature, soluble solids and pH conditions for the design of fruit processes. *International Journal of Food Microbiology* 51, 95–103.

Slaghuis, B.A., te Giffel, M.C., Beumer, R.R. & André, G. (1997) Effect of pasturing on the incid-

ence of *Bacillus cereus* spores in raw milk. *International Dairy Journal* 7, 201–205.

Sørhaug, T. & Stepaniak, L. (1997) Psychrotrophs and their enzymes in milk and dairy products: Quality aspects. *Trends in Food Science and Technology* 8, 35–41.

Splittstoesser, D.F., Churey, J.J. & Lee, C.Y. (1994) Growth characteristics of aciduric sporeforming bacilli isolated from fruit juices. *Journal of Food Protection* 57, 1080–1083.

Splittstoesser, D.F., Worobo, R.W. & Churey, J.J. (1998) Food safety and you: *Alicyclobacillus*: an emerging problem for New York's processors of fruit juices. *Venture* 1, http://www.nysaes.cornell.edu/fst/fvc/Venture/venture3_safety.html.

Stadhouders, J. (1992) Taxonomy of *Bacillus cereus*. *Bulletin of the International Dairy Federation* 275, 4–8.

Stadhouders, J. & Driessen, F.M. (1992) *Bacillus cereus* in liquid milk and other milk products: other milk products. *Bulletin of the International Dairy Federation* 275, 40–45.

Sutherland, A.D. & Murdoch, R. (1994) Seasonal occurrence of psychrotrophic *Bacillus* species in raw milk, and studies on the interactions with mesophilic *Bacillus* sp. *International Journal of Food Microbiology* 21, 279–292.

Tatzel, R., Ludwig, W., Schleifer, K.H. & Wallnöfer, P.R. (1994) Identification of *Bacillus* strains isolated from milk and cream with classical and nucleic acid hybridisation methods. *Journal of Dairy Research* 61, 539–535.

te Giffel, M.C., Beumer, R.R., Slaghuis, B.A. & Rombouts, F.M. (1995) Occurrence and characterization of (psychrotrophic) *Bacillus cereus* on farms in the Netherlands. *Netherlands Milk and Dairy Journal* 49, 125–138.

te Giffel, M.C., Beumer, R.R., Bonestroo, M.H. & Rombouts, F.M. (1996) Incidence and characterization of *Bacillus cereus* in two dairy processing plants. *Netherlands Milk and Dairy Journal* 50, 479–492.

te Giffel, M.C., Beumer, R.R., Klijn, N., Wagendorp, A. & Rombouts, F.M. (1997a) Discrimination between *Bacillus cereus* and *Bacillus thuringiensis* using specific DNA probes based on variable regions of 16S rRNA. *FEMS Microbiology Letters* 146, 47–51.

te Giffel, M.C., Beumer, R.R., Langeveld, L.P.M. & Rombouts, F.M. (1997b) The role of heat exchangers in the contamination of milk with *Bacillus cereus* in dairy processing plants. *International Journal of Dairy Technology* 50, 43–47.

Ternström, A., Lindberg, A.M. & Molin, G. (1993) Classification of the spoilage flora of raw and pasteurized bovine milk, with special reference to

Pseudomonas and *Bacillus*. *Journal of Applied Bacteriology* **75**, 25–34.

Vaerewijck, M., De Vos, P., Lebbe, L., Scheldeman, P., Hoste, B. & Heyndrickx, M. (2001) Occurrence of *Bacillus sporothermodurans* and other aerobic sporeforming species in feed concentrate for dairy cattle. *Journal of Applied Microbiology* **91**, 1074–1084.

Van Heddeghem, A. & Vlaemynck, G. (1992) Sources of contamination of milk with *B. cereus* on the farm and in the factory. *Bulletin of the International Dairy Federation* **275**, 19–22.

von Stetten, F., Francis, K.P., Lechner, S., Neuhaus, K. & Scherer, S. (1998) Rapid discrimination of psychrotolerant and mesophilic strains of the *Bacillus cereus* group by PCR targeting of 16S rDNA. *Journal of Microbiological Methods* **34**, 99–106.

von Wintzingerode, F., Rainey F.A., Kroppenstedt, R.M. & Stackebrandt, E. (1997) Identification of environmental strains of *Bacillus mycoides* by fatty acid analysis and species-specific 16S rDNA oligonucleotide probe. *FEMS Microbiology Ecology* **24**, 201–209.

Waes, G. (1976) Aerobic mesophilic spores in raw milk. *Milchwissenschaft* **31**, 521–525.

Washam, C.J., Olson, H.C. & Vedamuthu, E.R. (1977) Heat-resistant psychrotrophic bacteria isolated from pasteurized milk. *Journal of Food Protection* **40**, 101–108.

Westhoff, D.C. & Dougherty, S.L. (1981) Characterization of *Bacillus* species isolated from spoiled ultrahigh temperature processed milk. *Journal of Dairy Science* **64**, 572–580.

Wisotzkey, J.D., Jurtshuk, P., Fox, G.E., Deinhard, G. & Poralla, K. (1992) Comparative sequence analyses on the 16S rRNA (rDNA) of *Bacillus acidocaldarius*, *Bacillus acidoterrestris*, and *Bacillus cycloheptanicus* and proposal for creation of a new genus, *Alicyclobacillus* gen. nov. *International Journal of Systematic Bacteriology* **42**, 263–269.

Yamada, S., Ohashi, E., Agata, N. & Venkateswaran, K. (1999) Cloning and nucleotide sequence analysis of gyrB of *Bacillus cereus*, *B. thuringiensis*, *B. mycoides*, and *B. anthracis* and their application to the detection of *B. cereus* in rice. *Applied and Environmental Microbiology* **65**, 1483–1490.

Yamazaki, K., Teduka, H., Inoue, N. & Shinano, H. (1996) Specific primers for detection of *Alicyclobacillus acidoterrestris* by RT-PCR. *Letters in Applied Microbiology* **23**, 350–354.

Chapter 7

Moderately Halophilic and Halotolerant Species of *Bacillus* and Related Genera

David R. Arahal and Antonio Ventosa

Introduction

The origins of the study of halophilic microorganisms go back to the end of the nineteenth and the beginning of the twentieth centuries, mostly a response to the reddening-spoilage of salt-cured products. Hanzawa and Takeda (1931) reviewed previous publications dating from 1880 to 1911, and it is concluded that, besides the 'chromogenic microorganisms' causing the red coloration in cured cod fish, there is also a 'symbiotic life of microorganisms, each of which does not produce a red colour in itself', and some of the examples given are spore-forming rods. Two early reports about the occurrence of halophilic bacilli that produced endospores in habitats other than cured fish, are the studies of Hof (1935) and Volcani (1940). The bacilli isolated by Hof were obtained from salty mud from a solar saltern in Java and were able to grow in media containing up to 24% salt (Hof 1935). The strains studied by Volcani (1940) were isolated from Dead Sea water samples and tolerated up to 6 or 12% salt content (one strain was even able to grow in 24% salt), but they also grew in salt-free medium and therefore he referred to them as 'haloresistants'. He considered that they were contaminants that had entered the lake, carried with the inflow of freshwater from the Jordan river. A recent review of the diversity of moderately halophilic Gram-positive bacteria in hypersaline environments deals with this matter further (Ventosa *et al.* 1998a).

Halophilism and its degrees have been defined and classified in many fashions (Trüper & Galinski 1986; Vreeland 1987), employing different terms and concepts. The most commonly accepted of these definitions is that of Kushner and Kamekura (1988). According to it there are to date 18 recognized species of the genus *Bacillus* and related genera that could fit into the categories of moderate halophiles or halotolerants (table 7.1).

Moderately halophilic bacteria are those organisms growing best in the presence of 3–15% salt, while halotolerant bacteria are nonhalophilic microorganisms that can thrive in the absence of salts and can tolerate relatively high salt concentrations. Other categories of halophilic organisms included in this definition are the extreme halophiles and the slight halophiles (Kushner & Kamekura 1988). The former are well represented among the *Halobacteriaceae*, and the slight halophiles (which grow best in media with 1–3% salt) are typically represented by marine bacteria (Kushner & Kamekura 1988).

Table 7.1 Currently recognized moderately halophilic and halotolerant species of the genus *Bacillus* and related genera.

	Type strain	Reference	Habitat	16S rRNA accession number
Moderate halophiles				
Bacillus halophilus	DSMZ 4771	Ventosa *et al.* (1989)	Unknown (rotting wood from seawater)	AB021188
Bacillus haloalkaliphilus	DSMZ 5271	Fritze (1996)	Wadi Natrun	
Halobacillus halophilus	DSMZ 2266	Spring *et al.* (1996)	Salt marsh soil, saline soil, salterns	X62174
Halobacillus litoralis	DSMZ 10405	Spring *et al.* (1996)	Great Salt Lake sediment	X94558
Halobacillus trueperi	DSMZ 10404	Spring *et al.* (1996)	Great Salt Lake sediment	
Salibacillus salexigens	DSMZ 11483	Waino *et al.* (1999)	Solar saltern, hypersaline soil	Y11603
Salibacillus marismortui	DSMZ 12325	Arahal *et al.* (2000)	Dead Sea	AJ009793
Halotolerant				
Bacillus halodurans	ATCC 27557	Nielsen *et al.* (1995)	Soil	AB021187
Bacillus pseudofirmus	DSMZ 8715	Nielsen *et al.* (1995)	Soil and animal manure	X76439
Bacillus agaradhaerens	DSMZ 8721	Nielsen *et al.* (1995)	Soil and faeces	X76445
Bacillus clarkii	DSMZ 8720	Nielsen *et al.* (1995)	Soil	X76444
Bacillus pseudoalcaliphilus	DSMZ 8725	Nielsen *et al.* (1995)	Soil	X76449
Bacillus halodenitrificans	ATCC 49067	Denariaz *et al.* (1989)	Solar saltern	AB021186
Bacillus mojavensis	DSMZ 9205	Roberts *et al.* (1994)	Desert soil	AB021191
Bacillus vallismortis	DSMZ 11031	Roberts *et al.* (1996)	Death Valley	AB021198
Gracilibacillus halotolerans	DSMZ 11805	Waino *et al.* (1999)	Great Salt Lake	AF036922
Gracilibacillus dipsosauri	DSMZ 11125	Waino *et al.* (1999)	Iguana nasal cavity	X82436
Virgibacillus pantothenticus	ATCC 14576	Heyndrickx *et al.* (1998)	Soil	D16275

DSMZ, Deutsche Sammlung für Mikroorganismen und Zellkulturen; ATCC, American Type Culture Collection.

Studies based on the isolation of strains on culture media have shown that the most common moderately halophilic inhabitants of hypersaline media are Gram-negative rods (Ventosa *et al.* 1998b). Moderately halophilic Gram-positive endospore-forming rods are found in very low proportions in salterns (Rodriguez-Valera *et al.* 1985; Marquez *et al.* 1987; Bejar *et al.* 1992); however, in hypersaline soils they make up a large fraction of the population, in which most *Bacillus* spp. isolated were able to grow optimally at 5–10% salt (Quesada *et al.* 1982). Although in hypersaline environments most microbial inhabitants are members of the halophilic archaea or bacteria branches, there are also non-halophilic microorganisms that are able to survive and compete with them. The ecological role and contribution of the latter to the microbial processes is unknown. In fact, very few studies have been carried out with halotolerant Gram-positive endospore-forming rods from hypersaline environments. Marquez *et al.* (1987) reported the isolation of 150 nonhalophilic bacteria from a saltern located in southern Spain. Their numerical study showed that of the ten phena obtained, six were members of the genus *Bacillus*. In a similar study from water and sediment samples from hypersaline environments, a large collection of heavy-metal-tolerant nonhalophilic bacteria were isolated and most Gram-positive metal-tolerant strains were assigned to the genus *Bacillus* (Ríos *et al.* 1998). Recently, Garabito *et al.* (1998) studied several halotolerant *Bacillus* isolates in detail that were able to grow in media containing up to 20–25% salt, and they were assigned to the species *Virgibacillus pantothenticus*, *B. firmus*, *B. alcalophilus*, *B. megaterium* and *Brevibacillus laterosporus*.

Moderate halophiles

There are to date seven validly published species considered as moderately halophilic bacilli (table 7.1). The main phenotypic features useful for the differentiation of these species are shown in table 7.2. The phylogenetic interrelationships among the moderately halophilic species and with the other members of the *Bacillus* and relatives branch can be seen in the tree shown in figure 7.1. A simplified version of this, containing only moderately halophilic and halotolerant species, is shown in figure 7.2. It is apparent that most moderately halophilic species are placed in a monophyletic group that includes several halotolerant and nonhalophilic species. This group comprises many different genera, namely *Amphibacillus*, *Gracilibacillus*, *Halobacillus*, *Marinococcus*, *Salibacillus* and *Virgibacillus*, as well as some *Bacillus* species (figure 7.2).

Moderate halophiles within the genus *Bacillus*

The first halophilic species described in the genus *Bacillus* was *B. halophilus* (Ventosa *et al.* 1989), an organism able to produce an extracellular nuclease with RNase and DNase activities. This enzyme, of marked halophilic properties, has been purified and characterized (Onishi *et al.* 1983). *Bacillus halophilus* has a very broad salinity range for growth (from 3 to 30%) and a very high optimal

Table 7.2 The main characteristics that differentiate the moderately halophilic bacilli. Data from Arahal *et al.* (1999), Claus *et al.* (1983), Fritze (1996), Garabito *et al.* (1997), Spring *et al.* (1996), Ventosa *et al.* (1989) and Weisser and Trüper (1985).

Feature	Bacillus		Halobacillus			Salibacillus	
	B. halophilus	*B. haloalkaliphilus*	*H. halophilus*	*H. litoralis*	*H. trueperi*	*S. salexigens*	*S. marismortui*
Morphology	Rods	Rods	Coccoid	Rods	Rods	Rods	Rods
Size (µm)	0.5–1.0 × 2.5–9.0	0.3–0.5 × 3.0–8.0	1.0–2.0 × 2.0–3.0	0.7–1.1 × 2.0–4.5	0.7–1.1 × 2.0–4.5	0.3–0.6 × 1.5–3.5	0.5–0.7 × 2.0–3.6
Pigmentation	None	Cream white	Orange	Orange	Orange	None	Cream
Spore shape	E	S	S	E/S	E/S	E	E
Spore position	C	T	C/ST	C/ST	C/ST	ST/C	T/ST
Sporangium swollen	–	+	–	–	n.d.	+	+
Facultative anaerobe	–	n.d.	–	–	–	–	–
NaCl range (%)	3–30	1–25	2–20	0.5–25	0.5–30	7–20	3–20
NaCl optimum (%)	15	5	10	10	10	10	10
Temp. range (°C)	15–50	15–40	15–40	10–43	10–44	15–45	15–50
pH range	6.0–8.0	8.0–9.7	7.0–9.0	6.0–9.5	6.0–9.5	6–11	6–9
pH optimum	7.0	>8.5	n.d.	7.5	7.5	7.5	7.5
Acid produced from:							
Glucose	+	–	–	+	+	+	+
Mannitol	–	–	–	+	–	+	–
Trehalose	+	+	–	+	+	–	–
Hydrolysis of:							
Casein	–	–	+	–	–	+	+
Esculin	+	n.d.	–	–	–	+	n.d.
Gelatin	–	+	+	+	+	+	+
Starch	–	+	+	–	–	+	–
H₂S production	–	n.d.	n.d.	n.d.	n.d.	+	d
Nitrate reduction	–	–	–	–	–	–	+
Phosphatase	–	n.d.	d	–	–	d	+
Urease	+	–	–	–	–	n.d.	+
Menaquinone system	MK-7	MK-7	MK-7	n.d.	n.d.	MK-7	n.d.
Cell wall type	*m*-Dpm	*m*-Dpm	Orn-D-Asp	Orn-D-Asp	Orn-D-Asp	*m*-Dpm	*m*-Dpm
G+C content (mol%)	51.5	38	40.8	42	43	39.5	40.7

E, ellipsoidal; C, central; S, spherical; ST, subterminal; T, terminal. +, positive; –, negative; d, differs among strains; n.d, not determined. MK-7, menaquinone with seven isoprene units; *m*-Dpm, *meso*-diaminopimelic acid.

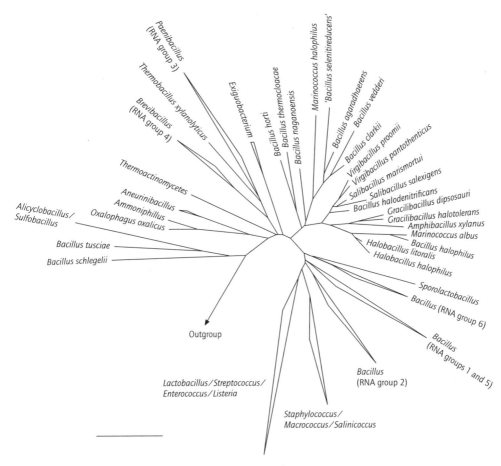

Figure 7.1 16S rRNA-based phylogenetic tree reflecting the interrelationships between the moderately halophilic and halotolerant bacilli and other members of the genus *Bacillus* and their relatives, corrected by maximum parsimony analysis. The sequences of *B. halodurans*, *B. pseudalcaliphilus* and *B. pseudofirmus* fall into the *Bacillus* rRNA group 6, and those of *B. mojavensis* and *B. vallismortis* into the *Bacillus* rRNA group 1. The bar indicates a 5% sequence divergence.

NaCl concentration (around 15% salt), enabling growth (Ventosa *et al.* 1989) that cannot be observed in any other bacilli (table 7.2).

The 16S rRNA of *B. halophilus* has been sequenced recently (Goto & Omura, unpubl. public databases; Arahal *et al.*, unpubl. data) and it has a close similarity to that of *Marinococcus albus* (figure 7.2), an organism probably misclassified if the low 16S rRNA similarity with *M. halophilus* (the type species of its genus) is taken into account. The taxonomic placement of *Bacillus halophilus* should be investigated and it should probably be reclassified in a new genus.

Weisser and Trüper (1985) isolated an obligately alkaliphilic and halophilic strain (WM13) from Lake Abu Gabara in the Wadi Natrun (Egypt), which served as a model for studies on osmoregulation. This strain failed to grow in media containing no salt and had an optimal growth at 5% salt content, being thus a

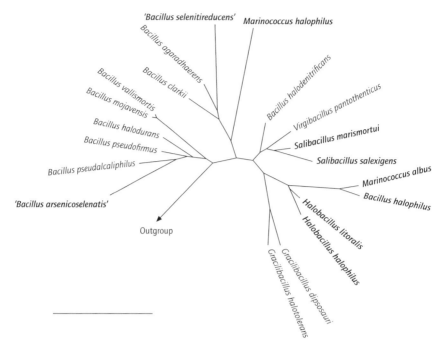

Figure 7.2 Simplified 16S rRNA-based phylogenetic tree of the moderately halophilic (in bold face) and halotolerant *Bacillus* species and related organisms, corrected by maximum parsimony analysis. The bar indicates a 5% sequence divergence.

moderate halophile. It was also not able to grow below pH 8 and grew optimally at pH 8.5–10; thus, it was regarded as a haloalkaliphile.

In 1992, brine and dried soil, mud and dung samples collected in various places in the Wadi Natrun permitted the isolation of ten haloalkaliphilic strains that were used, together with strain WM13, for the description of *B. haloalkaliphilus* (Fritze 1996). All of these strains produced round, terminally located spores in swollen sporangia. They were described as obligately alkaliphilic, extremely halotolerant isolates; however, because they do not grow in media without added NaCl, and are able to grow in media containing up to 25% NaCl and optimally at 5% NaCl they should be considered as moderate halophiles. They contain saturated branched-chain fatty acids (*iso*-$C_{15:0}$, 48%; *anteiso*-$C_{15:0}$, 11%; *anteiso*-$C_{17:0}$, 11%; and *iso*-$C_{17:0}$, 6%) as the major fatty acids and a minor proportion of unsaturated fatty acids (19%). The osmoregulation is achieved by ectoine synthesis that can be inhibited by the uptake and accumulation of betaine when this compatible solute is present in the medium (Fritze 1996).

The genus *Halobacillus*

The genus *Halobacillus* (Spring *et al.* 1996) was created to accommodate two new species, namely *H. litoralis* and *H. trueperi*, and the species *H. halophilus*, previously described as *Sporosarcina halophila* (Claus *et al.* 1983). *Halobacillus halophilus* is the type species of the genus and was isolated from salt marsh soils of the North Sea coasts of Germany (Claus *et al.* 1983) as well as from saline soils and water samples from salterns in Spain (Ventosa *et al.* 1983). *Halobacillus litoralis* and *H. trueperi* were obtained from surface sediments collected at the south arm (less concentrated than the hypersaline north arm) of the Great Salt Lake (Utah, USA) (Spring *et al.* 1996). *Sporosarcina halophila* was described as a coccoid, motile, spore-forming moderate halophile but phenotypic and phylogenetic features clearly support its placement in the genus *Halobacillus*, together with *H. litoralis* and *H. trueperi*. The main characteristics that differentiate members of the genus *Halobacillus* from *Bacillus* and other related genera are the peptidoglycan of the cell wall, which is of the Orn-D-Asp type and the G+C content of the DNA, ranging from 40 to 43 mol% (Spring *et al.* 1996) (see table 7.2).

The dependence of *H. halophilus* on chloride for growth has recently been estimated quantitatively (Roeßler & Müller 1998). The final yield was found to be strictly dependent on the Cl^- concentration, the optimal levels being between 0.5 and 2.0 M. The Cl^- ion could be replaced by Br^-, but not by any other anion tested (I^-, SO_4^{2-}, NO_2^-, SO_2^-, OCN^-, SCN^-, BO_2^- or BrO_3^-). In addition, growth was strictly dependent on Na^+. The latter could not be replaced by Li^+ or K^+. At low external Cl^- concentrations, Cl^- is excluded from the cytoplasm but it is taken up at higher external Cl^- concentrations decreasing the external to internal Cl^- gradient (Roeßler & Müller 1998), a phenomenon also observed in *B. haloalkaliphilus* (Weisser & Trüper 1985).

Further investigations of *H. halophilus* revealed that the Cl^- concentration was essential for the endospore germination in this organism (Dohrmann & Müller 1999), as was the case for the physiology of the vegetative cell. Again, Cl^- could only be replaced by Br^-. A new observation was that the Cl^- concentration also plays a role in the motility of the cells. It was shown that *H. halophilus* cultures grown in the absence of Cl^- lacked flagella, which nevertheless were restored by the addition of Cl^-, thus indicating that Cl^- is involved in the synthesis of flagella (Roeßler *et al.* 2000).

The genus *Salibacillus*

Salibacillus salexigens was originally described as *Bacillus salexigens* and was based on six strains isolated from different geographical areas of Spain (solar saltern ponds and hypersaline soils), that require at least 5–7% NaCl for growth (Garabito *et al.* 1997). It produces oval endospores at subterminal or central positions in swollen sporangia. It has DNase and protease activities but not amylase or lipase activities (Garabito *et al.* 1997). In a recent study, Wainö *et al.* (1999) proposed the transfer of *B. salexigens*, which is phylogenetically distant from *Bacillus* group 1, to the new genus *Salibacillus* as its single species, *Virgibacillus pantothenticus* being the closest relative to this new species.

Almost simultaneously to the study of Wainö *et al.* (1999), a new species, *Bacillus marismortui*, closely related to *Salibacilllus salexigens* in terms of phylogeny and phenotype, was described (Arahal *et al.* 1999). The description of this organism was based on a group of 91 strains, all of them having the same origin: a collection of 57-year-old enrichments from Dead Sea water samples taken in 1936 by B.E. Volcani during his studies of the microbiota of the Dead Sea (Volcani 1940). Despite their long storage time, the enrichments were shown to contain viable cells of different phylogenetic groups (Arahal *et al.* 1996; Ventosa *et al.* 1999).

The Dead Sea can be considered as one of the most extreme environments on Earth as a result of its very high salinity (close to 34% total dissolved salts) and its particular ionic composition (Mg^{2+} is the main cation). This is reflected in its low microbial diversity. As reported in a recent review by Ventosa and Arahal (1999), only a limited variety of microorganisms, not including any endospore-forming rods, has been demonstrated. *Bacillus marismortui* was, therefore, the first aerobic, Gram-positive, moderately halophilic endospore-forming bacterium isolated from the Dead Sea (Arahal *et al.* 1999).

DNA–DNA hybridization experiments between this species and *S. salexigens* showed that they constitute two separate species but considering their resemblance at the phenotypic level and their close phylogenetic relatedness (see figure 7.2), the transfer of *B. marismortui* to the genus *Salibacillus* has been recently proposed (Arahal *et al.* 2000).

Further moderately halophilic bacilli

So far, we have referred to species that have been validated by publication in the *International Journal of Systematic and Evolutionary Microbiology*, but now we will deal with organisms outside this category. Switzer Blum *et al.* (1998) isolated two Gram-positive anaerobic bacteria from the anoxic muds of Mono Lake (USA) – an alkaline, hypersaline and arsenic-rich lake – and named them *Bacillus arsenicoselenatis* and *Bacillus selenitireducens*. Both strains grow by dissimilatory reduction of As(V) to As(III) with the parallel oxidation of lactate to acetate plus carbon dioxide. *Bacillus arsenicoselenatis* is a spore-forming rod that also grows by dissimilatory reduction of Se(VI) to Se(IV), while *B. selenitireducens* is a short, asporulating rod that grows by dissimilatory reduction of Se(IV) to Se(0). The complete reduction of selenate [Se(VI)] to Se(0) can be achieved by co-culturization (Switzer Blum *et al.* 1998). Although both strains share similar pH requirements for growth (they are alkaliphiles and require pH values of 8.5–10 for optimal growth), they differ in their response to salts. *Bacillus arsenicoselenatis* has its salinity optimum at 6% salt and a salinity range for growth between 2 and 12% salts, while *B. selenitireducens* shows optimal growth in media containing between 2.4 and 6% salts and is able to grow in the range 2–22% salts (Switzer Blum *et al.* 1998). Phylogenetically they are placed in the *Bacillus* branch (figure 7.2).

Another organism that has not yet been validated as a new species is 'Halobacillus thailandensis', a single strain isolated from a fish sauce production line containing 25% NaCl and identified as a member of the genus *Halobacillus* according to its phenotypic and genotypic characteristics (Chaiyanan *et al.* 1999). Three extracellular proteases that are highly expressed in this strain have

been detected, two identified as serine proteases and the third as a metalloprotease. The strain has been used successfully to improve the fermentation process
in the industrial production of fish sauce in Thailand (Chaiyanan *et al.* 1999).

Halotolerant bacilli

The members of the genus *Bacillus* and relatives known to have a tolerant
response to salts are listed in table 7.1. Probably the list could be much longer, but
appropriate tests showing the response to salt (NaCl range and estimation of the
optimal NaCl concentration for growth) are lacking for many strains, especially
those that are not suspected to have halotolerance properties. For instance,
Virgibacillus pantothenticus, although it does not constitute a classical example
of a halophilic microorganism, is able to grow in the presence of 10% salt
(Heyndrickx *et al.* 1998).

Within this group a separation of the alkaliphilic and halotolerant strains and
the nonalkaliphilic ones is possible. Some differentiating features for the species
that can be considered more representative of both subgroups are shown in tables
7.3 and 7.4, respectively. Figure 7.1 shows the phylogenetic tree of the halotolerant species within the bacilli, while the phylogenetic relationships of halotolerant
species with the moderate halophiles are presented in figure 7.2.

Species within the genus *Bacillus*

Boyer *et al.* (1973) described a new subspecies of *Bacillus alcalophilus* that was
named *B. alcalophilus* ssp. *halodurans*, an organism that produced an extracellular alkaline amylase and was able to tolerate up to 12% NaCl, this being the reason for its subspecies status. Later, in a study dealing with alkaliphilic *Bacillus*
strains (Nielsen *et al.* 1995), it was proposed as a separate species under the name
B. halodurans. In that study, where eight other new species were proposed, it was
shown that in spite of the epithet *halodurans*, it only had a moderate tolerance to
NaCl in comparison to other groups. For example *B. pseudofirmus*, *B. agaradhaerens* and *B. clarkii* were able to tolerate up to 16% NaCl. For *Bacillus pseudalcaliphilus* the limit was lower (10% NaCl) but still considerable.

Thus, among the alkaliphilic bacilli a certain degree of halotolerance is not
unusual. In addition to the moderate halophile *B. haloalkaliphilus*, that we have
discussed earlier, there are many other alkaliphilic species that can be regarded as
halotolerant microorganisms. In the study of Nielsen *et al.* (1995), six groups out
of 11 studied were able to grow in contrations of 10% salt, and around 60% of
these strains still grew at 12% salts. It is interesting to note that most of these
strains were isolated from soils.

Another interesting species is the facultatively anaerobic *Bacillus halodenitrificans*, a halophilic denitrifier isolated from a solar saltern by enrichment culture in a liquid medium supplemented with 1.06 M (9%) $NaNO_3$ (Denariaz *et al.*
1989). Removal of nitrate from urban residues or industrial effluents is a major
concern in wastewater treatment plants. A very high concentration of nitrate and

Table 7.3 The main characteristics that differentiate the halotolerant alkaliphilic *Bacillus* species. Data from Boyer *et al.* (1973) and Nielsen *et al.* (1995). All species are rod shaped and produce ellipsoidal endospores.

Feature	B. halodurans	B. pseudofirmus	B. agaradhaerens	B. clarkii	B. pseudalcaliphilus
Size (μm)	0.9–1.0 × 3.0–4.0	0.6–0.8 × 3.0–6.0	0.5–0.6 × 2.0–5.0	0.6–0.7 × 2.0–5.0	0.5–0.6 × 2.0–4.0
Pigmentation	White	Yellow	White	Cream-white	White
Spore position	ST	C/ST	ST	ST	ST
Sporangium swollen	+	–	+	+	+
NaCl range (%)	0–12	0–16	0–16	0–16	0–10
Temperature range (°C)	15–55	10–45	10–45	15–45	10–40
pH optimum	9.0–10.0	9.0	10.0	>10.0	10.0
Hydrolysis of starch	+	+	+	–	+
Nitrate reduction	+	–	+	n.d.	–
Phenylalanine deaminase	–	+	–	n.d.	–
G+C content (mol%)	42.5	39–40.8	39.3–39.5	42.4–43	38.2–39

C, central; ST, subterminal. +, positive; –, negative; n.d., not determined.

Table 7.4 The main characteristics that differentiate the halotolerant, nonalkaliphilic bacilli. Data from Denariaz et al. (1989), Heyndrickx et al. (1999), Lawson et al. (1996), Roberts et al. (1994, 1996) and Wainö et al. (1999). All species are motile rods, oxidase-positive, hydrolyse starch and produce acid from glucose.

Feature	Bacillus			Gracilibacillus		Virgibacillus
	B. halodenitrificans	B. mojavensis	B. vallismortis	G. dipsosauri	G. halotolerans	V. pantothenticus
Size (µm)	0.6–0.8 × 2.5–4.0	0.5–1.0 × 2.0–4.0	0.8–1.0 × 2.0–4.0	n.d.	0.4–0.6 × 2.0–5.0	0.5–0.7 × 2.0–8.0
Pigmentation	Cream	n.d.	n.d.	White	Cream-white	Cream-grey
Spore shape	A	E	E	S	E	S/E
Spore position	A	C	C	T	T	T/ST
Sporangium swollen	A	–	–	+	+	+
Facultative anaerobe	+	–	–	+	–	+
NaCl range (%)	2–25	0–10	0–10	0–15	0–20	0–10
NaCl optimum (%)	3–8	n.d.	n.d.	n.d.	0	4
Temperature range (°C)	10–45	5–55	5–50	28–50	6–50	15–50
pH optimum	7.4	n.d.	n.d.	7.5	7.5	7
Hydrolysis of:						
Casein	+	+	+	–	–	+
Tween 80	–	–	+	n.d.	+	n.d.
Acid produced from mannitol	n.d.	+	+	+	+	–
H_2S production	–	–	–	–	+	–
Urease	–	–	–	–	+	–
Menaquinone system	MK-7	n.d.	n.d.	MK-7	MK-7	MK-7
Cell wall type	n.d.	n.d.	n.d.	m-Dpm	m-Dpm	m-Dpm
G+C content (mol%)	38	43	43	39.4	38	36.9

A, asporulated; C, central; E, ellipsoidal; S, spherical; ST, subterminal; T, terminal. +, positive; –, negative; n.d, not determined. MK-7, menaquinone with seven isoprene units, m-Dpm, meso-diaminopimelic acid.

sodium ions causes inhibition to many denitrifiers. In addition, the first intermediate of denitrification, nitrite, is often toxic. Nevertheless, *B. halodenitrificans*
showed a great flexibility in many senses: broad pH, salt and temperature ranges
for growth, and good tolerance to nitrate and nitrite (Denariaz *et al.* 1989). This
organism was never observed to produce spores. It has been reported that many
bacilli tested fail to sporulate under anaerobic conditions and that strains known
to produce spores seem to lose this ability after exposure to high concentrations
of nitrate (Couchot & Maier 1974), as is the case in the medium used for the isolation of *B. halodenitrificans* (Denariaz *et al.* 1989). Phylogenetically, the closest
relatives of *B. halodenitrificans* are the species of the genera *Salibacillus* and
Virgibacillus (figure 7.2).

Bacillus mojavensis (Roberts *et al.* 1994) was isolated from soil samples
collected in several deserts (Mojave, Gobi, Sahara and Tumamo Hill). It is a
halotolerant organism which grows well at 10% NaCl but it is not reported to be
alkaliphilic. This is also the case for *B. vallismortis* (Roberts *et al.* 1996) isolated
from Death Valley soil.

Bacillus marinus (Rüger 1983) is unable to grow without salts but it grows in
media with ≤3.5% NaCl and some strains grow weakly at 7% NaCl. Therefore,
according to the classification of Kushner and Kamekura (1988), this marine
organism is considered as a slight halophile rather than a moderately halophilic
or halotolerant microorganism.

Bacillus infernus (Boone *et al.* 1995), isolated from a depth of 2700 m in the
terrestrial subsurface, is a thermophilic, strictly anaerobic Fe(III) and Mn(IV)
reducer, also reported to be halotolerant. As it is able to grow at only up to 3.5%
NaCl, it too is a slight halophile.

The genus *Gracilibacillus*

This recently created genus (Wainø *et al.* 1999) is represented to date by two
halotolerant species: *Gracilibacillus halotolerans*, which is the type species and
was described at the same time as the genus, and *G. dipsosauri*, a species originally described as *Bacillus dipsosauri* (Lawson *et al.* 1996) and transferred to
the genus *Gracilibacillus* on the basis of chemotaxonomic features and its close
phylogenetic position to *G. halotolerans* (Wainø *et al.* 1999). The exotic hypersaline environment habitat from which the single strain *G. dipsosauri* was isolated was the salt glands in the nasal cavity of desert iguanas (*Dipsosaurus
dorsalis*). The halotolerant properties of this microorganism enabled it to grow in
media containing up to 15% salts.

Applications of moderately halophilic and halotolerant
species of the genus *Bacillus* and related genera

There are two recent reviews of the biotechnological applications of halophiles
(Ventosa & Nieto 1995; Ventosa *et al.* 1998b). Therefore, we only summarize
here some actual and potential applications of moderately halophilic and halotolerant bacilli.

Enzymes

Moderate halophiles produce extra- and intracellular enzymes, such as amylases, proteases, lipases, nucleases, etc., that are or may be of industrial interest. However, few studies have reported the features of these enzymes. Most work has been carried out in Japan and has been reviewed by Kamekura (1986). *Bacillus halophilus* produced an halophilic exonuclease that was active at 1.4–3.2 M NaCl or 2.3–3.2 M KCl. Also, it was stimulated by Mg^{2+} and Ca^{2+} ions (Onishi *et al.* 1983).

Kamekura and Onishi (1974) reported a moderate halophile, strain 21–1, isolated from unrefined solar salt that produces a protease. It was not studied taxonomically in detail but it was assigned to the genus *Bacillus*. More recently, it was reported that this organism grows better in media without salts and, thus, it should be considered to be halotolerant and not as a moderate halophile (Garabito *et al.* 1998).

Nagata *et al.* (1995) reported an halophilic and thermophilic isolate from sand of Tottori Dune that produces a thermostable and halophilic leucine dehydrogenase. The organism was identified as a new strain of *B. licheniformis*, and the enzyme, which consisted of eight subunits, was very similar to that of *B. stearothermophilus* (there was 84.6% identity in the amino acid sequence deduced from the nucleotide sequence of the gene). Indeed, both enzymes were shown to have very similar thermostability, substrate and coenzyme specificities, and stereospecificity for hydrogen transfer. Nevertheless, the *B. licheniformis* enzyme was much more stable in the presence of high salt concentrations.

Exopolysaccharides and lipopeptides

Microorganisms producing exopolymers may be potentially useful for *in situ* enhanced oil recovery (EOR) processes. They may be indigenous to the reservoir or injected from above ground and act as selective plugging agents. They are fed with a carbohydrate-based medium to stimulate growth and the *in situ* production of extracellular polymer. The purpose is to reduce the permeability of the high permeability regions, which improves the sweeping efficiency, by avoiding the bypassing of oil entrapped in the less permeable regions which remains uncovered when the injected fluid moves freely through the more permeable regions (Pfiffner *et al.* 1986). Microorganisms used in this way must grow and produce exopolymers under the environmental conditions that occur in the reservoir, which are often anaerobic and with high salinities and temperatures. One bacterium able to do this is strain SP018, a Gram-positive spore-forming rod that grows and produces a heteropolysaccharide at salinities up to 12% (optimally between 4 and 10%) and is also thermotolerant (Pfiffner *et al.* 1986). Recently, Bouchotroch *et al.* (1999) described the isolation and characterization of moderately halophilic bacteria that produced exopolysaccharides from several salterns in Morocco, six of them were Gram-positive endospore-forming and were tentatively placed in the genus *Bacillus*.

Another approach to EOR involves the use of biosurfactants. These are useful because they reduce the interfacial tension between the oil and the brine. *Bacillus licheniformis* strain BAS50 was isolated from a deep oil well and produces a lipopeptide surfactant (lichenysin A) at salinities up to 13% NaCl (optimally at 5% NaCl), both under aerobic and anaerobic conditions (Yakimov *et al.* 1995).

Compatible solutes

The intracellular ion concentration of moderately halophilic and halotolerant bacteria is generally insufficient to provide an osmotic balance with the external growth medium, and most of them accumulate organic intracellular solutes which are called 'compatible solutes' because they provide osmotic balance without interfering with the cellular metabolism (Brown 1976; Galinski 1995). *Bacillus halophilus, B. haloalkaliphilus* and *Halobacillus halophilus* synthesize ectoine (derived from N-acetylated diamino-butyrate) as the main compatible solute. Proline, N-acetylornithine and N-acetyllysine are also found in these moderate halophiles (Ventosa *et al.* 1998b). Ectoines (ectoine and hydroxyectoine) are considered to be compounds of great biotechnological importance, because they can be used as stabilizers of enzymes and whole cells. Currently, they are produced industrially for molecular biology kits and several other applications (Bitop, Germany) using *Marinococcus* as producer strain (Frings *et al.* 1995).

Other applications

Several applications have been suggested for moderate halophiles. They can be used for the production of antimicrobial and bioactive compounds, in the salted food industry, as fish and meat fermenting microorganisms, and for the degradation of toxic compounds (Ventosa & Nieto 1995; Ventosa *et al.* 1998b). Some industrial activities generate toxic industrial residues and these should be degraded before they are disposed of. As many of these residues have a high concentration of salts, the biological treatment of them requires the use of halotolerant or moderate halophiles that are able to degrade compounds under saline conditions. Organisms like *B. halodenitrificans* could be used for the anaerobic treatment of industrial effluents containing high concentrations of sodium nitrate (Denariaz *et al.* 1989). In recent years, we have isolated a large number of halotolerant and moderately halophilic bacteria that are able to grow over a wide range of salt concentrations and are tolerant to high concentrations of heavy metals (Ríos *et al.* 1998, unpubl. results). Further work is necessary to establish whether any of these organisms are of use in the treatment of saline waste polluted with heavy metals.

Acknowledgements

Work in our laboratory was supported by grants from the Ministerio de Educación y Cultura, Spain (grants 1FD97-1162 and PB98-1150), and Junta de Andalucía.

References

Arahal, D.R., Dewhirst, F.E., Paster, B.J., Volcani, B.E. & Ventosa, A. (1996) Phylogenetic analyses of some extremely halophilic Archaea isolated from Dead Sea water, determined on the basis of their 16S rRNA sequences. *Applied and Environmental Microbiology* 62, 3779–3786.

Arahal, D.R., Márquez, M.C., Volcani, B.E., Schleifer, K.H. & Ventosa, A. (1999) *Bacillus marismortui* sp. nov., a new moderately halophilic species from the Dead Sea. *International Journal of Systematic Bacteriology* 49, 521–530.

Arahal, D.R., Márquez, M.C., Volcani, B.E., Schleifer, K.H. & Ventosa, A. (2000) Reclassification of *Bacillus marismortui* as *Salibacillus marismortui* comb. nov. *International Journal of Systematic and Evolutionary Microbiology* 50, 1501–1503.

Bejar, V., Quesada, E., Gutierrez, M.C. *et al.* (1992) Taxonomic study of moderately halophilic Gram-positive endospore-forming rods. *Systematic and Applied Microbiology* 14, 223–228.

Boone, D.R., Liu, Y., Zhao, Z.J. *et al.* (1995) *Bacillus infernus* sp. nov., an Fe(III) and Mn(IV) reducing anaerobe from the deep terrestrial subsurface. *International Journal of Systematic Bacteriology* 45, 441–448.

Bouchotroch, S., Quesada, E., Del Moral, A. & Béjar, V. (1999) Taxonomic study of exopolysaccharide-producing, moderately halophilic bacteria isolated from hypersaline environments in Morocco. *Systematic and Applied Microbiology* 22, 412–419.

Boyer, E.E., Ingle, M.B. & Mercer, G.D. (1973) *Bacillus alcalophilus* subsp. *halodurans* subsp. nov.: an alkaline amylase producing alkalophilic organism. *International Journal of Systematic Bacteriology* 23, 238–242.

Brown, A.D. (1976) Microbial water stress. *Bacteriological Reviews* 40, 803–846.

Chaiyanan, S., Chaiyanan, S., Maugel, T., Huq, A., Robb, F.T. & Colwell, R.R. (1999) Polyphasic taxonomy of a novel *Halobacillus, Halobacillus thailandensis* sp. nov. isolated from fish sauce. *Systematic and Applied and Microbiology* 22, 360–365.

Claus, D., Fahmy, F., Rolf, H.J. & Tosunoglu, N. (1983) *Sporosarcina halophila* sp. nov., an obligate, slightly halophilic bacterium from salt marsh soils. *Systematic and Applied Microbiology* 4, 496–506.

Couchot, K.R. & Maier, S. (1974) Anaerobic sporulation in facultatively anaerobic species of the genus *Bacillus*. *Canadian Journal of Microbiology* 20, 1291–1296.

Denariaz, G., Payne, W.J. & Gall, J.L. (1989) A halophilic denitrifier, *Bacillus halodenitrificans* sp. nov. *International Journal of Systematic Bacteriology* 39, 147–151.

Dohrmann, A.B. & Müller, V. (1999) Chloride dependance of endospore germination in *Halobacillus halophilus*. *Archives of Microbiology* 172, 264–267.

Frings, E., Sauer, T. & Galinski, E.A. (1995) Production of hydroxyectoine: high cell-density cultivation and osmotic downshock of *Marinococcus* strain M52. *Journal of Biotechnology* 43, 53–61.

Fritze, D. (1996) *Bacillus haloalkaliphilus* sp. nov. *International Journal of Systematic Bacteriology* 46, 98–101.

Galinski, E.A. (1995) Osmoadaptation in bacteria. *Advances in Microbial Physiology* 37, 273–328.

Garabito, M.J., Arahal, D.R., Mellado, E., Márquez, M.C. & Ventosa, A. (1997) *Bacillus salexigens* sp. nov., a new moderately halophilic *Bacillus* species. *International Journal of Systematic Bacteriology* 47, 735–741.

Garabito, M.J., Márquez, M.C. & Ventosa, A. (1998) Halotolerant *Bacillus* diversity in hypersaline environments. *Canadian Journal of Microbiology* 44, 95–102.

Hanzawa, J. & Takeda, S. (1931) On the reddening of boned codfish. *Archiv für Mikrobiologie* 2, 1–22.

Heyndrickx, M., Lebbe, L., Kersters, K., De Vos, P., Forsyth, G. & Logan, N.A. (1998) *Virgibacillus*: a new genus to accomodate *Bacillus pantothenticus* (Proom and Knight 1950). Emended description of *Virgibacillus pantothenticus*. *International Journal of Systematic Bacteriology* 48, 99–106.

Heyndrickx, M., Lebbe, L., Kersters, K. *et al.* (1999) Proposal of *Virgibacillus proomii* sp. nov. and emended description of *Virgibacillus pantothenticus* (Proom and Knight 1950) Heyndrickx *et al.* 1998. *International Journal of Systematic Bacteriology* 49, 1083–1090.

Hof, T. (1935) Investigations concerning bacterial life in strong brines. *Recueil des Travaux Botaniques Néerlandais* 32, 92–171.

Kamekura, M. (1986) Production and function of enzymes of eubacterial halophiles. *FEMS Microbiology Reviews* 39, 145–150.

Kamekura, M. & Onishi, H. (1974) Protease formation by a moderately halophilic *Bacillus* strain. *Applied Microbiology* 27, 809–810.

Kushner, D.J. & Kamekura, M. (1988) Physiology of halophilic eubacteria. In: *Halophilic Bacteria* (ed. F. Rodríguez-Valera), Vol. 1, pp. 109–138. CRC Press, Boca Raton, FL.

Lawson, P.A., Deutch, C.E. & Collins, M.D. (1996) Phylogenetic characterization of a novel salt-tolerant *Bacillus* species: description of *Bacillus dipsosauri* sp. nov. *Journal of Applied Bacteriology* **81**, 109–112.

Marquez, M.C., Ventosa, A. & Ruiz-Berraquero, F. (1987) A taxonomic study of heterotrophic halophilic and non-halophilic bacteria from a solar saltern. *Journal of General Microbiology* **133**, 45–56.

Nagata, S., Bakthavatsalam, S., Galkin, A.G. *et al.* (1995) Gene cloning, purification, and characterization of a thermostable and halophilic leucine dehydrogenase from a halophilic thermophile, *Bacillus licheniformis* TSN9. *Applied Microbiology and Biotechnology* **44**, 432–438.

Nielsen, P., Fritze, D. & Priest, F.G. (1995) Phenetic diversity of alkaliphilic *Bacillus* strains: proposal for nine new species. *Microbiology* **141**, 1745–1761.

Onishi, H., Mori, T., Takeuchi, S., Tani, K., Kobayashi, T. & Kamekura, M. (1983) Halophilic nuclease of a moderately halophilic *Bacillus* sp.: production, purification, and characterization. *Applied and Environmental Microbiology* **45**, 24–30.

Pfiffner, S.M., McInerney, M.J., Jenneman, G.E. & Knapp, R.M. (1986) Isolation of halotolerant, thermotolerant, facultative polymer-producing bacteria and characterization of the exopolymer. *Applied and Environmental Microbiology* **51**, 1224–1229.

Quesada, E., Ventosa, A., Rodriguez-Valera, F. & Ramos-Cormenzana, F. (1982) Types and properties of some bacteria isolated from hypersaline soils. *Journal of Applied Bacteriology* **53**, 155–161.

Ríos, M., Nieto, J.J. & Ventosa, A. (1998) Numerical taxonomy of heavy metal-tolerant nonhalophilic bacteria isolated from hypersaline environments. *International Microbiology* **1**, 45–51.

Roberts, M.S., Nakamura, L.K. & Cohan, F.M. (1994) *Bacillus mojavensis* sp. nov., distinguishable from *Bacillus subtilis* by sexual isolation, divergence in DNA sequence, and differences in fatty acid composition. *International Journal of Systematic Bacteriology* **44**, 256–264.

Roberts, M.S., Nakamura, L.K. & Cohan, F.M. (1996) *Bacillus vallismortis* sp. nov., a close relative of *Bacillus subtilis*, isolated from soil in Death Valley, California. *International Journal of Systematic Bacteriology* **46**, 470–475.

Rodriguez-Valera, F., Ventosa, A., Juez, G. & Imhoff, J.F. (1985) Variation of environmental features and microbial populations with salt concentrations in a multi-pond saltern. *Microbial Ecology* **11**, 107–115.

Roeßler, M. & Müller, V. (1998) Quantitative and physiological analyses of chloride dependence of growth of *Halobacillus halophilus*. *Applied and Environmental Microbiology* **64**, 3813–3817.

Roeßler, M., Wanner, G. & Müller, V. (2000) Motility and flagellum synthesis in *Halobacillus halophilus* are chloride dependent. *Journal of Bacteriology* **182**, 532–535.

Rüger, H.-J. (1983) Differentiation of *Bacillus globisporus*, *Bacillus marinus* comb. nov., *Bacillus aminovorans*, and *Bacillus insolitus*. *International Journal of Systematic Bacteriology* **33**, 157–161.

Spring, S., Ludwig, W., Márquez, M.C., Ventosa, A. & Schleifer, K.-H. (1996) *Halobacillus* gen. nov., with description of *Halobacillus litoralis* sp. nov. and *Halobacillus trueperi* sp. nov., and transfer of *Sporosarcina halophila* to *Halobacillus halophilus* comb. nov. *International Journal of Systematic Bacteriology* **46**, 492–496.

Switzer Blum, J., Burns Bindi, A., Buzzeli, J., Stolz, J.F. & Oremland, R.S. (1998) *Bacillus arsenicoselenatis*, sp. nov., and *Bacillus selenitireducens*, sp. nov.: two haloalkaliphiles from Mono Lake, California that respire oxyanions of selenium and arsenic. *Archives of Microbiology* **171**, 19–30.

Trüper, H.G. & Galinski, E.A. (1986) Concentrated brines as habitats for microorganisms. *Experientia* **42**, 1182–1187.

Ventosa, A. & Arahal, D.R. (1999) Microbial life in the Dead Sea. In: *Enigmatic Microorganisms and Life in Extreme Environments* (ed. J. Seckbach), pp. 357–368. Kluwer, Dordrecht.

Ventosa, A. & Nieto, J.J. (1995) Biotechnological applications and potentialities of halophilic microorganisms. *World Journal of Microbiology and Biotechnology* **11**, 85–94.

Ventosa, A., Ramos-Cormenzana, A. & Kocur, M. (1983) Moderately halophilic Gram-positive cocci from hypersaline environments. *Systematic and Applied Microbiology* **4**, 564–570.

Ventosa, A., García, M.T., Kamekura, M., Onishi, H. & Ruiz-Berraquero, F. (1989) *Bacillus halophilus* sp. nov., a new moderately halophilic *Bacillus* species. *Systematic and Applied Microbiology* **12**, 162–166.

Ventosa, A., Márquez, M.C., Garabito, M.J. & Arahal, D.R. (1998a) Moderately halophilic gram-positive bacterial diversity in hypersaline environments. *Extremophiles* **2**, 297–304.

Ventosa, A., Nieto, J.J. & Oren, A. (1998b) Biology of moderately halophilic aerobic bacteria. *Microbiology and Molecular Biology Reviews* **62**, 504–544.

Ventosa, A., Arahal, D.R. & Volcani, B.E. (1999) Studies on the microbiota of the Dead Sea – 50 years later. In: *Microbiology and Biogeochemistry of Hypersaline Environments* (ed. A. Oren), pp. 139–147. CRC Press, Boca Raton, FL.

Volcani, B. (Elazari-Volcani, B.) (1940) *Studies on the microflora of the Dead Sea*. Ph.D. thesis, The Hebrew University of Jerusalem, Israel.

Vreeland, R.H. (1987) Mechanisms of halotolerance in microorganisms. *Critical Reviews in Microbiology* 14, 311–356.

Wainö, M., Tindall, B.J., Schumann, P. & Ingvorsen, K. (1999) Gracilibacillus gen. nov., with description of *Gracilibacillus halotolerans* gen. nov., sp. nov.; transfer of *Bacillus dipsosauri* to *Gracilibacillus dipsosauri* comb. nov., and *Bacillus salexigens* to the genus *Salibacillus* gen. nov., as *Salibacillus salexigens* comb. nov. *International Journal of Systematic Bacteriology* 49, 821–831.

Weisser, J. & Trüper, H.G. (1985) Osmoregulation in a new haloalkaliphilic *Bacillus* from the Wadi Natrun (Egypt). *Systematic and Applied Microbiology* 6, 7–11.

Yakimov, M.M., Timmis, K.N., Wray, V. & Fredrickson, H.L. (1995) Characterization of a new lipopeptide surfactant produced by thermotolerant and halotolerant subsurface *Bacillus licheniformis* BAS50. *Applied and Environmental Microbiology* 61, 1706–1713.

Chapter 8

Bacillus Identification – Traditional Approaches

Dagmar Fritze

What is a *Bacillus*?

This heading, posing a crucial question, is taken from Slepecky (1992), who presented a very useful overview of the history and taxonomic status of the genus in the context of the industrial applications of its members. The answer to this question cannot be short as only a few, salient characteristics are used to circumscribe this group. Thus, there is a multitude of quite different organisms within the *Bacillus* taxon.

The present definition, as it stands, says 'aerobic, endospore forming, Gram-positive, rod-shaped organisms'. However, several species have been allocated to *Bacillus* even though they are described as nonspore-formers, for example, *B. thermoamylovorans* (Combet-Blanc *et al.* 1995) and *B. halodenitrificans* (Denariaz *et al.* 1989). Similarly, organisms constantly staining Gram-negative or those staining only weakly positive, like *B. thermosphaericus* (Andersson *et al.* 1995), *B. horti* (Yumoto *et al.* 1998), *B. oleronius* (Kuhnigk *et al.* 1995a) and *B. azotoformans* (Pichinoty *et al.* 1983) have been included in the genus. In addition, when classification is based upon rRNA/DNA sequence analysis, nonspore-forming, strict anaerobes are also placed within the *Bacillus* taxon (*B. infernus*; Boone *et al.* 1995), as well as a coccus-shaped species *Halobacillus halophilus* (ex *Sporosarcina halophila*) and two rod-shaped species *H. trueperi* and *H. litoralis* in a neighbouring genus (Spring *et al.* 1996). Also, with nucleic acid sequences as the determining background, numerous traditional nonspore-forming species and genera overlap with the domain of the spore-formers (see Chapter 2). Meanwhile, several groups of species have been separated from the 'core' of *Bacillus* and described as new genera, and, conversely, there exist new genera which all fall more or less under the definition given above.

For the purpose of this chapter the term '*Bacillus* and related genera' is adopted and used for 'organisms that, with some exceptions, are able to produce endospores, are, with some exceptions, rod-shaped, are able to grow aerobically and which have a Gram-positive cell wall'. Thus, herein there is discussion neither of the nonspore-forming genera in the rRNA/DNA radiation of the genus *Bacillus* nor the obligately anaerobic species *B. infernus* (Boone *et al.* 1995). Another group that needs to be mentioned here because of its ability to form typical endospores, is the genus *Thermoactinomyces*. It also falls within the rRNA/DNA radiation of the genus *Bacillus*. However, because of its distinct appearance on solid media, where it forms aerial mycelia like the streptomycetes, the species of this genus are not covered further.

At the time of writing (April 2000) there were 16 validly published genera of aerobic, spore-forming, Gram-positive organisms (table 8.1). As can be seen, the organisms of some of these genera have previously been allocated to *Bacillus*. Others have been created to harbour new organisms. The definitions of some of these 'new out of old' genera (e.g. *Aneurinibacillus, Brevibacillus, Paenibacillus*) and of the ones comprising new isolates (e.g. *Thermobacillus*) have been based mainly on sequence results. With *Gracilibacillus* and *Salibacillus*, chemotaxonomic data have been considered also, and for *Halobacillus* the presence of an Orn-D-Asp-containing cell wall was established. *Ammoniphilus* has been described as a genus of ammonium dependent and obligately oxalotrophic organisms. *Sporolactobacillus* and *Amphibacillus* have been differentiated from strains of *Bacillus* species mainly through their response to oxygen, because optimal growth occurs under microaerophilic conditions. In the case of *Amphibacillus*, spore formation occurs not only under aerobic but also under anaerobic conditions. In addition, members of both genera lack catalase; it should be noted, however, that catalase is also absent in some *Bacillus* species or strains. The genus *Alicyclobacillus* has been formed to encompass 3-ω-alicyclic fatty acid-producing thermoacidophilic organisms, and in this it resembles the genus *Sulfobacillus* – the species of which have a slightly lower pH optimum and a clearly lower pH minimum than those of the genus *Alicyclobacillus*. Both genera are close neighbours on an rRNA branch on the lower end of which the species *B. tusciae* and *B. schlegelii* are found.

Despite the reported divergences – some bigger, some smaller – from the genus description, the key characteristics 'aerobic', 'spore formation' and 'Gram-positive' might still be meaningful, and might be used (on a higher level) to embrace all of the mentioned genera. However, as different techniques are used across the *Bacillus* taxon to test for these characteristics, attention turns in this chapter to these technical details.

Some technical details concerning spore formation and cell morphology, oxygen effect on spore formation and Gram reaction

Spore formation and cell morphology

Today, cellular morphology is often judged to be insufficiently clear or constant and therefore is considered to be of little importance. Seeing the quality of some of the published microphotographs, which are often very poor, this view can be understood (while not shared). When it comes to the question of spore formation and sporangium shape, researchers tend to forget that spore formation usually needs to be induced, for example through some kind of deficiency or suboptimal condition or through the addition of certain compounds to the growth medium. Often organisms are incubated in rich media with compounds which are easy to

Table 8.1 Validly published genera of the aerobic endospore-forming organisms (as of May 2000).

Original name	Correct name	No. of sp. or ssp.	Aerobic/facultatively anaerobic growth	Spore form	Gram reaction
Bacillus	Alicyclobacillus Wisotzkey et al. (1992)	4	+/+	e	+, v
–	Ammoniphilus Zaitsev et al. (1998)	2	+/–	e	v
–	Amphibacillus Niimura et al. (1990)	1	+/+[a]	e	+
Bacillus	Aneurinibacillus Shida et al. (1996)	3	+/–	e	+
Bacillus	Bacillus Cohn (1872)	73/2 (c. 80 serovar.)	+/+(+)	e, r, e/r, n.d.	+, v, –
Bacillus	Brevibacillus Shida et al. (1996)	10	+/–[b]	e	+, v
Bacillus	Gracilibacillus Wainö et al. (1999)	2	+/–	e	+
Bacillus/Sporosarcina	Halobacillus Spring et al. (1996)	3	+/–	e, r	+
Bacillus	Paenibacillus Ash et al. (1993, 1994)	24/2	+/+	e	+, v, –
Bacillus	Salibacillus Wainö et al. (1999)	1	+/–	e	+
Sporolactobacillus	Sporolactobacillus Kitahara & Suzuki (1963)	5/2	Microaerobic	e	+
Sporosarcina	Sporosarcina Kluyver & van Niel (1936)	1	+/–	r	+
Sulfobacillus	Sulfobacillus Golovacheva & Karavaiko (1978, 1991)	3	+/–	e	+, v
Thermoactinomyces	Thermoactinomyces Tsiklinsky (1899)	8	+/–	r	+
–	Thermobacillus Touzel et al. (2000)	1	+/–	e	–
Bacillus	Virgibacillus Heyndrickx et al. (1998)	2	+/+	e/r	+

[a] Spore formation has been observed also under anaerobic conditions.
[b] One species described as being facultatively anaerobic.
e, ellipsoid; r, round; e/r, both kinds in one culture; n.d., not detected; v, variable; (+), one obligately anaerobic species has been allocated to Bacillus.

assimilate, with the expectation that spore formation will nevertheless occur. Under such conditions cells are more likely to lyse than to form spores and the conclusion 'nonspore-former' is easily drawn. If spores are formed at all under 'rich' conditions, the very few spores detectable may have either already been liberated from the cell, which makes a decision on sporangial shape impossible, or they may be within collapsed cells, the appearance of which will, to the inexperienced observer, look like a nonswollen sporangium. It might, however, be a swollen one; from a collapsed cell no deduction can be drawn as to the shape of a sporangium. Poor microscopic methodology can also lead to incorrect conclusions. At low magnification (×40 objective), with badly adjusted beams or improper focusing, haloes around spores may easily lead to the false impression of swollen sporangia.

An interesting approach to the question of spore-forming ability has been published by Brill and Wiegel (1997), who described a PCR method based on certain sporulation genes to distinguish between bacteria containing sporulation genes and those which do not, regardless of whether sporulation is observed or not.

An arrangement of all validly published species (as of April 2000) of aerobic spore-forming organisms according to spore and sporangial shape is given in table 8.2.

Table 8.2 Morphological groups of aerobic endospore-forming bacteria.

Group I   Cells rod-shaped, spores ellipsoid, sporangia not swollen
SUBGROUP 1: CELL DIAMETER >1 μm

B. anthracis	B. fastidiosus	B. pseudomycoides
B. benzoevorans	B. megaterium	B. thuringiensis
B. cereus	B. mycoides	B. weihenstephanensis

SUBGROUP 2: CELL DIAMETER ≤1 μm

Amm. oxalaticus (also II)	B. firmus	B. pseudofirmus
Amm. oxalivorans (also II)	B. flexus	B. pumilus
	B. gibsonii	B. salexigens
B. alcalophilus	B. halmapalus	B. simplex
B. amyloliquefaciens	B. halophilus	B. smithii (also II)
B. atrophaeus	B. lentus	B. sporothermodurans
B. badius	B. licheniformis	B. subtilis (2 ssp.)
B. clausii (also II)	B. mojavensisB.	B. vallismortis
B. coagulans (also II)	mucilaginosus	
B. edaphicus	B. oleronius	H. litoralis (also II)
		H. trüperi (also II)

Group II   Cells rod-shaped, spores ellipsoid, sporangia swollen

Al. acidocaldarius	B. pallidus	P. azotofixans
Al. acidoterrestris	B. pseudalcaliphilus	P. campinasensis
Al. cycloheptanicus	B. psychrosaccharolyticus	P. chibensis
Al. hesperidum	B. smithii (also I)	P. chondroitinus
	B. stearothermophilus	P. curdlanolyticus
Amm. oxalaticus (also I)	B. thermoamylovorans	P. dendritiformis
Amm. oxalivorans (also I)	B. thermocatenulatus	P. glucanolyticus
	B. thermocloacae	P. illinoisensis

(continued on p. 104)

Table 8.2 *(cont'd)*

Amp. xylanus	*B. thermodenitrificans*	*P. kobensis*
	B. thermoglucosidasius	*P. larvae* (2 ssp.)
An. aneurinilyticus	*B. thermoleovorans*	*P. lautus*
An. migulanus	*B. tusciae*	*P. lentimorbus*
An. thermoaerophilus	*B. vedderi*	*P. macerans*
		P. macquariensis
B. agaradhaerens	*Br. agri*	*P. pabuli*
B. azotoformans	*Br. borstelensis*	*P. peoriae*
B. carboniphilus	*Br. brevis*	*P. polymyxa*
B. chitinolyticus	*Br. centrosporus*	*P. popilliae*
B. circulans	*Br. choshinensis*	*P. thiaminolyticus*
B. clarkii	*Br. formosus*	*P. validus*
B. clausii (also I)	*Br. laterosporus*	
B. coagulans (also I)	*Br. parabrevis*	*Sal. salexigens*
B. cohnii	*Br. reuszeri*	
B. ehimensis	*Br. thermoruber*	*Spl. inulinus*
B. fusiformis		*Spl. kofuensis*
B. halodurans	*G. halotolerans*	*Spl. lactosus*
B. horikoshii		*Spl. nakayamae* (2 ssp.)
B. horti	*H. litoralis* (also I)	*Spl. terrae*
B. kaustophilus	*H. trüperi* (also I)	
B. laevolacticus		*Sul. acidophilus*
B. marismortui	*P. alginolyticus*	*Sul. disulfidooxidans*
B. methanolicus	*P. alvei*	*Sul. thermosulfidooxidans*
B. naganoensis	*P. amylolyticus*	
B. niacini	*P. apiarius*	*Thb. xylanilyticus*

Group III Cells rod-shaped, spores round, sporangia not swollen
B. insolitus   *B. marinus*

Group IV Cells rod-shaped, spores round, sporangia swollen
B. globisporus   *B. psychrophilus*   *G. dipsosauri*
B. haloalkaliphilus   *B. schlegelii*
B. pasteurii   *B. silvestris*

Group V Cells rod-shaped, spores round and ellipsoid, sporangia swollen
B. sphaericus   *V. pantothenticus*
B. thermosphaericus   *V. proomii*

Species not allocated to above morphological groups
NO SPORES DEMONSTRATED
B. halodenitrificans   *B. infernus*   *B. thermoamylovorans*

ROUND CELLS, ROUND SPORES
H. halophilus   *Sps. ureae*
  (ex *Sps. halophila*)

Abbreviations of genera as they appear in the table: *B.*, *Bacillus*; *Amm.*, *Ammoniphilus*;
H., *Halobacillus*; *Al.*, *Alicyclobacillus*; *Amp.*, *Amphibacillus*; *An.*, *Aneurinibacillus*;
Br., *Brevibacillus*; *G.*, *Gracilibacillus*; *P.*, *Paenibacillus*; *Sal.*, *Salibacillus*; *Spl.*,
Sporolactobacillus; *Sul.*, *Sulfobacillus*; *Thb.*, *Thermobacillus*; *V.*, *Virgibacillus*; *Sps.*,
Sporosarcina.

Effect of oxygen on spore formation

In the context of *Bacillus*, the term aerobic is used in the sense that spore formation occurs only under aerobic conditions. This is in contrast to the genus
Clostridium which – by definition – is able to form spores only under anaerobic
conditions. *Bacillus* species growing facultatively anaerobically have long been
known. It is likely, however, that the propensity to form spores under anaerobic
conditions has not been tested with most of the presently recognized genera and
species of aerobic spore-forming organisms. The ability to sporulate under both
aerobic and anaerobic conditions has been described only for *Amphibacillus*.
With some species of the genus *Clostridium* aerobic spore formation has been
observed.

Gram reaction

The majority of all aerobic spore-forming organisms stain Gram-positive,
although not necessarily in all phases of their cell cycle. Often only very young
cultures will definitely stain Gram-positive. As demonstrated by Wiegel (1981),
using electron microscopy, the structure of the cell wall, the 'Gram type', is not
always the same as the Gram-staining reaction. He showed that the cell wall of a
Gram-negatively staining organism can have a true Gram-positive structure. A
Gram-negatively structured cell wall will always stain Gram-negative. However,
a Gram-positively structured cell wall, under certain circumstances, may stain
Gram-positive or Gram-negative.

For a number of species it has been recorded that they stain Gram-negative or,
as was described in the case of *B. horti* (Yumoto *et al.* 1998), lose their colour
more easily than the standard strain used for comparison. This phenomenon is
also known for lactobacilli. For some of these and also for some *Bacillus* strains it
has been shown by chemical analysis that decoloration is not the result of a
Gram-negative cell wall but instead because of a very thin peptidoglycan which
did not hold the dye completely.

The KOH reaction is often regarded as a substitute for Gram staining. As with
the Gram stain, however, a number of *Bacillus* strains have been found to be
lysed by KOH whereas the majority will not.

Advantages of morphological/
physiological approaches to identification

In the applied areas of microbiology it is not only the occurrence of a given organism that is of interest but also its numbers, the occurrence of physiologically similar organisms, the likely source of contamination, why the organism can multiply
in the environment from which it has been isolated, and so on. These correlating
questions can only be answered when knowledge of its physiological properties
is available. Morphological information such as the appearance of colonies
on certain substrates or of cells under the microscope can further accelerate

identification and details of the physiological properties of a given contaminant enable targeted measures to combat it. In particular, whether a given contaminant is a spore-former is of great importance for quality managers in industry.

When food or other substrates are spoilt by microorganisms, this spoilage is caused by organisms growing in or on the substrate. That means these organisms use certain compounds of the substrate as nutrients. They form colonies of certain shapes and colours, they cause typical changes in taste, odour, texture, etc. Only specific groups of organisms will develop in a given nutritional environment. In some cases, obvious spoilage occurs only when the number of organisms reaches a critical level. Here, physiological approaches like selective/elective cultivation, direct enumeration and diagnostic identification are indispensable and often the only way to obtain appropriate answers. The information obtained, for example, by PCR that a certain organism is indeed present may not suffice in this case. Physiological tests will always have their advantages, as the methodology is usually cheap and quick and offers a level of 'exactness' or perhaps better 'non-exactness' that is sometimes more appropiate for the problem than the highly specific molecular methods. Some examples of the practical impact of the aerobic spore-forming organisms are given in table 8.3.

Problems of standardization

Standardization of methodology in the traditional approaches for identification is as important as in any other area of identification. Unfortunately, the identification results of different authors are often difficult to compare because the methods applied were not thoroughly standardized. The physiological tests designed by Smith *et al.* (1952) and Gordon *et al.* (1973) specifically for *Bacillus* are described in detail in Claus and Berkeley (1986) and should be followed precisely. Modifications are often made without need. If adjustments need to be made for the specific requirements of a given organism, this should be clearly stated together with evidence that the standard technique was not applicable. Several attempts have been made to design dichotomous keys for the identification of aerobic spore-formers based on physiological characteristics (Slepecky 1992). At present, however, this approach is thwarted by the difficulties in comparing the published properties of species.

Indeed, with many recently proposed species requiring low or high pH (reported from 0.5 to 11), low or high temperatures (reported from $-5°C$ to $78°C$) or low or high NaCl concentrations (reported from 0% NaCl added to the medium to 30%) for growth, characterization results cannot and should not be directly compared without full background knowledge of the precise conditions used for a particular test. Only in this way can there be confidence about comparibility of the results obtained.

For many physiological test media, several recipes have been described and these will yield differing results. Problems often occur with reports of urease activity for example. This should be tested in a medium not containing peptone. Most *Bacillus* species are strongly proteolytic and so any pH change may be the

Table 8.3 Species of *Bacillus* and related genera of relevance to humans with respect to agriculture, health and industry.

Species	Risk group[a]	Application
B. thuringiensis, B. sphaericus	RG 1	Plant pest control
B. cereus	RG 2/RG 1[b]	Probioticum
B. subtilis	RG 1	Fermented food – 'natto'
B. subtilis, B. pumilus, B. licheniformis, B. benzoevorans and others	RG 1	Production of enzymes, antibiotic components, surfactants, degradation of xenobiotics, etc.

Species	Risk group	Pathogenic for
B. anthracis	RG 3	Humans and animals (anthrax)
B. cereus	RG 2	Cattle (mastitis), humans (diarrhoea/nausea/ vomiting)
B. sphaericus (certain strains)	RG 1	Mosquito larvae
B. thuringiensis	RG 1	*Lepidoptera* or *Diptera* larvae
B. weihenstephanensis	RG 2	As for B. cereus
Paenibacillus larvae ssp. larvae	RG 1	Honey bees (foulbrood)
P. popilliae	RG 1	Some *Coleoptera* larvae (milky disease)
P. lentimorbus	RG 1	Some *Coleoptera* larvae (milky disease)

Species	Risk group	Spoilage of
Alicyclobacillus spp.	RG 1	Fruit juices
B. coagulans, B. smithii and B. stearothermophilus	RG 1	Canned food
B. cereus	RG 2	Chilled food
Brevibacillus brevis, Bacillus sporothermodurans and B. kaustophilus	RG 1	Milk

[a] Risk group allocation (European Community 1990, 1993) is done with a view to pathogenicity to humans. Only one species of aerobic spore-forming organisms is allocated to RG 3. Only two species are allocated to RG 2.
[b] Strains of *B. cereus* used as probiotics must have been demonstrated to be nonpathogenic and would thus be classified as RG 1.
Occasionally, opportunistic infections of humans and animals may be caused also by otherwise harmless *Bacillus* species, notably *Paenibacillus alvei, P. laterosporus, B. licheniformis, B. subtilis* and *B. sphaericus*.

consequence of liberated ammonium. Similarly, the starch hydrolysis test will yield different results depending on whether iodine or 95% ethanol is used for revealing nonhydrolysed starch. With the nitrate reduction test, if a combined reagent is used instead of the two separate solutions, as described by Gordon *et al.* (1973), a false negative result may be recorded.

A number of commercial identification kits such as API, Biolog, Vitek, etc. are available for the rapid and standardized identification of *Bacillus*. These are all based on physiological tests. These systems are quite helpful within their abilities, but, again, results obtained with one kit are only comparable with those obtained with the same system, and not with those obtained with other test kits or the conventional test tube/Petri dish tests. Another disadvantage of the

commercial systems is that the range of genera of aerobic spore-forming organisms for which they can be used is relatively restricted. Furthermore, such systems
will nearly always produce a result, no matter what organism (pure or contaminated) has been tested and great experience is needed to be able to interpret the
data obtained.

Gordon *et al.* (1973): reliable as ever

Until the classical works of Smith *et al.* (1952) and Gordon *et al.* (1973) were
published, the taxonomic situation of the aerobic spore-forming organisms
was – at best – chaotic. After recharacterization of over one thousand strains they
reorganized the genus *Bacillus* according to practical and pragmatic viewpoints.
The beauty of their work is that because of the great care taken in the performance of the tests, the large number of strains involved and their critical interpretation of the results, modern researchers can still have confidence in their results.
They were, purposefully and carefully, 'lumpers' of species rather than 'splitters',
knowing that, when better methods became available, the 'lumps' would be
taken apart to form 'good species'. And indeed, there is a high correlation of their
groupings with those based on recent results. Species of aerobic spore-forming
organisms included in the Approved Lists (Skerman *et al.* 1980) and their name
changes up to May 2000 are presented in table 8.4. All other species of aerobic
spore-forming organisms validly published since 1980 up to May 2000 are listed
in table 8.5.

A general approach for the identification of pure cultures of aerobic endospore-
forming organisms today could start with some sort of differential cultivation.
The principle applied would be to combine several physiological properties to
select for certain groups of organisms. A list of all validly published species of
aerobic spore-forming organisms expected to grow under specific temperature or
pH conditions is given in table 8.6. However, it should be noted that, especially
with the recently described new species, ability to grow at low temperatures has
often not been examined, and therefore many additional psychrotolerant species
may exist.

Differential step 1 would be selection for spores by pasteurization or ethanol
treatment (Koransky *et al.* 1978). Ethanol treatment is preferable because not all
spores exhibit the necessary heat resistance to survive the former differential process. Differential step 2 would use the combination of three different incubation
temperatures with three different pH levels.

For thermophiles a temperature of 55°C would be chosen and media at pH 4.5,
pH 7–7.2 and pH 9 prepared and inoculated. For mesophiles a temperature of
30°C would be chosen and media with the same pH values (pH 4.5, pH 7–7.2,
pH 9) inoculated. For psychrophiles a temperature of 5°C would be chosen
and media prepared as above. Through this approach a first subdivision of the
aerobic spore-forming organisms is achieved, which facilitates further taxonomic
differentiation.

A medium which, after appropriate adjustment of pH, allows growth of most
species of aerobic spore-formers (including the halophilic ones) under the con-

Table 8.4 Species of aerobic endospore-forming bacteria as listed in the Approved Lists (Skerman *et al.* 1980) and their name changes up to May 2000.

Approved Lists 1980		Name changes up to May 2000
Bacillus acidocaldarius Darland & Brock (1971)	>>	*Alicyclobacillus acidocaldarius* Wisotzkey *et al.* (1992)
B. alcalophilus Vedder (1934)		–
B. alvei Cheshire & Cheyne (1885)	>>	*Paenibacillus alvei* Ash *et al.* (1993, 1995)
B. anthracis Cohn (1872)		–
B. badius Batchelor (1919)		–
B. brevis Migula (1900)	>>	*Brevibacillus brevis* Shida *et al.* (1996)
B. cereus Frankland & Frankland (1887)		–
B. circulans Jordan (1890)		–
B. coagulans Hammer (1915)		–
B. fastidiosus den Dooren de Jong (1929)		–
B. firmus Bredemann & Werner (1933)		–
B. globisporus Larkin & Stokes (1967)		–
B. insolitus Larkin & Stokes (1967)		–
B. larvae White (1906)	>>	*Paenibacillus larvae* (2 ssp.) Ash *et al.* (1993, 1995)
B. laterosporus Laubach (1916)	>>	*Brevibacillus laterosporus* Shida *et al.* (1996)
B. lentimorbus Dutky (1940)	>>	*Paenibacillus lentimorbus* Petterson *et al.* (1999)
B. lentus Gibson (1935)		–
B. licheniformis Chester (1901)		–
B. macerans Schardinger (1905)	>>	*Paenibacillus macerans* Ash *et al.* (1993, 1995)
B. macquariensis Marshall & Ohye (1966)	>>	*Paenibacillus macquariensis* Ash *et al.* (1993, 1995)
B. megaterium de Bary (1884)		–
B. mycoides Flügge (1886)		–
B. pantothenticus Proom & Knight (1950)	>>	*Virgibacillus pantothenticus* Heyndrickx *et al.* (1998)
B. pasteurii Chester (1898)		–
B. polymyxa Macé (1889)	>>	*Paenibacillus polymyxa* Ash *et al.* (1993, 1994)
B. popilliae Dutky (1940)	>>	*Paenibacillus popilliae* Petterson *et al.* (1999)
B. pumilus Meyer & Gottheil (1901)		–
B. sphaericus Meyer & Neide (1904)		–
B. stearothermophilus Donk (1920)		–
B. subtilis Cohn (1872)		(2 ssp.)
B. thuringiensis Berliner (1915)		–
Sporolactobacillus inulinus Kitahara & Lai (1967)		–
Sporosarcina ureae Kluyver & van Niel (1936)		–
Thermoactinomyces candidus Kurup *et al.* (1975)		–
T. dichotomicus Cross & Goodfellow (1973)		–
T. peptonophilus Nonomura & Ohara (1971)		–
T. sacchari Lacey (1971)		–
T. vulgaris Tsiklinsky (1899)		–

ditions given above is brain heart infusion (Difco), which may also be used with agar as a solid medium. Only very few species such as *B. fastidiosus*, *B. pasteurii* and probably the organisms of the genus *Ammoniphilus* will not grow.

Further diagnostic or identification tests should be applied as appropriate. It should be noted that some of the species listed in table 8.6 (or some strains thereof) are quite tolerant in the medium ranges of pH or temperature and may equally grow under several conditions. As far as possible the standardized tests of Gordon *et al.* (1973) as described in Claus and Berkeley (1986) should be followed. When standard methods cannot be applied, the original literature relating to similar organisms should be consulted.

Table 8.5 Newly described genera and species and new combinations of names of aerobic endospore-forming bacteria since 1980, not including those of table 8.4 or invalidated species (as of May 2000).

Alicyclobacillus acidoterrestris Wisotzkey *et al.* (1992)
Al. cycloheptanicus Wisotzkey *et al.* (1992)
Al. hesperidum Albuquerque *et al.* (2000)
Ammoniphilus oxalaticus Zaitsev *et al.* (1998)
Amm. oxalivorans Zaitsev *et al.* (1998)
Amphibacillus xylanus Niimura *et al.* (1990)
Aneurinibacillus aneurinilyticus Shida *et al.* (1996)
An. migulanus Shida *et al.* (1996)
An. thermoaerophilus Heyndrickx *et al.* (1997)
Bacillus agaradhaerens Nielsen *et al.* (1995a,b)
B. amyloliquefaciens Priest *et al.* (1987)
B. atrophaeus Nakamura (1989)
B. azotoformans Pichinoty *et al.* (1983)
B. carboniphilus Fujita *et al.* (1996)
B. chitinolyticus Kuroshima *et al.* (1996)
B. clarkii Nielsen *et al.* (1995a,b)
B. clausii Nielsen *et al.* (1995a,b)
B. cohnii Spanka & Fritze (1993)
B. edaphicus Shelobolina *et al.* (1997, 1998)
B. ehimensis Kuroshima *et al.* (1996)
B. flexus Priest *et al.* (1988, 1989)
B. fusiformis Priest *et al.* (1988, 1989)
B. gibsonii Nielsen *et al.* (1995a,b)
B. halmapalus Nielsen *et al.* (1995a,b)
B. haloalkaliphilus Fritze (1996)
B. halodenitrificans Denariaz *et al.* (1998)
B. halodurans Nielsen *et al.* (1995a,b)
B. halophilus Ventosa *et al.* (1989a,b)
B. horikoshii Nielsen *et al.* (1995a,b)
B. horti Yumoto *et al.* (1998)
B. kaustophilus Priest *et al.* (1988, 1989)
B. laevolacticus Andersch *et al.* (1994)
B. marinus Rüger (1983)
B. marismortui Arahal *et al.* (1999)
B. methanolicus Arfman *et al.* (1992)
B. mojavensis Roberts *et al.* (1994)
B. mucilaginosus Avakyan *et al.* (1986, 1998)
B. naganoensis Tomimura *et al.* (1990)
B. niacini Nagel & Andreesen (1991)
B. oleronius Kuhnigk *et al.* (1995a,b)
B. pallidus Scholz *et al.* (1987, 1988)
B. pseudalcaliphilus Nielsen *et al.* (1995a,b)
B. pseudofirmus Nielsen *et al.* (1995a,b)
B. pseudomycoides Nakamura (1998)
B. psychrophilus Nakamura (1984)
B. psychrosaccharolyticus Priest *et al.* (1988, 1989)
B. schlegelii Schenk & Aragno (1979, 1981)
B. silvestris Rheims *et al.* (1999)
B. simplex Priest *et al.* (1988, 1989)
B. smithii Nakamura *et al.* (1988)
B. sporothermodurans Petterson *et al.* (1996)
B. thermoamylovorans Combet-Blanc *et al.* (1995)
B. thermocatenulatus Golovacheva *et al.* (1975, 1991)

B. thermocloacae Demharter & Hensel (1989a,b)
B. thermodenitrificans Manachini *et al.* (2000)
B. thermoglucosidasius Suzuki *et al.* (1983, 1984)
B. thermoleovorans Zarilla & Perry (1987, 1988)
B. thermosphaericus Andersson *et al.* (1995, 1996)
B. tusciae Bonjour & Aragno (1984, 1985)
B. vallismortis Roberts *et al.* (1996)
B. vedderi Agnew *et al.* (1995, 1996)
B. weihenstephanensis Lechner *et al.* (1998)
Brevibacillus agri Shida *et al.* (1996)
Br. borstelensis Shida *et al.* (1996)
Br. centrosporus Shida *et al.* (1996)
Br. choshinensis Shida *et al.* (1996)
Br. formosus Shida *et al.* (1996)
Br. parabrevis Shida *et al.* (1996)
Br. reuszeri Shida *et al.* (1996)
Br. thermoruber Shida *et al.* (1996)
Gracilibacillus dipsosauri Wainö *et al.* (1999)
G. halotolerans Wainö *et al.* (1999)
Halobacillus halophilus Spring *et al.* (1996)
H. litoralis Spring *et al.* (1996)
H. trueperi Spring *et al.* (1996)
Paenibacillus alginolyticus Shida *et al.* (1997a)
P. amylolyticus Ash *et al.* (1993, 1995)
P. apiarius Nakamura (1996)
P. azotofixans Ash *et al.* (1993, 1995)
P. campinasensis Yoon *et al.* (1998)
P. chibensis Shida *et al.* (1997b)
P. chondroitinus Shida *et al.* (1997a)
P. curdlanolyticus Shida *et al.* (1997a)
P. dendritiformis Tcherpakov *et al.* (1999)
P. glucanolyticus Shida *et al.* (1997a)
P. illinoisensis Shida *et al.* (1997b)
P. kobensis Shida *et al.* (1997a)
P. lautus Heyndrickx *et al.* (1996)
P. pabuli Ash *et al.* (1993, 1995)
P. peoriae Heyndrickx *et al.* (1996)
P. thiaminolyticus Shida *et al.* (1997a)
P. validus Ash *et al.* (1993, 1995)
Salibacillus salexigens Wainö *et al.* (1999)
Sporolactobacillus kofuensis Yanagida *et al.* (1997)
Sp. lactosus Yanagida *et al.* (1997)
Sp. nakayamae (2 ssp.) Yanagida *et al.* (1997)
Sp. terrae Yanagida *et al.* (1997)
Sulfobacillus acidophilus Norris *et al.* (1996a,b)
Sul. disulfidooxidans Dufresne *et al.* (1996)
Sul. thermosulfidooxidans Golovacheva & Karavaiko (1978, 1991)
Thermoactinomyces intermedius Kurup *et al.* (1980, 1981)
T. putidus Lacey & Cross (1989a,b)
T. thalpophilus Lacey & Cross (1989a,b)
Thermobacillus xylanilyticus Touzel *et al.* (2000)
Virgibacillus proomii Heyndrickx *et al.* (1999)

Table 8.6 Physiological groups within the aerobic endospore-forming organisms.

Group		Organism
Temperature tolerance	T_{max}	
Psychrophilic:	25/30°C	*Bacillus insolitus*
growth from around	25/30°C	*B. globisporus*
−5/−2°C to	25/30°C	*B. marinus*
	25/30°C	*B. psychrophilus*
	35°C	*B. psychrosaccharolyticus*
	25°C	*Paenibacillus macquariensis*
Psychrotolerant:	40°C	*B. weihenstephanensis*
growth from about 7°C to		(ex *B. cereus* and *B. mycoides* strains)
Mesophilic:	40–45°C	*B. cereus*
growth from about 10/15°C to	45–50°C	*B. subtilis*
		and all other species not listed in the other groups
Thermotolerant or moderately	55°C	*Alicyclobacillus acidoterrestris*
thermophilic:	53°C	*Al. cycloheptanicus*
growth from ≤30°C to	~58°C	*Al. hesperidum*
	60°C	*B. coagulans*
	55°C	*B. licheniformis*
	60°C	*B. methanolicus*
	70°C	*B. pallidus*
	65°C	*B. smithii*
	58°C	*B. thermoamylovorans*
	64°C	*B. thermosphaericus*
	58°C	*Brevibacillus thermoruber*
	55°C	*Sulfobacillus thermosulfidooxidans*
	63°C	*Thermobacillus xylanilyticus*
Thermophilic:	70°C	*Al. acidocaldarius*
growth from about 37/40°C to	60°C	*Aneurinibacillus thermoaerophilus*
	75°C	*B. kaustophilus*
	70°C	*B. pallidus*
	~75°C	*B. schlegelii*
	78°C	*B. thermocatenulatus*
	70°C	*B. thermocloacae*
	70°C	*B. thermodenitrificans*
	69°C	*B. thermoglucosidasius*
	70°C	*B. thermoleovorans*
	~60°C	*B. tusciae*
	70°C	*B. stearothermophilus*
pH tolerance	pH range	
Acidophilic:	2–6	*Al. acidocaldarius*
no growth above pH 6	2.2–5.8	*Al. acidoterrestris*
	3.0–5.5	*Al. cycloheptanicus*
	2.5–5.5	*Al. hesperidum*
	4–6	*B. naganoensis*
	~5.6	*B. tusciae*
	2–3	*Sulfobacillus acidophilus*
	0.5–6.0	*Sul. disulfidooxidans*
	1.5–5.5	*Sul. thermosulfidooxidans*
Acidotolerant	4.5–7.7	*B. coagulans*
	4.5–7.7	*B. leavolacticus*
Neutrophilic	~7	*B. cereus*
		B. megaterium
		B. subtilis
		and all other species not listed in the other groups

(*continued on p. 112*)

Table 8.6 (*cont'd*)

Group		Organism
Alkalitolerant or moderately	6.8–9.5	*Ammoniphilus oxalaticus*
alkaliphilic	6.8–9.5	*Amm. oxalivorans*
	5.0–9.0	*Aneurinibacillus aneurinolyticus*
	5.5–9.0	*An. migulanus*
	7–8	*An. thermoaerophilus*
	7–8	*B. clausii*
	7–9	*B. cohnii*
	7–8	*B. gibsonii*
	7–8	*B. halmapalus*
	7–10	*B. halodurans*
	7–8	*B. horikoshii*
	7–10	*B. horti*
	7–11	*B. niacini*
	6.5–8.5	*B. thermoglucosidasius*
	6.5–8.5	*Thermobacillus xylanilyticus*
	pH_{max}	
Alkaliphilic:	10	*Amphibacillus xylanus*
growth above pH 8 up to	>10	*B. agaradhaerens*
	10	*B. alcalophilus*
	>10	*B. clarkii*
	10	*B. haloalkaliphilus*
	10	*B. halodurans*
	8.5	*B. pallidus*
	10	*B. pseudalcaliphilus*
	> 9	*B. pseudofirmus*
	9	*B. thermocloacae*
	10	*B. vedderi*
	>10	*Paenibacillus campinasensis*
NaCl tolerance	Max. conc.	
Halotolerant:	16%	*B. agaradhaerens*
growth from without NaCl up to	16%	*B. clarkii*
	12%	*B. halodurans*
	10%	*B. horti*
	10%	*B. pasteurii*
	10%	*B. pseudalcaliphilus*
	16%	*B. pseudofirmus*
	15%	*Gracilibacillus dipsosauri*
	16%	*Salibacillus salexigens*
	10%	*Virgibacillus pantothenticus*
	10%	*V. proomii*
Halophilic:	20%	*B. haloalkaliphilus*
growth from about 0.5% NaCl to	30%	*B. halophilus*
	25%	*B. marismortui*
	20%	*Gracilibacillus halotolerans*
	24%	*Halobacillus halophilus*
	25%	*H. litoralis*
	30%	*H. trüperi*
Specialized nutrition		
C-compound urate or allantoin		*B. fastidiosus*
Facultatively chemolithotrophic		*B. schlegelii*
		B. tusciae
N-compound ammonia		*B. pasteurii*
Obligately oxalotrophic		*Amm. oxalaticus*
		Amm. oxalivorans

Traditional physiological and morphological groupings of aerobic endospore-forming bacteria

In what follows, some groups which comprise the most commonly occurring and studied species of aerobic spore-forming organisms are described briefly. All grow satisfactorily in the moderate ranges of temperature and pH and have no specific nutritional requirements.

The '*cereus* group'

Cells are >1 μm diameter, their sporangia are not swollen, spores are ellipsoidal and, in principle, they are mesophilic and neutrophilic. Organisms of this group are *B. anthracis*, *B. cereus*, *B. mycoides*, *B. pseudomycoides*, *B. thuringiensis* and *B. weihenstephanensis*.

Bacillus cereus and *B. mycoides* may occur as psychrotolerant strains (= *B. weihenstephanensis*). All of these organisms are placed in 16S rRNA/DNA group 1. Physiologically the members of this group are extremely difficult to distinguish from each other. Nonpathogenic *B. anthracis*, nonrhizoid *B. mycoides* and nontoxin-crystal-carrying *B. thuringiensis* may be judged as *B. cereus* and vice versa. Species with similar cell diameters (*B. megaterium*, *B. benzoevorans*, *B. fastidiosus*) are easily to distinguish physiologically. DNA–DNA hybridizations differentiate clearly between *B. cereus*, *B. mycoides*, *B. pseudomycoides* and *B. thuringiensis* (Nakamura 1994, 1998; Nakamura & Jackson 1995). The group have in common an inability to produce acid from mannitol and being lecithinase-positive. They can be discriminated using tests for toxin crystal formation, motility, haemolysis, penicillin resistance and gamma phage susceptibility (table 8.7).

The '*megaterium* group'

Cells of *B. megaterium* are >1 μm diameter whereas those of *B. flexus* and *B. simplex* are <1 μm. Their sporangia are not swollen, spores are ellipsoidal, and they are mesophilic and neutrophilic. Organisms of this group have been separated from *B. megaterium* on the grounds of morphological and physiological traits and DNA–DNA hybridizations. All species are placed in rRNA/DNA group 1.

The '*subtilis* group'

Cells are <1 μm diameter, their sporangia are not swollen, spores are ellipsoidal and, in principle, they are mesophilic and neutrophilic. Organisms of this group are *B. amyloliquefaciens*, *B. atrophaeus*, *B. licheniformis*, *B. mojavensis*, *B. pumilus*, *B. subtilis* ssp. *subtilis*, *B. subtilis* ssp. *spizizenii* and *B. vallismortis*. All of these organisms are placed in 16S rRNA/DNA group 1 and are physiologically very similar. *Bacillus amyloliquefaciens* is described as distinguishable from

Table 8.7 Some characteristic and differential properties of the *Bacillus cereus* group.

	B. anthracis	B. cereus	B. mycoides	B. pseudomycoides	B. thuringiensis	B. weihenstephanensis
Mannitol	–	–	–	–	–	n.d.[a]
Lecithinase	w	+	+	+	+	n.d.[a]
Rhizoid growth	–	–	+	+	–	–
Toxin crystal	–	+	–	–	+	n.d.[a]
Motility	–	+	–	–	+	n.d.[a]
Haemolysis	–	+	+	n.d.	+	n.d.[a]
Penicillin resistance	–	+	+	n.d.	+	n.d.[a]
Lysed by gamma phage	+	–	n.d.	n.d.	–	n.d.[a]

[a] According to Lechner *et al.* (1998) this species is similar to *B. cereus* in cell morphology and substrate utilization.
–, negative; +, positive; w, weak; n.d., not determined.

B. subtilis by its faster acid production from lactose and slower gluconate utilization. Pigment formation on tyrosine medium differentiates *B. atrophaeus* from *B. subtilis*, from which it is otherwise not distinguishable. Thus far it is impossible to distinguish physiologically between *B. subtilis*, *B. mojavensis* and *B. vallismortis*. All five species are, however, separated using DNA–DNA hybridization. *Bacillus pumilus* is starch-negative and hippurate-positive, *B. licheniformis* is propionate-positive, grows up to 55°C and is facultatively anaerobe. *Bacillus sporothermodurans* may be loosely attached to this group.

The '*circulans* group'

Cells are <1 µm diameter, with swollen sporangia and ellipsoidal spores. They are mesophilic and neutrophilic. Organisms of this group or 'complex' have been separated from *B. circulans* on physiological grounds and DNA–DNA hybridizations. Sequence analysis placed *B. circulans* in rRNA/DNA group 1, whereas all other species descending from *B. circulans* were placed in the new genus *Paenibacillus* (rRNA/DNA group 3) in which several other phenotypically related species also have been placed (e.g. *P. alvei*, *P. macerans*, *P. polymyxa*). At present, the group comprising decendants from the parent species *B. circulans* includes *B. circulans*, *P. alginolyticus*, *P. amylolyticus*, *P. chondroitinus*, *P. glucanolyticus*, *P. illinoisensis*, *P. lautus*, *P. pabuli* and *P. validus*.

The '*brevis* group'

Cells are <1 µm diameter, with swollen sporangia and ellipsoidal spores. They are mesophilic and neutrophilic. Organisms of this group have been separated from '*B. brevis*' on the grounds of physiological data and DNA–DNA hybridization. Sequence analysis effected transfer of all species to the newly formed genus *Brevibacillus* (rRNA/DNA group 4). At present, this group comprises *Br. agri*, *Br. brevis*, *Br. centrosporus*, *Br. choshinensis*, *Br. parabrevis*, *Br. formosus*, *Br. reuszeri* and *Br. borstelensis*.

Concluding remarks

Several other 'lumped' species are being and will be taken apart similarly in future. For example, *B. sphaericus* has been shown to have strains which belong to at least five DNA/DNA homology groups (Krych *et al.* 1980) and three close neighbours have been already described to date: *B. fusiformis*, *B. thermosphaericus* and *B. silvestris*. Another example is *B. stearothermophilus*. Up to now, at least five physiologically related thermophilic or thermotolerant species have been validly described, and the heterogeneity of the strains allocated to *B. stearothermophilus* indicates the variety of species still enclosed therein.

In this context, it should be clarified once more that the choice of species definition protocol determines the degree of overlap between species defined according to different methods. Hasty application of new methods may not be helpful for solving taxonomic problems. At the species level, DNA–DNA hybridizations seem to offer good resolution and are widely accepted as well as recommended (Murray *et al.* 1990; Stackebrandt & Goebel 1994). However, for a recently described species (*B. weihenstephanensis*), where hybridization results do not support this species, results of other methods have been ranked higher.

At the genus level, it is agreed that phylogenetic as well as phenotypic markers are essential for differentiation. Unfortunately, for only a few of the new genera, for example for *Alicyclobacillus*, have common phenotypic properties been determined that distinguish them from *Bacillus* and the other new genera. For others, e.g. *Paenibacillus* or *Brevibacillus* or some of the new genera not consisting of previous *Bacillus* species, differentiation seems less clear. A reason for this might be that genus borderlines, now set mainly by rRNA sequence similarities, have often been decided upon when only a few strains had been sequenced and distances were convincingly long. However, these distances shrink as more and more strains are sequenced and species added. The main points of branching become less obvious as 'bushes' develop instead of 'trees'. The individual decision of a taxonomist on where to cut off a genus may add to the difficulty in finding correlating phenotypic characteristics.

It is obvious that the phenotypic description 'aerobic, Gram-positive, spore-forming bacteria' is applicable not only to the genus *Bacillus sensu stricto* but also to '*Bacillus* and related genera'. As it seems clear that the group will be split up even further, with a view to the needs of the applied sciences, it becomes urgently necessary to establish for the genus *Bacillus* and its related genera more definite and consistent phenotypic descriptions.

References

Agnew, M.D., Koval, S.F. & Jarrell, K.F. (1995) Isolation and characterization of novel alkaliphiles from bauxite-processing waste and description of *Bacillus vedderi* sp. nov., a new obligate alkaliphile. *Systematic and Applied Microbiology* 18, 221–230.

Agnew, M.D., Koval, S.F. & Jarrell, K.F. (1996) *Bacillus vedderi* new species. In: *Validation of the publication of new names and new combinations previously effectively published outside the IJSB*. List no. 56. *International Journal of Systematic Bacteriology* 46, 362.

Albuquerque, L., Rainey, F.A., Chung, A.P. *et al.* (2000) *Alicyclobacillus hesperidum* sp. nov. and a related genomic species from solfataric soils of Sao Miguel in the Azores. *International Journal of Systematic Evolutionary Microbiology* 50, 451–457.

Andersch, I., Pianka, S., Fritze, D. & Claus, D. (1994) Description of *Bacillus laevolacticus* (ex Nakayama and Yanoshi 1967) sp. nov., nom. rev. *International Journal of Systematic Bacteriology* 44, 659–664.

Andersson, M., Laukkanen, M., Nurmiaho-Lassila, E.-L., Rainey, F.A., Niemelä, S.I. & Salkinoja-Salonen, M. (1995) *Bacillus thermosphaericus* sp. nov. a new thermophilic ureolytic *Bacillus* isolated from air. *Systematic and Applied Microbiology* 18, 203–220.

Andersson, M., Laukkanen, M., Nurmiaho-Lassila, E.-L., Rainey, F.A., Niemelä, S.I. & Salkinoja-Salonen, M. (1996) *Bacillus thermosphaericus* new species. In: *Validation of the publication of new names and new combinations previously effectively published outside the IJSB*. List no. 56. *International Journal of Systematic Bacteriology* 46, 362.

Arahal, D.R., Márquez, M.C., Volcani, B.E., Schleifer, K.H. & Ventosa, A. (1999) *Bacillus marismortui* sp. nov., a new moderately halophilic species from the Dead Sea. *International Journal of Systematic Bacteriology* 49, 521–530.

Arfman, N., Dijkhuizen, L., Kirchhof, G. *et al.* (1992) *Bacillus methanolicus* sp., a new species of thermotolerant, methanol-utilizing, endospore-forming bacteria. *International Journal of Systematic Bacteriology* **42**, 439–445.

Ash, C., Priest, F.G. & Collins, M.D. (1993) Molecular identification of rRNA group 3 bacilli (Ash, Farrow, Wallbanks and Collins) using a PCR probe test. *Antonie van Leeuwenhoek* **64**, 253–260.

Ash, C., Priest, F.G. & Collins, M.D. (1994) *Paenibacillus polymyxa* comb. nov. In: *Validation of the publication of new names and new combinations previously effectively published outside the IJSB*. List no. 51. *International Journal of Systematic Bacteriology* **44**, 852.

Ash, C., Priest, F.G. & Collins, M.D. (1995) *Paenibacillus alvei, Paenibacillus amylolyticus, Paenibacillus azotofixans, Paenibacillus gordonae, Paenibacillus larvae, Paenibacillus macerans, Paenibacillus macquariensis, Paenibacillus pabuli, Paenibacillus pulvifaciens, Paenibacillus validus* new combinations. In: *Validation of the publication of new names and new combinations previously effectively published outside the IJSB*. List no. 52. *International Journal of Systematic Bacteriology* **45**, 197–198.

Avakyan, Z.A., Pivovarova, T.A. & Karavaiko, G.I. (1986) Properties of a new species, *Bacillus mucilaginosus. Mikrobiologiya* **55**, 477–482.

Avakyan, Z.A., Pivovarova, T.A. & Karavaiko, G.I. (1998) *Bacillus mucilaginosus* new species. In: *Validation of the publication of new names and new combinations previously effectively published outside the IJSB*. List no. 66. *International Journal of Systematic Bacteriology* **48**, 631–632.

Batchelor, M.D. (1919) Aerobic spore-bearing bacteria in the intestinal tract of children. *Journal of Bacteriology* **4**, 23–24.

Berliner, E. (1915) Über die Schlaffsucht der Mehlmottenraupe (*Ephestia kühniella* Zell) und ihren Erreger *Bacillus thuringiensis n. sp. Zeitschrift für angewandte Entomologie Berlin* **2**, 29–56.

Bonjour, F. & Aragno, M. (1984) *Bacillus tusciae*, a new species of thermoacidophilic, facultatively chemolithoautotrophic, hydrogen oxidizing spore-former from a geothermal area. *Archives of Microbiology* **139**, 397–401.

Bonjour, F. & Aragno, M. (1985) *Bacillus tusciae* new species. In: *Validation of the publication of new names and new combinations previously effectively published outside the IJSB*. List no. 17. *International Journal of Systematic Bacteriology* **35**, 223.

Boone, D.R., Liu, Y., Zhao, Z.-J. *et al.* (1995) *Bacillus infernus* sp. nov., an Fe(III)- and Mn(IV)-reducing anearobe from the deep terrestrial subsurface. *International Journal of Systematic Bacteriology* **45**, 441–448.

Bredemann, G. & Werner, W. (1933) In: Werner, W. *Zentralblatt für Bakteriologie, Parasitenkunde, Infektionskrankheiten und Hygiene. Abteilung II.* **87**, 446–475.

Brill, J.A. & Wiegel, J. (1997) Differentiation between spore-forming and asporogenic bacteria using a PCR and Southern hybridization based method. *Journal of Microbiological Methods* **31**, 29–36.

Cheshire, F.R. & Cheyne, W.W. (1885) *Journal of the Royal Microscopic Society, Series II* **5**, 581–601.

Chester, F.D. (1898) Report of the mycologist: Bacteriological work. *Delaware Experimental Station Annual Report* **10**, 47–137.

Chester, F.D. (1901) *A Manual of Determinative Bacteriology*. Macmillan, New York.

Claus, D. & Berkeley, R.C.W. (1986) Genus *Bacillus* Cohn 1872. In: *Bergey's Manual of Systematic Bacteriology* (eds P.H.A. Sneath *et al.*), Vol. 2, pp. 1105–1139. Williams & Wilkins, Baltimore, MD.

Cohn, F. (1872) Untersuchungen über Bakterien. *Beiträge zur Biologie der Pflanzen* **1**, 127–224.

Combet-Blanc, Y., Ollivier, B., Streicher, C. *et al.* (1995) *Bacillus thermoamylovorans* sp. nov., a moderately thermophilic and amylolytic bacterium. *International Journal of Systematic Bacteriology* **45**, 9–16.

Cross, T. & Goodfellow, M. (1973) Taxonomy and classification of the actinomycetes. In: *Actinomycetales: Characteristics and Practical Importance*, (eds I.K. Sykes & F.A. Skinner), pp. 11–112. Academic Press, London.

Darland, G. & Brock, T.D. (1971) *Bacillus acidocaldarius* sp. nov., an acidophilic thermophilic spore-forming bacterium. *Journal of General Microbiology* **67**, 9–15.

de Bary, A. (1884) *Vergleichende Morphologie und Biologie der Pilze, Mycetozoen und Bacterien*. Wilhelm Engelmann, Leipzig.

Demharter, W. & Hensel, R. (1989a) *Bacillus thermocloaceae* sp. nov., a new thermophilic species from sewage sludge. *Systematic and Applied Microbiology* **11**, 272–276.

Demharter, W. & Hensel, R. (1989b) *Bacillus thermocloaceae* new species. In: *Validation of the publication of new names and new combinations previously effectively published outside the IJSB*. List no. 31. *International Journal of Systematic Bacteriology* **39**, 495.

den Dooren de Jong, L.E. (1929) Über *Bacillus fastidiosus. Zentralblatt für Bakteriologie, Parasitenkunde, Infektionskrankheiten und Hygiene. Abteilung II* **79**, 344–353.

Denariaz, G., Payne, W.J. & Gall, J.L. (1989) A halophilic denitrifier, *Bacillus halodenitrificans* sp. nov. *International Journal of Systematic Bacteriology* **39**, 145–151.

Donk, P.J. (1920) A highly resistant thermophilic organism. *Journal of Bacteriology* **5**, 373–374.

Dufresne, S., Bousquet, J., Boissinot, M. & Guay, R. (1996) *Sulfobacillus disulfidooxidans* sp. nov., a new acidophilic, disulfide-oxidizing, Gram-positive, spore-forming bacterium. *International Journal of Systematic Bacteriology* **46**, 1056–1064.

Dutky, S.R. (1940) Two new spore-forming bacteria causing milky diseases of Japanese beetle larvae. *Journal of Agricultural Research* **61**, 57–68.

European Community (1990) Council Directive 90/679/EEC on the protection of workers from risks related to exposure to biological agents at work. *OJL 374 of 31.12.1990*.

European Community (1993) Council Directive 93/88/EEC amending Directive 90/679/EEC on the protection of workers from risks related to exposure to biological agents at work. *OJL 268 of 29.10.1993*.

Flügge, C. (1886) *Die Mikroorganismen*. F.C.W. Vogel, Leipzig.

Frankland, G.C. & Frankland, P.F. (1887) Studies on some new microorganisms obtained from air. *Royal Society London, Philosophical Transactions, Series B: Biological Sciences* **178**, 257–287.

Fritze, D. (1996) *Bacillus haloalkaliphilus* sp. nov. *International Journal of Systematic Bacteriology* **46**, 98–101.

Fujita, T., Shida, O., Takagi, H., Kunugita, K., Pankrushina, A.N. & Matsuhashi, M. (1996) Description of *Bacillus carboniphilus* sp. nov. *International Journal of Systematic Bacteriology* **46**, 116–118.

Gibson, T. (1935) The urea-decomposing microflora of soils. I. Description and classification of the organisms. *Zentralblatt für Bakteriologie, Parasitenkunde, Infektionskrankheiten und Hygiene. Abteilung II* **92**, 364–380.

Golovacheva, R.S. & Karavaiko, G.I. (1978) A new genus of thermophilic spore-forming bacteria, *Sulfobacillus*. *Microbiology* (Engl. translation of *Mikrobiologiya*) **47**, 658–664.

Golovacheva, R.S. & Karavaiko, G.I. (1991) *Sulfobacillus* new genus, *Sulfobacillus thermosulfidooxidans* new species. In: *Validation of the publication of new names and new combinations previously effectively published outside the IJSB*. List no. 36. *International Journal of Systematic Bacteriology* **41**, 179.

Golovacheva, R.S., Loginova, L.G., Salikhov, T.A., Kolesnikov, A.A. & Zaitseva, G.N. (1975) A new thermophilic species, *Bacillus thermocatenulatus* nov. spec. *Mikrobiologiya* **44**, 265–268.

Golovacheva, R.S., Loginova, L.G., Salikhov, T.A., Kolesnikov, A.A. & Zaitseva, G.N. (1991) *Bacillus thermocatenulatus* new species In: *Validation of the publication of new names and new combinations previously effectively published outside the IJSB*. List no. 36. *International Journal of Systematic Bacteriology* **41**, 178.

Gordon, R.E., Haynes W.C. & Pang, C.H. (1973) *The genus* Bacillus. Handbook 427. U.S. Department of Agriculture, Washington, DC.

Hammer, B.W. (1915) Bacteriological studies on the coagulation of evaporated milk. *Research Bulletin of the Iowa Agricultual Experimental Station* **19**, 119–131.

Heyndrickx, M., Vandermeulebroecke, K., Kersters, K. *et al.* (1996) A polyphasic reassessment of the genus *Paenibacillus*, reclassification of *Bacillus lautus* (Nakamura 1984) as *Paenibacillus lautus* comb. nov. and of *Bacillus peoriae* (Montefusco *et al.* 1993) as *Paenibacillus peoriae* comb. nov., and emended description of *P. lautus* and of *P. peoriae*. *International Journal of Systematic Bacteriology* **46**, 988–1003.

Heyndrickx, M., Lebbe, L., Vancanneyt, M. *et al.* (1997) A polyphasic reassessment of the genus *Aneurinobacillus*, reclassification *Bacillus thermoaerophilus* (Meier-Stauffer *et al.* 1996) as *Aneurinobacillus thermoaerophilus* comb. nov., and emended description of *A. aneurinilyticus* (corrig.), *A. migulanus*, and *A. thermoaerophilus*. *International Journal of Systematic Bacteriology* **47**, 808–817.

Heyndrickx, M., Lebbe, L., Kersters, K., De Vos, P., Forsyth, G. & Logan, N.A. (1998) *Virgibacillus*: a new genus to accommodate *Bacillus pantothenticus* (Proom and Knight 1950). Emended description of *Virgibacillus pantothenticus*. *International Journal of Systematic Bacteriology* **48**, 99–106.

Heyndrickx, M., Lebbe, L., Kersters, K. *et al.* (1999) Proposal of *Virgibacillus proomii* sp. nov. and emended description of *Virgibacillus pantothenticus* (Proom and Knight 1950) Heyndrickx *et al.* 1998. *International Journal of Systematic Bacteriology* **49**, 1083–1090.

Jordan, E.O. (1890) A report on certain species of bacteria observed in sewage. In: *A Report of the Biological Work of the Lawrence Experiment Station, including an Account of Methods Employed and Results Obtained in the Microscopical and Bacteriological Investigation of Sewage and Water* (ed. W.T. Sedgewick). *Report of the Massachusetts Board of Public Health* **II**, 821–844.

Kitahara, K. & Lai, C.L. (1967) On the spore formation of *Sporolactobacillus inulinus*. *Journal of General Applied Microbiology* **13**, 197–203.

Kitahara, K. & Suzuki, J. (1963) *Sporolactobacillus* nov. subgen. *Journal of General and Applied Microbiology* 9, 59–71.

Kluyver, A.J. & van Niel, C.B. (1936) Prospects for a natural classification of bacteria. *Zentralblatt für Bakteriologie, Parasitenkunde, Infektionskrankheiten und Hygiene. Abteilung II* 94, 369–403.

Koransky, J.R., Allen, S.D. & Dowell, V.R. (1978) Use of ethanol for selective isolation of sporeforming microorganisms. *Applied and Environmental Microbiology* 35, 762–765.

Krych, V.A., Johnson, J.L. & Yousten, A.A. (1980) Deoxyribonucleic acid homologies among strains of *Bacillus sphaericus*. *International Journal of Systematic Bacteriology* 30, 476–484.

Kuhnigk, T., Borst, E.-M., Breunig, A. *et al.* (1995a) *Bacillus oleronius* sp. nov., a member of hindgut flora of the termite *Reticulitermes santonensis* (Feytand). *Canadian Journal Microbiology* 41, 699–706.

Kuhnigk, T., Borst, E.-M., Breunig, A. *et al.* (1995b) *Bacillus oleronius* new species. In: *Validation of the publication of new names and new combinations previously effectively published outside the IJSB.* List no. 57. *International Journal of Systematic Bacteriology* 46, 625.

Kuroshima, K.-I., Sakane, T., Takata, R. & Yokota, A. (1996) *Bacillus ehimensis* sp. nov. and *Bacillus chitinolyticus* sp. nov., new chitinolytic members of the genus *Bacillus*. *International Journal of Systematic Bacteriology* 46, 76–80.

Kurup, V.P., Barboriak, J.J., Fink, J.N. & Lechevalier, M.P. (1975) *Thermoactinomyces candidus*, a new species of thermophilic actinomyces. *International Journal of Systematic Bacteriology* 25, 150–154.

Kurup, V.P., Hollick, G.E. & Pagan, E.F. (1980) *Thermoactinomyces intermedius*, a new species of amylase negative thermophilic actinomycetes. *Science-Scientia. Bolletin de Ciencias del Sur* 7, 104–108.

Kurup, V.P., Hollick, G.E. & Pagan, E.F. (1981) *Thermoactinomyces intermedius* new species. In: *Validation of the publication of new names and new combinations previously effectively published outside the IJSB.* List no. 6. *International Journal of Systematic Bacteriology* 31, 216.

Lacey, J. (1971) *Thermoactinomyces sacchari* sp. nov., a thermophilic actinomycete causing bagassosis. *Journal General Microbiology* 66, 327–338.

Lacey, J. & Cross, T. (1989a) Genus *Thermoactinomyces* Tsiklinsky 1899, 501[AL]. In: *Bergey's Manual of Systematic Bacteriology* (eds S.T. Williams *et al.*), Vol. 4, pp. 2574–2585. Williams & Wilkins, Baltimore, MD.

Lacey, J. & Cross, T. (1989b) *Thermoactinomyces putidus* new species and *Thermoactinomyces thalpophilus* revived name. In: *Validation of the publication of new names and new combinations previously effectively published outside the IJSB.* List no. 31. *International Journal of Systematic Bacteriology* 39, 496.

Larkin, J.M. & Stokes, J.L. (1967) Taxonomy of psychrophilic strains of *Bacillus*. *Journal of Bacteriology* 94, 889–895.

Laubach, C.A. (1916) Studies on aerobic spore-bearing non-pathogenic bacteria. Spore-bearing organisms in water. *Journal of Bacteriology* 1, 505–512.

Lechner, S., Mayr, R., Francis, K.P. *et al.* (1998) *Bacillus weihenstephanensis* sp. nov. is a new psychrotolerant species of the *Bacillus cereus* group. *International Journal of Systematic Bacteriology* 48, 1373–1382.

Macé, E. (1889) *Traité pratique de Bactériologie*, 1st edn. J.-B. Ballière, Paris.

Manachini, P.L., Mora, D., Nicastro, G. *et al.* (2000) *Bacillus thermodenitrificans* sp. nov., nom. rev. *International Journal of Systematic and Evolutionary Microbiology* 50, 1331–1337.

Marshall, B.J. & Ohye, D.F. (1966) *Bacillus macquariensis* n. sp., a psychrotrophic bacterium from sub-antarctic soil. *Journal of General Microbiology* 44, 41–46.

Meyer, A. & Gottheil, O. (1901) In: *Botanische Beschreibung einiger Bodenbakterien. Beiträge zur Methode der Speciesbestimmung und Vorarbeit für die Entscheidung der Frage nach der Bedeutung der Bodenbakterien für die Landwirtschaft* (general author O. Gottheil). *Zentralblatt für Bakteriologie, Parasitenkunde, Infektionskrankheiten und Hygiene. Abteilung II* 7, 680–691.

Meyer, A. & Neide, E. (1904) In: *Botanische Beschreibung einiger sporenbildenden Bakterien* (general author E. Neide). *Zentralblatt für Bakteriologie, Parasitenkunde, Infektionskrankheiten und Hygiene. Abteilung II* 12, 337–352.

Migula, W. (1900) *System der Bakterien*, Vol. 2. Gustav Fischer, Jena.

Murray, R.G.E., Brenner, D.J., Colwell, R.R. *et al.* (1990) Report of the Ad Hoc Committee on Approaches to Taxonomy within the Proteobacteria. *International Journal of Systematic Bacteriology* 40, 213–215.

Nagel, M. & Andreesen, J.R. (1991) *Bacillus niacini* sp. nov., a nicotinate-metabolizing mesophile isolated from soil. *International Journal of Systematic Bacteriology* 41, 134–139.

Nakamura, L.K. (1984) *Bacillus psychrophilus* sp. nov., nom. rev. *International Journal of Systematic Bacteriology* 34, 121–123.

Nakamura, L.K. (1989) Taxonomic relationship of black-pigmented *Bacillus subtilis* strains and a proposal for *Bacillus atrophaeus* sp. nov. *International Journal of Systematic Bacteriology* **39**, 295–300.

Nakamura, L.K. (1994) DNA relatedness among *Bacillus thuringiensis* serovars. *International Journal of Systematic Bacteriology* **44**, 125–129.

Nakamura, L.K. (1996) *Paenibacillus apiarius* sp. nov. *International Journal of Systematic Bacteriology* **46**, 688–693.

Nakamura, L.K. (1998) *Bacillus pseudomycoides* sp. nov. *International Journal of Systematic Bacteriology* **48**, 1031–1035.

Nakamura, L.K., Blumenstock, I. & Claus, D. (1988) Taxonomic study of *Bacillus coagulans* Hammer 1915 with a proposal for *Bacillus smithii* sp. nov. *International Journal of Systematic Bacteriology* **38**, 63–73.

Nakamura, L.K. & Jackson, M.A. (1995) Clarification of the taxonomy of *Bacillus mycoides*. *International Journal of Systematic Bacteriology* **45**, 46–49.

Nielsen, P., Fritze, D. & Priest, F.G. (1995a) Phenetic diversity of alkaliphilic *Bacillus* strains: proposal for nine new species. *Microbiology* **141**, 1745–1761.

Nielsen, P., Fritze, D. & Priest, F.G. (1995b) *Bacillus agaradhaerens, Bacillus clarkii, Bacillus clausii, Bacillus gibsonii, Bacillus halmapalus, Bacillus halodurans, Bacillus horikoshii, Bacillus pseudalcaliphilus, Bacillus pseudofirmus* new species. In: *Validation of the publication of new names and new combinations previously effectively published outside the IJSB*. List no. 55. *International Journal of Systematic Bacteriology* **45**, 879–880.

Niimura, Y., Koh, E., Yanagida, F., Suzuki, K.-I., Komagata, K. & Kozaki, M. (1990) *Amphibacillus xylanus* gen. nov., sp. nov., a facultatively anaerobic sporeforming xylan-digesting bacterium which lacks cytochrome, quinone, and catalase. *International Journal of Systematic Bacteriology* **40**, 297–301.

Nonomura, H. & Ohara, Y. (1971) Distribution of actinomycetes in soil X. New genus and species of monosporic actinomycetes. *Journal of Fermentation Technology* **49**, 895–903.

Norris, P.R., Clark, D.A., Owen, J.P. & Waterhouse, S. (1996a) Characteristics of *Sulfobacillus acidophilus* sp. nov. and other moderately thermophilic mineral-sulphide-oxidizing bacteria. *Microbiology* **142**, 775–783.

Norris, P.R., Clark, D.A., Owen, J.P. & Waterhouse, S. (1996b) *Sulfobacillus acidophilus* new species. In: *Validation of the publication of new names and new combinations previously effectively published outside the IJSB*. List no. 59. *International Journal of Systematic Bacteriology* **46**, 1189.

Pettersson, B., Lembke, F., Hammer, P., Stackebrandt, E. & Priest, F.G. (1996) *Bacillus sporothermodurans*, a new species produces highly heat-resistent endospores. *International Journal of Systematic Bacteriology* **46**, 759–764.

Pettersson, B., Rippere, K.E., Yousten, A.A. & Priest, F.G. (1999) Transfer of *Bacillus lentimorbus* and *Bacillus popilliae* to the genus *Paenibacillus* with emended descriptions of *Paenibacillus lentimorbus* comb. nov. and *Paenibacillus popilliae* comb. nov. *International Journal of Systematic Bacteriology* **49**, 531–540.

Pichinoty, F., de Barjac, H., Mandel, M. & Asselineau, J. (1983) Description of *Bacillus azotoformans* sp. nov. *International Journal of Systematic Bacteriology* **33**, 660–662.

Priest, F.G., Goodfellow, M., Shute, L.A. & Berkeley, R.C.W. (1987) *Bacillus amyloliquefaciens* sp. nov., nom. rev. *International Journal of Systematic Bacteriology* **37**, 69–71.

Priest, F.G., Goodfellow, M. & Todd, C. (1988) A numerical classification of the genus *Bacillus*. *Journal of General Microbiology* **134**, 1847–1882.

Priest, F.G., Goodfellow, M. & Todd, C. (1989) *Bacillus flexus, Bacillus fusiformis, Bacillus kaustophilus, Bacillus psychrosaccharolyticus, Bacillus simplex* new species. In: *Validation of the publication of new names and new combinations previously effectively published outside the IJSB*. List no. 28. *International Journal of Systematic Bacteriology* **39**, 93–94.

Proom, H. & Knight, B.C.J.G. (1950) *Bacillus pantothenticus* (n.sp.). *Journal of General Microbiology* **4**, 539–541.

Rheims, H., Frühling, A., Schumann, P., Rohde, M. & Stackebrandt, E. (1999) *Bacillus silvestris* sp. nov., a new member of the genus *Bacillus* that contains lysine in its cell wall. *International Journal of Systematic Bacteriology* **49**, 795–802.

Roberts, M.S., Nakamura, L.K. & Cohan, F.M. (1994) *Bacillus mojavensis* sp. nov., distinguishable from *Bacillus subtilis* by sexual isolation, divergence in DNA sequence, and differences in fatty acid composition. *International Journal of Systematic Bacteriology* **44**, 256–264.

Roberts, M.S., Nakamura, L.K. & Cohan, F.M. (1996) *Bacillus vallismortis* sp. nov., a close relative of *Bacillus subtilis*, isolated from soil in Death Valley, California. *International Journal of Systematic Bacteriology* **46**, 470–475.

Rüger, H.-J. (1983) Differentiation of *Bacillus globisporus, Bacillus marinus* comb. nov., *Bacillus aminovorans*, and *Bacillus insolitus*. *International Journal of Systematic Bacteriology* **33**, 157–161.

Schardinger, F. (1905) *Bacillus macerans*, ein Aceton bildender Rottebacillus. *Zentralblatt für Bakteriologie, Parasitenkunde, Infektionskrankheiten und Hygiene. Abteilung II* **14**, 772–781.

Schenk, A. & Aragno, M. (1979) *Bacillus schlegelii*, a new species of thermophilic facultatively chemo-lithoautotrophic bacterium oxidizing molecular hydrogen. *Journal of General Microbiology* **115**, 333–341.

Schenk, A. & Aragno, M. (1981) *Bacillus schlegelii* new species. In: *Validation of the publication of new names and new combinations previously effectively published outside the IJSB*. List no. 6. *International Journal of Systematic Bacteriology* **31**, 215.

Scholz, T., Demharter, W., Hensel, R. & Kandler, O. (1987) *Bacillus pallidus* sp. nov., a new thermophilic species from sewage. *Systematic and Applied Microbiology* **9**, 91–96.

Scholz, T., Demharter, W., Hensel, R. & Kandler, O. (1988) *Bacillus pallidus* new species. In: *Validation of the publication of new names and new combinations previously effectively published outside the IJSB*. List no. 24. *International Journal of Systematic Bacteriology* **38**, 136–137.

Shelobolina, E.S., Avakyan, Z.A., Bulygina, E.S. *et al.* (1997) Description of a new species of mucilaginous bacteria, *Bacillus edaphicus* sp. nov., and confirmation of the taxonomic status of *Bacillus mucilaginosus* Avakyan *et al.* 1986 based on data from phenotypic and genotypic analysis. *Mikrobiologiya* **66**, 813–822.

Shelobolina, E.S., Avakyan, Z.A., Bulygina, E.S. *et al.* (1998) *Bacillus edaphicus* new species. In: *Validation of the publication of new names and new combinations previously effectively published outside the IJSB*. List no. 66. *International Journal of Systematic Bacteriology* **48**, 631.

Shida, O., Takagi, H., Kadowaki, K. & Komagata, K. (1996) Proposal for two new genera, *Brevibacillus* gen. nov. and *Aneurinibacillus* gen. nov. *International Journal of Systematic Bacteriology* **46**, 939–946.

Shida, O., Takagi, H., Kadowaki, K., Nakamura, L.K. & Komagata, K. (1997a) Transfer of *Bacillus alginolyticus*, *Bacillus chondroitinus*, *Bacillus curdlanolyticus*, *Bacillus glucanolyticus*, *Bacillus kobensis*, and *Bacillus thiaminolyticus* to the genus *Paenibacillus* and emended description of the genus *Paenibacillus*. *International Journal of Systematic Bacteriology* **47**, 289–298.

Shida, O., Takagi, H., Kadowaki, K., Nakamura, L.K. & Komagata, K. (1997b) Emended description of *Paenibacillus amylolyticus* and description of *Paenibacillus illinoisensis* sp. nov. and *Paenibacillus chibensis* sp. nov. *International Journal of Systematic Bacteriology* **47**, 299–306.

Skerman, V.B.D., McGowan, V. & Sneath, P.H.A. (eds) (1980) Approved Lists of Bacterial Names. *International Journal of Systematic Bacteriology* **30**, 225–420.

Slepecky, R.A. (1992) What is a *Bacillus*? In: *Biology of Bacilli: Applications to Industry* (eds R.H. Doi & M. McGloghlin), pp. 1–21. Butterworth-Heinemann, Boston, MA.

Smith, N.R., Gordon, R.E. & Clark, F.E. (1952) *Aerobic Sporeforming Bacteria*. Monograph 16. U.S. Department of Agriculture, Washington, DC.

Spanka, R. & Fritze, D. (1993) *Bacillus cohnii* sp. nov., a new, obligately alkaliphilic, oval-spore-forming *Bacillus* species with ornithine and aspartic acid instead of diaminopimelic acid in the cell wall. *International Journal of Systematic Bacteriology* **43**, 150–156.

Spring, S., Ludwig, W., Marquez, M.C., Ventosa, A. & Schleifer, K.-H. (1996) *Halobacillus* gen. nov., with descriptions of *Halobacillus litoralis* sp. nov. and *Halobacillus trueperi* sp. nov., and transfer of *Sporosarcina halophila* to *Halobacillus halophilus* comb. nov. *International Journal of Systematic Bacteriology* **46**, 492–496.

Stackebrandt, E. & Goebel, B.M. (1994) Taxonomic note: a place for DNA-DNA reassociation and 16S rRNA sequence analysis in the present apecies definition in bacteriology. *International Journal of Systematic Bacteriology* **44**, 846–849.

Suzuki, Y., Kishigami, T., Inoue, K. *et al.* (1983) *Bacillus thermoglucosidasius* sp. nov., a new species of obligately thermophilic bacilli. *Systematic and Applied Microbiology* **4**, 487–495.

Suzuki, Y., Kishigami, T., Inoue, K. *et al.* (1984) *Bacillus thermoglucosidasius* new species. In: *Validation of the publication of new names and new combinations previously effectively published outside the IJSB*. List no. 14. *International Journal of Systematic Bacteriology* **34**, 270.

Tcherpakov, M., Ben-Jacob, E. & Gutnick, D.L. (1999) *Paenibacillus dendritiformis* sp. nov., proposal for a new pattern-forming species and its localization within a phylogenetic cluster. *International Journal of Systematic Bacteriology* **49**, 239–246.

Tomimura, E., Zeman, N.W., Frankiewicz, J.R. & Teague, W.M. (1990) Description of *Bacillus naganoensis* sp. nov. *International Journal of Systematic Bacteriology* **40**, 123–125.

Touzel, J.P., O'Donohue, M., Debeire, P., Samain, E. & Breton, C. (2000) *Thermobacillus xylanilyticus* gen. nov., sp. nov., a new aerobic thermophilic xylan-degrading bacterium isolated from farm soil. *International Journal of Systematic and Evolutionary Microbiology* **50**, 315–320.

Tsiklinsky, P. (1899) On the thermophilic molds (in French). *Annales de L'Institut Pasteur* **13**, 500–505.

Vedder, A. (1934) *Bacillus alcalophilus* n. sp.; benevens enkele ervaringen met sterk alcalische voedingsbodems. *Antonie van Leeuwenhoek* **1**, 143–147.

Ventosa, A., Garcia, M.T., Kamekura, M., Onishi, H. & Ruiz-Berraquero, F. (1989a) *Bacillus halophilus* sp. nov., a moderately halophilic *Bacillus* species. *Systematic and Applied Microbiology* **12**, 162–166.

Ventosa, A., Garcia, M.T., Kamekura, M., Onishi, H. & Ruiz-Berraquero, F. (1989b) *Bacillus halophilus* new species. In: *Validation of the publication of new names and new combinations previously effectively published outside the IJSB*. List no. 32. *International Journal of Systematic Bacteriology* **40**, 105.

Wainö, M., Tindall, B.J., Schumann, P. & Ingvorsen, K. (1999) *Gracilibacillus* gen. nov., with description of *Gracilibacillus halotolerans* gen. nov., sp. nov.; transfer of *Bacillus dipsosauri* to *Gracilibacillus dipsosauri* comb. nov., and *Bacillus salexigens* to the genus *Salibacillus* gen. nov., as *Salibacillus salexigens* comb. nov. *International Journal of Systematic Bacteriology* **49**, 821–831.

White, G.F. (1906) *The Bacteria of the Apiary, with Special Reference to Bee Diseases*. Technical Series 14. United States Department of Agriculture, Bureau of Entomology.

Wiegel, J. (1981) Distinction between the Gram reaction and the Gram type of bacteria. *International Journal of Systematic Bacteriology* **31**, 88.

Wisotzkey, J.D., Jurtshuk, P., Jr, Fox, G.E., Deinhard, G. & Poralla, K. (1992) Comparative sequences analyses on the 16S rRNA (rDNA) of *Bacillus acidocaldarius*, *Bacillus acidoterrestris*, and *Bacillus cycloheptanicus* and proposal for creation of a new genus, *Alicyclobacillus* gen. nov. *International Journal of Systematic Bacteriology* **42**, 263–269.

Yanagida, F., Suzuki, K.-I., Kozaki, M. & Komagata, K. (1997) Proposal of *Sporolactobacillus nakayamae* ssp. *nakayamae* sp. nov., ssp. nov., *Sporolactobacillus nakayamae* ssp. *racemicus* ssp. nov. *Sporolactobacillus terrae* sp. nov., *Sporolactobacillus kofuensis* sp. nov., *Sporolactobacillus lactosus* sp. nov. *International Journal of Systematic Bacteriology* **47**, 499–504.

Yoon, J.-H., Yim, D.K., Lee, J.-S. *et al.* (1998) *Paenibacillus campinasensis* sp. nov., a cyclodextrin-producing bacterium isolated in Brazil. *International Journal of Systematic Bacteriology* **48**, 833–837

Yumoto, I., Yamazaki, K., Sawabe, T. *et al.* (1998) *Bacillus horti* sp. nov., a new Gram-negative alkaliphilic bacillus. *International Journal of Systematic Bacteriology* **48**, 565–571.

Zaitsev, G., Tsitko, I.V., Rainey, F.A. *et al.* (1998) New aerobic ammonium-dependent obligately oxalotrophic bacteria: description of *Ammoniphilus oxalaticus* gen. nov., sp. nov. and *Ammoniphilus oxalivorans* gen. nov., sp. nov. *International Journal of Systematic Bacteriology* **48**, 151–163.

Zarilla, K.A. & Perry, J.J. (1987) *Bacillus thermoleovorans*, sp. nov., a species of obligately thermophilic hydrocarbon utilizing endospore-forming bacteria. *Systematic and Applied Microbiology* **9**, 258–264.

Zarilla, K.A. & Perry, J.J. (1988) *Bacillus thermoleovorans* new species. In: *Validation of the publication of new names and new combinations previously effectively published outside the IJSB*. List no. 25. *International Journal of Systematic Bacteriology* **38**, 220.

Chapter 9

Modern Methods for Identification

Niall A. Logan

Introduction

Most aerobic endospore-forming species are saprophytes distributed widely in the natural environment (in soils of all kinds, ranging from acid to alkaline, hot to cold, and fertile to desert, and in the water columns and bottom deposits of fresh and marine waters), but some species are opportunistic or obligate pathogens of animals, including humans, other mammals and insects. Their persistent spores readily survive distribution from natural environments to a wide variety of other habitats. They are nearly everywhere, and their huge diversity not only reflects their ubiquity, but also presents a huge diagnostic challenge.

In the widely used diagnostic scheme for *Bacillus* species developed by Smith *et al.* (1952) and updated by Gordon *et al.* (1973), 18 species were split into three groups according to their sporangial morphologies, and then further divided by biochemical and physiological tests. Although this approach was effective for some years, *Bacillus* identification was still generally perceived as complicated, the chief difficulties being the need for special media, and between-strain variation. Much of the latter was a reflection of unsatisfactory taxonomy, but, as the studies of Logan and Berkeley (1981) revealed, test inconsistency exacerbated the problems. Logan and Berkeley (1984) addressed these problems with a large database for 38 clearly defined taxa (species) using miniaturized tests in the API 20E and 50CHB Systems (bioMérieux, Marcy l'Etoile, France), and this scheme, with updates, remains in common use.

The problems of *Bacillus* identification, however, far from being solved, have worsened: the groupings used in the phenotypically-based schemes of Smith and Gordon and of Logan and Berkeley do not always correlate with current, phylogenetically led classifications, and since 1973 over 90 new species and subspecies of aerobic endospore-formers have been proposed. These make a grand total of 123 species in *Bacillus sensu lato*. In the same period, only four proposals for merging species have been made (Shida *et al.* 1994b; Heyndrickx *et al.* 1995, 1996a; Rosado *et al.* 1997).

In addition to the establishment of numerous new species, phylogenetic studies (see also Chapter 2) based upon 16S rDNA sequence comparisons have led to proposals for seven new genera to accommodate species formerly assigned to *Bacillus*: *Alicyclobacillus* (Wisotzkey *et al.* 1992), *Paenibacillus* (Ash *et al.* 1993; validated Collins *et al.* 1994), *Brevibacillus* (Shida *et al.* 1996), *Aneurinibacillus*

(Shida *et al.* 1996), *Virgibacillus* (Heyndrickx *et al.* 1998) and *Gracilibacillus* and *Salibacillus* (Wainö *et al.* 1999). Furthermore, the motile, spore-forming coccus *Sporosarcina ureae* is phylogenetically close to *B. sphaericus*, while a more distantly related species, *Sporosarcina halophila*, has been generically reclassified as *Halobacillus* (Spring *et al.* 1996). Other new genera include: *Sulfobacillus* (Golovacheva & Karavaiko 1978), which now contains two species of facultative autotrophs (Dufresne *et al.* 1996); *Amphibacillus*, which was proposed for a species comprising three strains of catalase-negative, facultatively anaerobic endospore-formers (Niimura *et al.* 1990); *Ammoniphilus* (Zaitsev *et al.* 1998), with two species of ammonium-dependent and obligately oxalotrophic organisms; and most recently the genus *Thermobacillus* which was proposed on the basis of a single isolate (Touzel *et al.* 2000).

Therefore, the genus *Bacillus sensu stricto* continues to accommodate the best-known species, such as *B. subtilis* (the type species), *B. anthracis*, *B. cereus*, *B. licheniformis*, *B. megaterium*, *B. pumilus*, *B. sphaericus* and *B. thuringiensis*, and with 76 species and subspecies it remains a large genus. Many of its species fall into several apparently distinct rDNA sequence groups such as the '*B. subtilis* group', the '*B. cereus* group' and the '*B. sphaericus* group'. Therefore, while it is likely that further rearrangements at the generic and species levels will be proposed, phenotypically and/or phylogenetically intermediate organisms may make satisfactory subdivision difficult.

Identification problems

Unfortunately, taxonomic progress has not revealed readily determinable features characteristic of each genus. Also, many recently described species represent genomic groups disclosed by DNA–DNA pairing experiments (Nakamura 1984, 1993), and routine phenotypic characters for distinguishing between several of them are very few and of unproven value. Furthermore, several recently described species have been proposed on the basis of very few strains so that the within-species diversities of such taxa, and so their true boundaries, remain unknown. It is clear, therefore, that the identification of aerobic endospore-formers has become increasingly difficult since the publication of the scheme of Gordon *et al.* (1973), a period during which demands to identify such organisms have increased greatly, especially in the medical and biotechnological fields.

From the point of view of the routine diagnostic laboratory, the aerobic endospore-formers comprise two groups: the reactive ones, which will give positive results in various routine biochemical tests (and which are therefore more amenable to identification by traditional methods and modern developments of such approaches), and the nonreactive ones, which give few if any positive results in such tests. These nonreactive isolates – which are often members of the genus *Brevibacillus* – figure prominently in the identification requests sent to reference laboratories.

When faced with aerobic endospore-former contamination problems, bacteriologists in biotechnological industries frequently ask: 'Is it *Bacillus cereus* or *B.*

anthracis, or is it another (and by implication, nonpathogenic) species?'. It is probably fair to say that most routine laboratories that have the confidence to attempt the identification of an aerobic endospore-former are well able to recognize members of the *B. subtilis* group and the *B. cereus* group. Once faced with a strain of *Brevibacillus* or *Paenibacillus*, however, it is a more difficult matter. Species of the former are usually inactive in routine biochemical tests – be they carbohydrate utilizations, or assimilations of carbohydrates, amino acids or organic acids as sole carbon sources – so that there are too few characters available to distinguish between the ten species, most of which are very closely related. With *Paenibacillus*, the opposite problem arises; many of the species are highly active in the routine tests so that distinctions between them are confined to rather few variable characters.

Databases

It is the experience of *Bacillus* taxonomists that collections of aerobic endospore-formers in laboratories around the world harbour many misnamed strains. This is not necessarily a reflection on the competence of those assembling the collections. It is a symptom of the unsatisfactory state of the classification of these organisms, which underlies the difficulty that many bacteriologists frequently encounter with their identification.

Unfortunately, several of the groups whose taxonomies are the most complex are the ones whose members are frequently submitted to reference laboratories. Such organisms are frequently included in the databases of commercial kits, but it can be difficult to obtain sufficient authentic strains of some these species to allow a satisfactory database entry to be made. Ideally, an entry in the database should reflect at least ten representative strains of the species, but for some taxa, and particularly for the new ones which have been based upon just one strain or very few strains, this can be impossible. It can also be a problem for some of the older-established species, as the representative strains found in culture collections around the world are sometimes the widely dispersed subcultures of a few original isolates. These problems emphasize the importance of basing proposals for new taxa on adequate numbers of strains to reflect the diversities of the taxa.

A further problem has emerged with the splitting of many well-established (although that is not to say homogeneous!) species or groups into large numbers of new taxa over a short period. Examples are *B. circulans* and *B.* (now *Brevibacillus*) *brevis*. While it is true that both taxa were very heterogeneous and no longer represented true species (*B. circulans* was often referred to as a complex rather than a species), the revisions of their taxonomies and consequent proposals for several new species to be split off from each have led to difficulties in identification. Although the proposals were based mostly upon polyphasic taxonomic studies, initial recognition of many of the new taxa depended largely upon DNA relatedness data or 16S rDNA sequence comparisons.

For example, a DNA homology study of *B. circulans* strains yielded *B. circulans sensu stricto*, *B. amylolyticus*, *B. lautus*, *B. pabuli* and *B. validus* and evidence for the existence of five other species (Nakamura & Swezey 1983; Nakamura 1984). Then, on the basis of 16S rDNA sequencing studies of *Bacillus* type strains, Ash *et al.* (1993) proposed the new genus *Paenibacillus* to accommodate *B. polymyxa* and *B. macerans* (which had long been regarded as close relatives of *B. circulans* on account of their phenotypic similarities) as well as, among others, *B. amylolyticus*, *B. pabuli* and *B. validus*. In fact, *B. lautus* also belongs in *Paenibacillus* (Heyndrickx *et al.* 1996b), but the single representative of this species that Ash *et al.* (1991) originally studied was apparently a contaminant. In the seven years since *Paenibacillus* was proposed with 11 species, expansion has been rapid: a further 11 species have been transferred to the genus, and five new species proposals made, yet only two mergers of species have occurred. The validity of the new genus is not in doubt, but many of its members, delineated as they were by DNA homology data or 16S rDNA sequence comparisons, are difficult to distinguish using routine phenotypic tests.

Following DNA relatedness studies of *B. brevis* strains, nine new species were proposed: *B. agri* and *B. centrosporus* (Nakamura 1993); *B. borstelensis*, *B. formosus* and *B. reuszeri* (Shida *et al.* 1995); *B. choshinensis*, *B. galactophilus*, *B. migulanus* and *B. parabrevis* (Takagi *et al.* 1993) [but *B. galactophilus* was later recognized as a synonym of *B. agri* (Shida *et al.* 1994b)]. In addition, *B. aneurinolyticus*, which is phenotypically similar to *B. brevis*, has been revived (Shida *et al.* 1994a). Subsequently, the two new genera *Brevibacillus* and *Aneurinobacillus* [*sic*] were proposed to accommodate all of these species (Shida *et al.* 1996), with the latter genus containing *A. aneurinolyticus* and *A. migulanus*.

These radical taxonomic revisions have left many culture collections worldwide with few representatives of both of the original species, *Br. brevis* and *B. circulans sensu stricto*, yet with numerous misnamed strains of these species, which may or may not belong to the newly proposed taxa. The curators will normally not be able to know which are which without considerable expenditure in scholarship and experimental work, and in many cases a collection will hold only one authentic strain, the type strain, of a species – be it an old or a new species.

For anyone attempting to construct an identification scheme for aerobic, endospore-forming bacteria the implication of such rapid taxonomic progress is huge. Accessing authentic strains of many species, even well-established ones, may require much time and effort, and for several of them the strains available may be too few to allow the diversities of the taxa to be adequately reflected in the identification scheme. Smith *et al.* (1952) and Gordon *et al.* (1973) showed commendable restraint in their concept of a bacterial species, saying in the latter monograph: 'When only a few strains of a group are available, as often happens, their species descriptions must remain tentative until verified by the study of more strains'. Just as taxonomists can only be as good as their culture collections, so identification systems can only be as good as their databases.

Although many new characterization methods have been developed over the last 30 years, the principle of identification remains the same: strictly speaking, identifications cannot be achieved – the best that can be done is to seek the taxon to which the unknown strain probably belongs. The outcome is expressed as a probability and, as with the classification upon which the scheme is based, the answer cannot be final.

Modern approaches to identification

Current schemes for identifying aerobic endospore-formers may be roughly divided into three categories according to the kinds of characters they use: (i) miniaturized versions of traditional biochemical tests (API kits, Vitek cards and Biolog plates), (ii) chemotaxonomic characters [such as fatty acid methyl ester (FAME) profiles, and pyrolysis mass spectrometry], and (iii) genomic characters (ribotyping, and nucleic acid probes). Whatever the method used, however, it is appropriate to establish at the outset that the isolate in question is really an aerobic endospore-former. This remark may seem uncalled-for, but it is not uncommon for a Gram-positive, aerobic, rod-shaped organism to be submitted for identification even though spores have not been demonstrated, and many routine laboratories appear to be reluctant to use microscopy.

Another frequently encountered error is the assumption that, because traditional approaches for identifying these organisms are perceived as being difficult and unreliable, any newer approach is likely to be superior regardless of the size and quality of its database. This takes us to the very heart of the problem: no matter what characterization method is used, considerable amounts of time, money and expertise need to be invested in the construction of reliable and detailed databases, which must be founded upon wide diversities of authentic reference strains. As already noted, most if not all sizeable collections of aerobic endospore-formers contain appreciable numbers of misnamed strains; misnamed either because of earlier identification difficulties, or because of the nomenclatural changes that have followed taxonomic revisions. Until such misnamed strains have been characterized and classified, the parent collection is not going to bear a reliable database. When Logan and Berkeley (1981, 1984) set about developing a *Bacillus* identification scheme based upon the miniaturized biochemical tests of the API System, it was necessary to undertake a taxonomic study first, in order to test the validities of their strains and of the taxa they represented. Some of this work, such as with the *B. subtilis* group, was polyphasic (O'Donnell *et al.* 1980). This kind of approach is even more desirable today, given both the ever-increasing number of species of aerobic endospore-formers, and the need for new polyphasic taxonomic studies that are set within phylogenetic frameworks so as to yield characters useful for routine identification.

Characterization

From as early as the work of Smith *et al.* (1952) it has been increasingly clear that no one phenotypic technique would be suitable for identifying all *Bacillus* species – for example, the insect pathogen *B. pulvifaciens* did not grow on their carbohydrate fermentation test medium – but with the small number of species covered by their scheme this caused little difficulty. However, as further species were isolated from extreme environments, the problems soon mounted up, and Logan and Berkeley (1984) acknowledged that their scheme, based upon miniaturized versions of traditional biochemical tests, could not embrace fastidious organisms, acidophiles or alkaliphiles, although thermophiles and psychrophiles could be accommodated by appropriate incubation. Unreactive species (such as members of the *B. brevis* and *B. sphaericus* group) were few at that time, and could be identified by their growth requirements and microscopic morphologies.

With the many further species of extremophiles subsequently proposed, however, the routine phenotypic test approaches of Smith *et al.* (1952), Gordon *et al.* (1973) and Logan and Berkeley (1984) appeared to have become untenable, and over the same period the potentials of chemotaxonomic and serological methods were investigated. The sections that follow will outline these approaches and summarize their current contributions to identification for *Bacillus* and its relatives. However, it is impossible to devise standard conditions to accommodate the growth of strains of all species for chemotaxonomic work, and it remains unknown to the taxonomist if differences between taxa are consequences of genetic or environmental factors.

The need to substantiate each characterization method by other techniques (be they phenotypic or genotypic) has become increasingly important as new techniques emerge. This need is satisfied by the polyphasic approach now usual for the better classification studies. This approach has the merit that it might be expected to reveal new taxonomic characters of diagnostic value. Another important advantage of the polyphasic approach is that the resolution of different taxonomic levels is possible. It is therefore astonishing that over 50 years after Smith *et al.* (1946) published their first identification scheme, the most widely used, commercially available methods for identifying members of the genus *Bacillus* and its relatives are still based upon miniaturized developments of traditional, routine biochemical tests.

Chemotaxonomic characters

Chemotaxonomic fingerprinting techniques applied to aerobic endospore-formers include FAME profiling, polyacrylamide gel electrophoresis (PAGE) analysis, pyrolysis mass spectrometry, and Fourier-transform infrared spectroscopy, all of which are covered at length in other chapters of this book. Only one of these approaches, FAME analysis, is supported by a commercially available database for routine identification.

As Kämpfer shows in Chapter 18, fatty acid analyses can play very useful parts in polyphasic taxonomic studies of *Bacillus* and its relatives, and certain groups such as *Alicyclobacillus* produce characteristic lipids valuable for identification. However, fatty acid profiles across the aerobic endospore-forming genera do not, given frequent and considerable within-species heterogeneity, form the basis of a reliable, stand-alone identification scheme (Kämpfer 1994). A further difficulty is the need for a standardized incubation temperature for preparing isolates for FAME analysis, making databases for psychrophiles, mesophiles and thermophiles incompatible.

The commercially available Microbial Identification System software (Microbial ID, Inc., Newark, DE) includes a FAME database for the identification of aerobic endospore-formers. Although it cannot be expected to give an accurate or reliable identification with every isolate, it is certainly a valuable screening tool when used with caution.

Serology

Attempts at developing simple serological differentiation systems for *Bacillus* species are commonly plagued by problems of cross-reacting antigens and, in the case of spores, hydrophobic surface properties leading to autoagglutination. A monoclonal antibody test for the distinction of *B. anthracis* from the vegetative cells or spores of other closely related *Bacillus* species is highly desirable, but it is not available.

Nonetheless, serology is routinely useful in two other parts of the *B. cereus* group. A strain differentiation system for *B. cereus* based on flagellar (H) antigens is available at the Food Hygiene Laboratory, Central Public Health Laboratory, Colindale, London, UK, for investigations of food-poisoning outbreaks or other *B. cereus*-associated clinical problems (Kramer & Gilbert 1992). The classification of *B. thuringiensis* strains on the basis of H-antigens and antigenic subfactors is well established, and superior to traditional phenotypic test schemes. Eighty-two serovars are currently recognized (Lecadet *et al.* 1999), and typing services are available at the Institut Pasteur, Paris, France, and at Abbott Laboratories, North Chicago, IL.

Genotypic methods

As with other groups of bacteria, studies of 16S rDNA and of DNA have very valuable applications in the classification of aerobic endospore-formers. This is discussed by De Vos in Chapter 10.

Nucleic acid fingerprinting techniques are also of great potential for typing work, of course. A good example is the differentiation of *Bacillus anthracis* strains by amplified fragment length polymorphism (AFLP) analysis (Keim *et al.* 1997), as the distinction of isolates of this species for epidemiological or strategic purposes has long been a challenge. AFLP also shows promise for the epidemiological typing of *B. cereus* (Ripabelli *et al.* 2000).

At present, however, nucleic acid analyses are not entirely suitable for the routine identification of aerobic endospore-formers; their value in classification does not necessarily make them suitable as routine diagnostic tools at the species level. Amplified ribosomal DNA restriction analysis (ARDRA), for example, has been and continues to be exceptionally effective in the classification of *Bacillus* and relatives (Heyndrickx *et al.* 1996c; Logan *et al.* 2000). It is a very powerful technique for recognizing new taxa and can be used to screen large numbers of strains much faster than is reasonably possible with 16S rDNA sequencing, but it is not always capable of distinguishing closely related species.

Other fingerprinting methods such as ribotyping, which is commercially available, are presently limited by the appropriateness of the restriction enzymes they use, and by the sizes of the databases available to those developing them – both in terms of the numbers of species included and of the numbers of authentic strains representing those species.

Miniaturized biochemical test systems

API System

The API System (bioMérieux, Marcy l'Etoile, France) represented a real advance in bacterial identification when it was introduced in 1970. The system offered miniaturization and standardization of conventional biochemical tests that had often proven to be difficult to perform and interpret consistently by traditional methods. It was also the first identification system to be supported by a computerized database.

Logan and Berkeley (1981) initially used two existing API systems, API 20E and API 50E, both originally developed for the identification of members of the Enterobacteriaceae. A pilot study indicated that the characters in these two kits not only showed promise for distinguishing between a number of well-established *Bacillus* species, but that they could also recognize biotypes within the *B. cereus* group (Logan *et al.* 1979). Consequently they compared test consistencies with those of the conventional tests described by Gordon *et al.* (1973) in international reproducibility trials employing code-numbered *Bacillus* strains. The conventional tests previously had been standardized by the *Bacillus* Sub-Committee (BSC) of the International Committee on Systematic Bacteriology.

As the overall findings of the trials showed that better test reproducibility could be achieved with API tests, further work was carried out to assess the applicability of API tests to *Bacillus* taxonomy. Six hundred cultures were collected and studied, including 137 fresh isolates from clinical specimens, foods, beverages and food-borne illness outbreaks. Twenty morphological and physiological characters, which were convenient and highly reproducible, were chosen from the list of standardized tests prepared by the BSC in 1975, and these were carried out alongside tests (table 9.1) in the API 20E strip and the API 50CHB gallery (the API 50E gallery having been discontinued). The coefficient of Gower (1971) was used to compute similarities, and cluster analysis was then used to split the strains into smaller sized groups for principal co-ordinate analysis (Logan & Berkeley 1981). The results were in general agreement with the taxonomic schemes of Smith *et al.* (1952) and Gordon *et al.* (1973). Moreover, the API System-based taxonomies were also consistent with DNA relatedness data from studies by Somerville and Jones (1972), Seki *et al.* (1975, 1978), Kaneko *et al.* (1978) and O'Donnell *et al.* (1980). It was therefore concluded that the tests in the API System provided a good basis for taxonomic work with *Bacillus*.

Logan and Berkeley (1984) continued their study to include 1075 *Bacillus* strains, and generated a database from 49 carbohydrate tests in the 50CHB gallery, 12 tests from the API 20E strip and 14 morphological and supplementary tests for the identification of 26 valid species and 12 further biotypes and unvalidated species. As available commercially, this system currently allows the recognition of only 19 species, but an enhanced database is in preparation, following the

Table 9.1 A listing of the substrates included in commercially available kits used for the characterization of aerobic endospore-forming bacteria.

Substrates	Test panel (number of tests in panel)			
	API 50CHB & 20E (61)[a]	API Biotype 100 (99)[b]	Vitek (29)[c]	Biolog (95)[d]
Adonitol	+	+		
Aesculin	+	+	+	
Amygdalin	+		+	+
Amylopectin			+	
D-Arabinose	+			
L-Arabinose	+	+	+	+
D-Arabitol	+	+	+	+
L-Arabitol	+	+		
Arbutin	+			+
2,3-Butanediol				+
Cellobiose	+	+		+
α-cyclodextrin				+
β-cyclodextrin				+
Dextrin				+
Dulcitol	+	+		
Erythritol	+	+		
Fructose	+	+		+
D-Fucose	+			
L-Fucose	+	+		+
Galactose	+	+	+	+
Gentiobiose	+	+.		+
Glucose	+	+	+	+
Glycerol	+	+		+
Glycogen	+			+
Hydroxyquinoline-β-glucuronide		+		
myo-Inositol	+	+	+	+
Inulin	+		+	+
Lactose	+	+		+
Lactulose		+		+
Lyxose	+	+		
Maltitol		+		
Maltose	+	+	+	+
Maltotriose		+		+
Mannan				+
Mannitol	+	+	+	+
Mannose	+	+		+
Melezitose	+	+		+
Melibiose	+	+		+
1-0-Methyl-α-galactopyranoside		+		
1-0-Methyl-β-galactopyranoside		+		
α-Methyl-D-galactoside				+
β-Methyl-D-galactoside				+
3-0-Methyl-D-glucopyranose		+		
1-0-Methyl-α-D-glucopyranoside		+		
1-0-Methyl-β-D-glucopyranoside		+		
3-Methyl-glucose				+
α-Methyl-D-glucoside	+			+
β-Methyl-D-glucoside				+

(*continued on p. 132*)

Table 9.1 *(cont'd)*

Substrates	Test panel (number of tests in panel)			
	API 50CHB & 20E (61)[a]	API Biotype 100 (99)[b]	Vitek (29)[c]	Biolog (95)[d]
α-Methyl-D-mannoside	+			+
β-Methyl-D-xyloside	+			
o-Nitrophenol-β-D-galactopyranoside	+			
Palatinose		+	+	+
Psicose				+
Raffinose	+	+	+	+
Rhamnose	+	+		+
Ribose	+	+	+	+
Salicin	+		+	+
Sedoheptulosan				+
Sorbitol	+	+	+	+
Sorbose	+	+		
Stachyose				+
Starch	+			
Sucrose	+	+	+	+
Tagatose	+	+	+	+
Trehalose	+	+	+	+
Turanose	+	+		+
Xylitol	+	+		+
D-Xylose	+	+	+	+
L-Xylose	+			
Acetate			+	+
N-Acetyl-L-glutamate				+
cis-Aconitate		+		
trans-Aconitate		+		
4-Aminobutyrate		+		
5-Aminovalerate		+		
Benzoate		+		
Caprate		+		
Caprylate		+		
Citrate	+	+		
m-Coumarate		+		
Fumarate		+		
Galacturonate		+		+
Gentisate		+		
Gluconate	+	+		+
Glucuronate		+		
Glutarate		+		
DL-Glycerate		+		
3-Hydroxybenzoate		+		
4-Hydroxybenzoate		+		
α-Hydroxybutyrate				+
β-Hydroxybutyrate		+		+
γ-Hydroxybutyrate				+
p-Hydroxyphenyl acetate				+
Itaconate		+		
2-Ketogluconate	+	+		
5-Ketogluconate	+	+		

(continued)

Table 9.1 *(cont'd)*

Substrates	Test panel (number of tests in panel)			
	API 50CHB & 20E (61)[a]	API Biotype 100 (99)[b]	Vitek (29)[c]	Biolog (95)[d]
α-Ketoglutarate		+		+
α-Ketovalerate				+
D-Lactic acid methyl ester				+
DL-Lactate		+		
L-Lactate				+
D-Malate		+		+
L-Malate		+		+
Malonate		+		
Methylpyruvate				+
Monomethyl succinate				+
Mucate		+		
Phenylacetate		+		
3-Phenylpropionate		+		
Propionate		+		+
Protocatechuate		+		
Pyruvate	+[e]			+
Quinate		+		
Saccharate		+		
Succinamate				+
Succinate		+		+
D-Tartrate		+		
L-Tartrate		+		
meso-Tartrate		+		
Tricarballylate		+		
N-Acetyl-D-glucosamine	+	+	+	+
N-Acetyl-D-mannosamine				+
Alaninamide				+
D-Alanine		+		+
L-Alanine		+		+
L-Alanyl-glycine				+
Arginine	+[f]			
Asparagine				+
Aspartate		+		
Betain		+		
Ethanolamine		+		
Glucosamine		+		
L-Glutamate		+		+
Glycyl-L-glutamate				+
Histamine		+		
L-Histidine		+		
Lactamide				+
Lysine	+[g]			
Ornithine	+[h]			
Proline		+		
Putrescine		+		+
L-Pyroglutamate				+
L-Serine		+		+
Trigonelline		+		

(continued on p. 134)

Table 9.1 (*cont'd*)

Substrates	API 50CHB & 20E (61)[a]	API Biotype 100 (99)[b]	Vitek (29)[c]	Biolog (95)[d]
Tryptamine		+		
Tryptophan	+[i]	+		
L-Tyrosine		+		
Adenosine				+
Adenosine-5'-monophosphate				+
2-Deoxyadenosine				+
Fructose-6-phosphate				+
Glucose-1-phosphate				+
Glucose-6-phosphate				+
DL-α-Glycerolphosphate				+
Inosine				+
Thymidine				+
Thymidine-5'-monophosphate				+
Urea	+			
Uridine				+
Uridine-5'-monophosphate				+
Gelatin (hydrolysis)	+			
Mandelic acid			+	
Nalidixic acid			+	
Nitrate (reduction)	+			
Oleandomycin			+	
Polyamidohygrostrepin			+	
Potassium thiocyanate			+	
Sodium chloride 7%			+	
Sodium thiosulphate	+[j]			
Tetrazolium red			+	
Tween 40				+
Tween 80				+

[a] The API 50CHB gallery tests for acid production from a range of carbohydrates, and 12 miscellaneous biochemical tests in the API 20E strip are used in conjunction with it.
[b] The API Biotype 100 gallery tests for assimilation of carbohydrates, organic acids and amino acids.
[c] The Vitek BAC card includes tests for: acid production from carbohydrates, substrate utilization, growth in the presence of inhibitory agents, and dye reduction.
[d] The Biolog MicroPlate tests for ability to utilize or oxidise various carbon sources.
[e] As substrate for the Voges–Proskauer test.
[f] As substrate for arginine dihydrolase.
[g] As substrate for lysine decarboxylase.
[h] As substrate for ornithine decarboxylase.
[i] As substrate for the indole and tryptophan deaminase tests.
[j] As substrate for H_2S production.

study of a large number of strains which have been carefully authenticated by polyphasic taxonomic study.

Biotype 100

The species of the genera *Brevibacillus* and *Aneurinibacillus*, and in the *B. sphaericus* group are largely unreactive in the carbohydrate utilization tests of the API 50CHB gallery, and insufficiently variable in the API 20E and supplementary tests, so that some 20 species are largely inseparable by this means. The API Biotype 100 gallery, which was developed as a research product for differentiating enterobacteria, contains 99 tests for the assimilation of carbohydrates, organic acids and amino acids (table 9.1), and one control tube. It is inoculated with a suspension in one of two semisolid media, which differ in the number of growth factors they contain, and after incubation the tubes are examined for turbidity. This kit has proven to be of great value in differentiating species in *Brevibacillus* and *Aneurinibacillus* (Heyndrickx *et al*. 1997) and shows promise with other unreactive species.

Vitek

In the 1960s, the National Aeronautics and Space Administration (NASA) sponsored the McDonnell Douglas Astronautics Company in the development of an instrumental system for detecting specific microorganisms in a space-craft environment. The resulting machine – the Microbial Load Monitor (MLM) – was then developed further for clinical application and led to the Auto Microbial System (AMS) for the direct identification of microbes from urine samples (Aldridge *et al*. 1977). Their results showed overall agreement of around 90% between AMS and established manual culture methods. The early influence of the space industry was reflected in the use of a disposable, miniaturized, plastic specimen-handling system, and solid-state optics with a minicomputer for control and processing. The optical system monitored changes in light transmission through micro-cuvettes placed inside an incubator/reader. The micro-cuvette contained dehydrated selective growth media that rehydrated on inoculation with the sample.

This whole project formed the concept of what was to become the world's leading automated instrumentation system for microbiology. The AMS was developed into the Vitek system (bioMérieux Vitek, Hazelwood, MO). The Vitek system is now used routinely for identification and for determination of antimicrobial susceptibility in thousands of laboratories worldwide, and has the capability to identify over 300 species of bacteria and yeasts, in both clinical and industrial environments. The automated and standardized inoculation of the test cards, and the automatic and regular screening of test reactions improves safety, eliminates many repetitive manual operations, and allows large numbers of samples to be analysed rapidly. Although the present Vitek *Bacillus* test card (table 9.1), coupled with the database, can identify only 17 species with confidence, the database is curently being expanded to recognize 34 well-defined species. A more highly automated system, the Vitek II, is now available for several clinically important bacterial groups and a similar system is being developed for aerobic endospore-formers.

Biolog

The Biolog system (Biolog, Hayward, CA), like the API Biotype system, is based on carbon source utilization patterns, using a plate that is inoculated with a standardized suspension of a pure culture and incubated as appropriate; there the similarity ends. The Biolog substrates are presented in a 96-well MicroPlate, one well of which is a negative control. Each substrate that supports metabolic activity is indicated by the reduction of a tetrazolium dye, so that the system is based on the process of metabolism itself rather than the release of metabolic by-products such as acid. The plate can be read using a conventional plate reader and the results compared with an extensive computerized database. It can also be read by eye, as with the API System kits, which is of great importance in allowing the less wealthy to use such products.

For *Bacillus* identification, the first release of the Biolog Gram-positive panel was dogged by problems of false positive results (Baillie *et al.* 1995). The recently released version of the Gram-positive aerobic panel (table 9.1) has addressed this problem and uses a more viscous suspension medium to reduce flocculation and pellicle formation. The database currently contains 23 *Bacillus* species (two of which are thermophiles, and therefore already part-identified by their growth temperature requirements), *Brevibacillus brevis* and six *Paenibacillus* species.

Probabilistic identification

One of the great advantages of using identification systems such as API 50CHB/20E, Vitek and Biolog is the support given by access to the manufacturers' ever-expanding databases. These databases are larger and more reliable than those found in most laboratories, excepting specialist reference facilities, and they offer identification using conditional probability or likelihood models (for explanation, see Logan 1994).

As these models work on the basis of multiplying probabilities together, they are best used with characterization tests that show low variation within the species they are used to distinguish – otherwise unacceptably low (say <90%) identification probabilities result. They do not necessarily lend themselves well to identifying aerobic endospore-formers, the species of which often show considerable within-species variation in many tests. This variation illustrates the comment made by Cowan and Steel (1965): 'The different kinds of bacteria are not separated by sharp divisions but by slight and subtle differences in characters so that they blend into each other and resemble a spectrum'. Sneath (1968) divided total dissimilarity between organisms into components of vigour (reflecting growth rate or times of reading tests) and pattern (representing differences between equally vigorous organisms), and both of these can be important in identifying aerobic endospore-formers. Differences of pattern are what we use to define our taxa, but in several species, particularly the more frequently encountered and, thus, better represented ones such as *B. subtilis*, appreciable within-species pattern differences can occur.

The current identification kits and databases for the aerobic endospore-formers reflect these problems in two ways. First, the large number of tests included in the API System and the Biolog MicroPlate mean that the influence of any single character on the result of an identification is less than would be the case with a smaller panel of tests. Secondly, the Vitek and Biolog databases can consider the development of the test results over time, as the times at which different tests become positive can be sufficiently characteristic of organisms to allow early identification. With Biolog, the MicroPlates can be read after 4–6 h of incubation, then reincubated for a further 12–18 h if an acceptable similarity threshold has not been reached. As the Vitek card is incubated within the test reader, it remains on-line, and therefore identifications are automatically attempted hourly between 6 h of incubation and the final reading at 15 h (for mesophiles). Each organism has an identification threshold established in the database, against which identification matches are screened for acceptability; any organism not passing a threshold by 15 h is then considered unidentified.

Conclusion

At the beginning of this chapter it was noted that there are 123 valid species in *Bacillus* and the seven new genera derived from it, yet the most up-to-date databases available (or soon to become available) with the three commercial identification systems only allow the recognition of 30–34 species. This shortfall reflects the obstacles – both practical and commercial – that hinder the achievement of a comprehensive routine identification scheme for these bacteria: (i) the group contains extremophiles and fastidious organisms; (ii) some species are very rarely encountered and may only be known from exotic environments; (iii) new species established on the basis of very few isolates are too poorly represented for satisfactory database construction; and (iv) other new species recognized from DNA relatedness studies may not be distinguished easily using phenotypic characters.

The foregoing well illustrates the statement made by Gordon *et al.* in their exemplary monograph of 1973: 'No matter how hard he struggles toward this goal [of building a system capable of identifying all kinds of bacteria that exist], the taxonomist is always painfully aware of the shortcomings of his efforts and of the enormous amount of work remaining to be done. At best he can only contribute to a progress report'.

References

Aldridge, C., Jones, P.W., Gibson, S. *et al.* (1977) Automated microbiological detection/identification system. *Journal of Clinical Microbiology* 6, 406–413.

Ash, C., Farrow, J.A.E., Wallbanks, S., & Collins, M.D. (1991) Phylogenetic heterogeneity of the genus *Bacillus* revealed by comparative analyses of small subunit-ribosomal RNA sequences. *Letters in Applied Microbiology* 13, 202–206.

Ash, C., Priest, F.G. & Collins, M.D. (1993) Molecular identification of rRNA group 3 bacilli (Ash, Farrow, Wallbanks and Collins) using a PCR probe test. *Antonie van Leeuwenhoek* **64**, 253–260.

Baillie, L.W.J., Jones, M.N., Turnbull, P.C.B. & Manchee, R.J. (1995) Evaluation of the Biolog system for the identification of *Bacillus anthracis*. *Letters in Applied Microbiology* **20**, 209–211.

Collins, M.D., Lawson, P.A., Willems, A. *et al.* (1994) The phylogeny of the genus *Clostridium*: proposal of five new genera and eleven new species combinations. *International Journal of Systematic Bacteriology* **44**, 812–826.

Cowan, S.T. & Steel, K.J. (1965) *Manual for the Identification of Medical Bacteria*. Cambridge University Press, Cambridge.

Dufresne, S., Bousquet, J., Boissinot, M. & Guay, R. (1996) *Sulfobacillus disulfidooxidans* sp. nov., a new acidophilic, disulfide-oxidising, Gram-positive, spore-forming bacterium. *International Journal of Systematic Bacteriology* **46**, 1056–1064.

Golovacheva, R.S. & Karavaiko, G.I. (1978) A new genus of thermophilic spore-forming bacteria. *Microbiology* **47**, 658–665.

Gordon, R.E., Haynes, W.C. & Pang, C.H.-N. (1973) *The Genus* Bacillus. Agriculture Handbook 427. US Department of Agriculture, Washington, DC.

Gower, J.C. (1971) A general coefficient of similarity and some of its properties. *Biometrics* **27**, 857–874.

Heyndrickx, M., Vandemeulebroeke, K., Scheldeman, P. *et al.* (1995) *Paenibacillus* (formerly *Bacillus*) *gordonae* (Pichinoty *et al.* 1986) Ash *et al.* 1994 is a later subjective synonym of *Paenibacillus* (formerly *Bacillus*) *validus* (Nakamura 1984) Ash *et al.* 1994: emended description of *P. validus*. *International Journal of Systematic Bacteriology* **45**, 661–669.

Heyndrickx, M., Vandemeulebroecke, K., Hoste, B. *et al.* (1996a) Reclassification of *Paenibacillus* (formerly *Bacillus*) *pulvifaciens* (Nakamura 1984) Ash *et al.* 1994, a later subjective synonym of *Paenibacillus* (formerly *Bacillus*) *larvae* (White 1906) Ash *et al.* 1994, as a subspecies of *P. larvae*. Emended description of *P. larvae* with *P. larvae* subsp. *larvae* and *P. larvae* subsp. *pulvifaciens*. *International Journal of Systematic Bacteriology* **46**, 270–279.

Heyndrickx, M., Vandemeulebroeke, K., Scheldeman, P. *et al.* (1996b) A polyphasic reassessment of the genus *Paenibacillus*, reclassification of *Bacillus lautus* (Nakamura 1984) as *Paenibacillus lautus* comb. nov. and of *Bacillus peoriae* (Montefusco *et al.* 1993) as *Paenibacillus peoriae* comb. nov., and emended descriptions of *P. lautus* and *P. peoriae*. *International Journal of Systematic Bacteriology* **46**, 988–1003.

Heyndrickx, M., Vauterin, L., Vandamme, P., Kersters, K. & De Vos, P. (1996c) Applicability of combined amplified 16S rDNA restriction analysis (ARDRA) patterns in bacterial phylogeny and taxonomy. *Journal of Microbiological Methods* **26**, 247–259.

Heyndrickx, M., Lebbe, L., Vancanneyt, M. *et al.* (1997) A polyphasic reassessment of the genus *Aneurinibacillus*, reclassification of *Bacillus thermoaerophilus* (Meier-Stauffer *et al.* 1996) as *Aneurinibacillus thermoaerophilus* comb. nov., and emended descriptions of *A. aneurinilyticus* corrig., *A. migulanus*, and *A. thermoaerophilus*. *International Journal of Systematic Bacteriology* **47**, 808–817.

Heyndrickx, M., Lebbe, L., Kersters, K. *et al.* (1998) *Virgibacillus*: a new genus to accommodate *Bacillus pantothenticus* (Proom and Knight 1950). Emended description of *Virgibacillus pantothenticus*. *International Journal of Systematic Bacteriology* **48**, 99–106.

Kämpfer, P. (1994) Limits and possibilities of total fatty acid analysis for classification and identification of *Bacillus* species. *Systematic and Applied Microbiology* **17**, 86–98.

Kaneko, T., Nozaki, R. & Aizawa, K. (1978) De-oxyribonucleic acid relatedness between *Bacillus anthracis*, *Bacillus cereus* and *Bacillus thuringiensis*. *Microbiology and Immunology* **22**, 639–641.

Keim, P., Kalif, A., Schupp, J. *et al.* (1997) Molecular evolution and diversity in *Bacillus anthracis* as detected by amplified fragment length polymorphism markers. *Journal of Bacteriology* **179**, 818–824.

Kramer, J.M. & Gilbert, R.J. (1992) *Bacillus cereus* gastroenteritis. In: *Food Poisoning: Handbook of Natural Toxins* (ed. A.T. Tu), Vol. 7, pp. 119–153. Marcel Dekker, New York.

Lecadet, M.-M., Frachon, E., Cosmao-Dumanoir, V. *et al.* (1999) Updating the H-antigen classification of *Bacillus thuringiensis*. *Journal of Applied Microbiology* **86**, 660–672.

Logan, N.A. (1994) *Bacterial Systematics*. Blackwell Scientific Publications, Oxford.

Logan, N.A. & Berkeley, R.C.W. (1981) Classification and identification of members of the genus *Bacillus*. In: *The Aerobic Endospore-Forming Bacteria* (eds R.C.W. Berkeley & M. Goodfellow), pp. 105–140. Academic Press, London.

Logan, N.A. & Berkeley, R.C.W. (1984) Identification of *Bacillus* strains using the API system.

Journal of General Microbiology **130**, 1871–1882.

Logan, N.A., Capel, B.J., Melling, J. & Berkeley, R.C.W. (1979) Distinction between emetic and other strains of *Bacillus cereus* using the API system and numerical methods. *FEMS Microbiology Letters* **5**, 373–375.

Logan, N.A., Lebbe, L., Hoste, B. *et al.* (2000) Aerobic endospore-forming bacteria from geothermal environments in northern Victoria Land, Antarctica, and Candlemas Island, South Sandwich archipelago, and proposal of *Bacillus fumarioli* sp. nov. *International Journal of Systematic and Evolutionary Microbiology* **50**, 1741–1753.

Nakamura, L.K. (1984) *Bacillus amylolyticus* sp. nov., nom. rev., *Bacillus lautus* sp. nov., nom. rev., *Bacillus pabuli* sp. nov., nom. rev., *Bacillus validus* sp. nov., nom. rev. *International Journal of Systematic Bacteriology* **34**, 224–226.

Nakamura, L.K. (1993) DNA relatedness of *Bacillus brevis* Migula 1900 strains and proposal of *Bacillus agri* sp. nov., nom. rev., and *Bacillus centrosporus* sp. nov., nom. rev. *International Journal of Systematic Bacteriology* **41**, 510–515.

Nakamura, L.K. & Swezey, J. (1983) Deoxyribonucleic acid study of *Bacillus circulans* Jordan 1890. *International Journal of Systematic Bacteriology* **33**, 703–708.

Niimura, Y., Koh, E., Yanagida, F., Suzuki, K.-I., Komagata, K. & Kozaki, M. (1990) *Amphibacillus xylanus* gen. nov., sp. nov., a facultatively anaerobic sporeforming xylan-digesting bacterium which lacks cytochrome, quinone, and catalase. *International Journal of Systematic Bacteriology* **40**, 297–301.

O'Donnell, A.G., Norris, J.R., Berkeley, R.C.W. *et al.* (1980) Characterisation of *Bacillus subtilis*, *Bacillus pumilus*, *Bacillus licheniformis* and *Bacillus amyloliquefaciens*, by pyrolysis gas-liquid chromatography: characterisation tested by using DNA-DNA hybridisation, biochemical tests and API systems. *International Journal of Systematic Bacteriology* **30**, 448–459.

Ripabelli, G., McLauchlin, J., Mithani, V. & Threlfall, E.J. (2000) Epidemiological typing of *Bacillus cereus* by amplified fragment length polymorphism. *Letters in Applied Microbiology* **30**, 358–363.

Rosado, A.S., van Elsas, J.D. & Seldin, L. (1997) Reclassification of *Paenibacillus durum* (formerly *Clostridium durum* Smith and Cato 1974) Collins *et al.* 1994 as a member of the species *P. azotoformans* (formerly *Bacillus azotoformans* Seldin et al. (1984) Ash *et al.* 1994. *International Journal of Systematic Bacteriology* **47**, 569–572.

Seki, T., Oshima, T. & Oshima, Y. (1975) Taxonomic study of *Bacillus* by deoxyribonucleic acid-deoxyribonucleic acid hybridisation and interspecific transformation. *International Journal of Systematic Bacteriology* **25**, 258–270.

Seki, T., Chung, C-K., Mikami, H., & Oshima, Y. (1978) Deoxyribonucleic acid homology and taxonomy of the genus *Bacillus*. *International Journal of Systematic Bacteriology* **28**, 182–189.

Shida, O., Takagi, H., Kadowaki, K. *et al.* (1994a) *Bacillus aneurinolyticus* sp. nov., nom. rev. *International Journal of Systematic Bacteriology* **44**, 143–150.

Shida, O., Takagi, H., Kadowaki, K., Udaka, S. & Komagata, K. (1994b) *Bacillus galactophilus* is a later subjective synonym of *Bacillus agri*. *International Journal of Systematic Bacteriology* **44**, 172–173.

Shida, O., Takagi, H., Kadowaki, K., Udaka, S., Nakamura, K. & Komagata, K. (1995) Proposal of *Bacillus reuszeri* sp. nov., *Bacillus formosus* sp. nov., nom. rev., and *Bacillus borstelensis* sp. nov., nom. rev. *International Journal of Systematic Bacteriology* **45**, 93–100.

Shida, O., Takagi, H., Kadowaki, K. & Komagata, K. (1996) Proposal for two new genera, *Brevibacillus* gen. nov. and *Aneurinibacillus* gen. nov. *International Journal of Systematic Bacteriology* **46**, 939–946.

Smith, N.R., Gordon, R.E. & Clark, F.E. (1946) *Aerobic Mesophilic Spore-Forming Bacteria*. Miscellaneous Publication 559. US Department of Agriculture, Washington, DC.

Smith, N.R., Gordon, R.E. & Clark, F.E. (1952) *Aerobic Spore-Forming Bacteria*. Monograph 16. US Department of Agriculture, Washington, DC.

Sneath, P.H.A. (1968) Vigour and pattern in taxonomy. *Journal of General Microbiology* **54**, 1–11.

Sommerville, H.J. & Jones, M.L. (1972) DNA competition studies within the *Bacillus cereus* group of bacilli. *Journal of General Microbiology* **73**, 257–265.

Spring, S., Ludwig, W., Marquez, M.C., Ventosa, A. & Schleifer, K.-H. (1996) *Halobacillus* gen. nov., with descriptions of *Halobacillus litoralis* sp. nov. and *Halobacillus trueperi* sp. nov., and transfer of *Sporosarcina halophila* to *Halobacillus halophilus* comb. nov. *International Journal of Systematic Bacteriology* **46**, 492–496.

Tagaki, H., Shida, O., Kadowaki, K., Komagata, K. & Udaka, S. (1993) Characterization of *Bacillus brevis* with descriptions of *Bacillus migulanus* sp. nov., *Bacillus choshinensis* sp. nov., *Bacillus parabrevis* sp. nov., and *Bacillus galactophilus* sp. nov. *International Journal of Systematic Bacteriology* **43**, 221–231.

Touzel, J.P., O'Donohue, M., Debeire, P., Samain, E. & Breton, C. (2000) *Thermobacilus xylanilyticus* gen. nov., sp. nov., a new aerobic thermophilic xylan-degrading bacterium isolated from soil. *International Journal of Systematic and Evolutionary Microbiology* 50, 315–320.

Wainö, M., Tindall, B.J., Schumann, P. & Ingvorsen, K. (1999) *Gracilibacillus* gen. nov., with description of *Gracilibacillus halotolerans* gen. nov., sp. nov.; transfer of *Bacillus dipsosauri* to *Gracilibacillus dipsosauri* comb. nov., and *Bacillus salexigens* to the genus *Salibacillus* gen. nov., as *Salibacillus salexigens* comb. nov. *International Journal of Systematic Bacteriology* 49, 821–831.

Wisotzkey, J.D., Jurtshuk, P., Jr, Fox, G.E., Deinhard, G. & Poralla, K. (1992) Comparative sequence analyses on the 16S rRNA (rDNA) of *Bacillus acidocaldarius, Bacillus acidoterrestris,* and *Bacillus cycloheptanicus* and proposal for creation of a new genus, *Alicyclobacillus* gen. nov. *International Journal of Systematic Bacteriology* 42, 263–269.

Zaitsev, G.M., Tsiko, I.V., Rainey, F.A. *et al.* (1998) New aerobic ammonium-dependent obligately oxalotrophic bacteria: description of *Ammoniphilus oxalaticus* gen. nov., sp. nov. and *Ammoniphilus oxalivorans* gen. nov., sp. nov. *International Journal of Systematic Bacteriology* 42, 151–163.

Chapter 10

Nucleic Acid Analysis and SDS-PAGE of Whole-cell Proteins in *Bacillus* Taxonomy

Paul De Vos

Bacterial taxonomy concerns (i) the classification of organisms into groups (taxa) on the basis of their similarities, (ii) the nomenclature of the taxa as defined in (i), and (iii) the identification of newly discovered organisms. The latter contains the process that in theory should reveal whether a new organism belongs in one of the existing taxa, or whether it belongs to a new taxon that has then to be named and reported to the scientific community. This type of definition holds for prokaryotes as well as for eukaryotes. In bacterial taxonomy it is generally accepted that taxa should be defined polyphasically. Polyphasic taxonomy can be considered as a concept for delineation of taxa through a consensus based on both phenotypic and genotypic information. A clear-cut general protocol or scheme cannot be given with which to perform a polyphasic taxonomic study of a given group of bacteria. The methods to be applied depend, of course, on the number of strains, the characteristics of the bacterial group under study, etc. (for a review see Vandamme *et al.* 1996). This chapter aims to give summaries of two rather different approaches that are currently used in polyphasic bacterial taxonomy in general and in taxonomic studies of *Bacillus* in particular.

Analysis of nucleic acids covers a wide range of taxonomic levels supported by a variety of methods – both simple and complex – that study the genotypic characteristics of the bacteria. Generally, the most reliable methods are expensive and demand special skills for manipulation and computerized interpretation of data.

Sodium-dodecylsulfate-polyacrylamide gel electrophoresis (SDS-PAGE) of whole-cell proteins results in a phenotypic fingerprint of the bacteria and can be used to discriminate at the species and often at the subspecies levels. The methodology is simple and cheap, and hence well suited for an initial study of large groups of strains. However, it does impose stringent standardized conditions of growth and technical manipulation.

In the following the two approaches will be discussed separately.

NUCLEIC ACID ANALYSIS IN *BACILLUS* TAXONOMY

Introduction

Since molecular methods have been introduced to biology, the nucleic acids of bacteria have been studied with growing interest. As a spin-off from these studies, the methodology in bacterial taxonomy has changed greatly. Indeed, it has been shown that bacterial genome characteristics reflect bacterial evolution, and it is now generally accepted that the taxonomy of living organisms should reflect this natural evolution. As a logical consequence of these progressive developments in molecular techniques, we are confronted with methods that allow us to determine the nucleic acid sequences of whole genomes, both prokaryotic and eukaryotic. The sequence data as such will not only offer new possibilities for tracing genotypic similarities between bacteria, but also provide information and new insights into the genetic organization of bacterial genomes, including metabolic diversity and the mechanisms of evolution itself. The sequence data revealed by previous genetic studies are about to be overruled if more entire bacterial genomes are sequenced. Indeed, the complete sequence of *Bacillus subtilis* (Kunst *et al.* 1997) revealed that at least one third of the open reading frames (ORFs) code for unknown products (Wipat & Harwood 1999). It is still too early, however, for a useful discussion on comparative analysis of overall genome sequences in relation to taxonomic implications.

In addition, the enormous quantity of genetic data on different aspects of biochemical pathways, and the morphological and physiological characteristics of *Bacillus* and related genera that might be of interest for their taxonomy will not be discussed, and the reader is referred to the vast number of reports in this field. Some of them can be found in the recent mini-review of Wipat and Harwood (1999).

This part of the chapter aims to give an overview of a variety of methods that are currently applied to the study of the nucleic acid composition of members of the genus *Bacillus* and relatives (the aerobic endospore-forming bacteria) in relation to their taxonomy and phylogeny. It concerns a wide phylogenetic group of bacteria, as can be deduced from the mol % G+C range of 43–68% (Claus & Berkeley 1986). This phylogenetic heterogeneity is demonstrated clearly by the 16S rRNA and rDNA sequence analyses that have been reported over the last two decades. The impact of 16S rDNA sequence analysis is discussed in Chapter 2 of this book. I will only mention here the application of ARDRA (amplified rDNA restriction analysis) of the 16S rDNA part of the operon as an alternative to direct sequence analysis. Furthermore, different (mostly indirect) methods applied in comparative studies of genomes of *Bacillus* and relatives will be discussed briefly, as well as their taxonomic relevance and resolution.

Methods to study nucleic acids in relation to the taxonomy of *Bacillus sensu lato*

rRNA- and rDNA-linked methods

Amplified rDNA restriction analysis (ARDRA)

16S rDNA (or rRNA) sequence analysis (either directly or indirectly measured) has formed the framework for modern bacterial taxonomy ever since it was shown by Woese (1987) and many others that characteristics of the rDNA operon can be considered as a biological clock to reveal the natural relationships between bacteria. Direct 16S rDNA sequencing of the type strains of many *Bacillus* species resulted in a new classification of the genus *Bacillus sensu* Claus and Berkeley (1986) – hereinafter referred to as *Bacillus sensu lato* (*s.l.*) – at the generic level. It has been shown that several rRNA groups deserve separate generic status, although the generic cut-off level is not fixed and varies between about 87 and 95% in 16S rDNA sequence similarity. On the basis of these rDNA sequence data, the genus *Bacillus s.l.* was split into *Bacillus sensu stricto* (containing organisms from rRNA groups 1 and 2 of Ash *et al.* 1993), *Paenibacillus* (i.e. rRNA group 3 of Ash *et al.* 1991, 1993), *Aneurinibacillus* and *Brevibacillus* (i.e. both rRNA group 4 of Ash *et al.* 1991; Shida *et al.* 1996), *Alicyclobacillus* (Wisotzkey *et al.* 1992), *Virgibacillus* (Heyndrickx *et al.* 1998), *Halobacillus* (Spring *et al.* 1996), *Gracilibacillus* and *Salibacillus* (Wainö *et al.* 1999). The genus *Amphibacillus* (Niimura *et al.* 1990) was delineated on the basis of 16S rDNA sequencing and also belongs to the *Bacillus s.l.* lineage as well as *Sporosarcina* (Farrow *et al.* 1992) and the nonsporulating genera *Planococcus*, *Kurthia*, *Caryophanon* and *Exiguobacterium* (Farrow *et al.* 1994).

However, many incompletely characterized aerobic spore-forming isolates are still erroneously assigned to species of *Bacillus s.l.*, and they need to be classified correctly[1]. The reclassification of these isolates will allow important taxonomic emendations of *Bacillus* and relatives at the genus and species levels and hence more reliable identifications. As direct sequencing of 16S rDNA is still expensive and not available to all microbiologists, indirect, fast and less expensive methods such as ARDRA, to characterize the 16S rDNA part of the ribosomal operon via restriction analysis, may offer a good alternative for sequencing (Heyndrickx *et al.* 1996b). This method also has the advantages that computerized interpretation of the data and database construction are posssible.

PRINCIPLES OF THE METHOD

A theoretical approach (via a computer program, e.g. GeneCompar, Applied Maths, Belgium) of a restriction analysis of 16S rDNA amplicons of well-chosen *Bacillus* reference strains of which the complete 16S rDNA sequence was known, resulted in a set of five restriction enzymes (*Hae*III, *Dpn*II, *Rsa*I, *Bfa*I and *Tru*9) that allowed maximal taxonomic discrimination at the species level. A computer combination of the five experimentally revealed restriction banding patterns resulted in a more complex banding pattern (20–30 bands) that can be used for numerical analysis (figure 10.1). From table 10.1, it can be deduced that the experimentally determined lengths of the restriction bands correspond closely to the theoretical lengths as deduced *a posteriori* from the sequence which needed to be determined for the type strain for the description of *B. fumarioli* (Logan *et al.* 2000).

Finally, the ARDRA method has been adapted for application on a nucleic acid sequencer (Pukall *et al.* 1998), although this application has not been used yet in taxonomic studies of *Bacillus*.

---

[1] Since this chapter was written, many adaptations concerning the taxonomy of *Bacillus s.l.* have been published.

Table 10.1 Distribution of restriction fragments of *Bacillus fumarioli* LMG 17489T obtained after experimental restriction and theoretical restriction using GeneCompar software. The slightly higher length of the experimentally revealed fragments can be at least partly explained by the length of the primers (about 30 bp). For the higher molecular weight bands, a slight change in the band marking affects the molecular weight significantly (e. g. the highest band in *Bfa*I). Sometimes (*Hae*III) the smaller fragments were not recovered on the gel. Sometimes (faint) false bands (resulting from incomplete restriction activity) are seen (a band of 630 bp in *Tru*9). As a result, the reproducibility of the method is about 92% similarity (Heyndrickx *et al.* 1996b).

Restriction enzyme	Restriction fragment distribution	
	Experimental	Theoretical
*Hae*III	497	479
	471	459
	231	212
	119	97
	82	85
	74	78
	–	34
	–	22
*Dpn*II	856	815
	238	238
	164	148
	152	136
	139	129
*Rsa*I	493	455
	424	406
	369	357
	147	146
	114	91
	–	11
*Bfa*I	832	763
	350	335
	252	228
	167	140
*Tru*9	–	630
	526	550
	393	432
	278	285
	136	138
	86	85
	47	–

EVALUATION OF THE DATABASE AND TAXONOMIC
RESOLUTION OF THE METHOD

Application of the ARDRA approach to *Bacillus* reference strains revealed a generally similar grouping of the strains to the one based on 16S rDNA sequence analysis (figure 10.2). However, the overall topology of the ARDRA clustering showed significant differences, indicating the lower reliability of the grouping at lower taxonomic levels. Comparative analysis of the ARDRA patterning allowed

Figure 10.1 Computer patterns of restriction analysis of 16S rDNA amplicons of the type strain of *Bacillus fumarioli* LMG 17489. Lanes 1 to 5 show the patterns obtained with the tetra cutters *Hae*III, *Dpn*II, *Rsa*I, *Bfa*I and *Tru*9, respectively. For the molecular weight marker in the left-hand lane, the number of base pairs are shown. The marker is used in the normalization process that is needed for computer-assisted interpretation of the similarities and for construction of a database. In the right-hand lane (6), a combined pattern of the five restriction patterns is shown. The bands of the individual restrictions are not always well separated on the figure, owing to compression and limitations in the drawing.

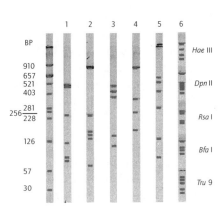

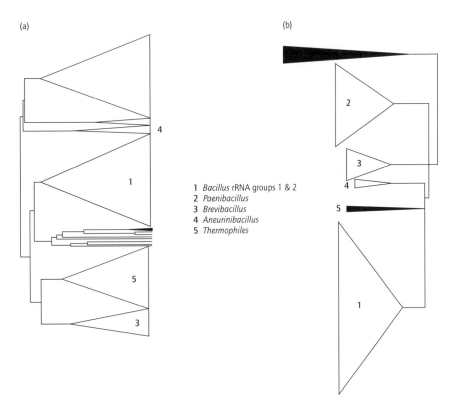

1 *Bacillus* rRNA groups 1 & 2
2 *Paenibacillus*
3 *Brevibacillus*
4 *Aneurinibacillus*
5 *Thermophiles*

Figure 10.2 Comparison between the groupings of *Bacillus s.l.* members based on sequence analysis. The clustering was obtained using (b) Treecon (Van De Peer & De Wachter 1994) and ARDRA clustering with (a) GelCompar (Heyndrickx *et al.* 1996b) software.

incompletely characterized isolates to be allocated at the genus level and often also at the species level. Furthermore, ARDRA indicated several possible sequence errors (e.g. *B. lautus*; see Heyndrickx *et al.* 1996b) that were later confirmed by repeated sequence analyses of the strains in question. As a result,

ARDRA analysis may be used as an initial approach to delineate new taxa, e.g.
Virgibacillus (Heyndrickx *et al.* 1998) and *B. fumarioli* (Logan *et al.* 2000), as
well as for the emendation of existing taxa, leading to better descriptions and
more reliable identifications (Heyndrickx *et al.* 1996a, 1997). A variant of the
ARDRA method as applied by the group of the Ghent University has been used
for the phylogenetic grouping of *B. licheniformis*-like isolates (Manachini *et al.*
1998).

Ribotyping

Originally, this method was developed to characterize Gram-negative organisms
belonging to the *Enterobacteriaceae* (Grimont & Grimont 1986). The method
is now available in an automated form using the Riboprinter™ (Qualicon,
Wilmington, DE).

PRINCIPLE OF THE METHOD

Genomic DNA is cut with restriction enzymes such as *Eco*RI or *Pvu*II. After separation on an agarose gel, the fragments are denatured and blotted on a filter. A
final hybridization step with a labelled fragment of the rRNA operon that can be
obtained (after a cloning step) from *Escherichia coli*, will reveal the distribution
of the complement of the probe over the restriction pattern on the filter. The
hybridization step allows reduction of the number of visualized bands, in order to
avoid the unworkable smear of banding that is produced if general staining with
(e.g.) ethidium bromide is used.

It is not the place to discuss all details of this method, but the choice of the
restriction enzyme and the choice of the fragment of the rRNA operon used as
probe is of decisive importance for the final resolution (for the effect of the probe
see, e.g., Lucchini & Altwegg 1992). Also, the number and the distribution of the
rRNA operons over the bacterial genome is very important. Indeed, bacteria with
only one or two rRNA operons will most probably show a very limited number of
bands in their ribopattern.

It is clear from the considerations outlined above that a universal protocol for
ribotyping resulting in a pattern with similar taxonomic resolution over a wide
variety of bacterial groups, is not realistic. Therefore, ribotyping of a group of
bacteria needs a preliminary investigation of the optimal conditions in order to
obtain the taxonomic resolution desired.

APPLICATION TO *BACILLUS*

Because *Bacillus s.l.* members contain 9–12 rRNA operons (e.g. Okamoto *et al.*
1993; Johansen *et al.* 1996; Nübel *et al.* 1996), ribotyping of the aerobic spore-
formers can be considered as an interesting approach to unravel their taxonomic
structure. Automation (via the riboprinter) makes the method much more user-
friendly, and reporting of new ribotyping patterns such as given in the description
of *Bacillus thermodenitrificans* (Manachini *et al.* 2000) can therefore be expected
in the future. At present, the number of studies in which ribotyping is used to
characterize members of *Bacillus s.l.* is rather limited. Furthermore, the existing
reports on ribotyping of *Bacillus* deal with intraspecific variation. In these studies

the investigators try to find a correlation between the intraspecific distribution of ribopatterning in correlation with, for example: (i) food poisoning with *Bacillus licheniformis* (Salkinoja-Salonen *et al.* 1999), (ii) toxin production by members of the *Bacillus cereus* group, including strains of *B. cereus* from food-poisoning incidents (Pittijarvi *et al.* 1999), (iii) tracing of certain *B. thuringiensis* types (Akhurst *et al.* 1997), and (iv) differentiation between toxic and nontoxic *Bacillus sphaericus* strains (De Muro *et al.* 1992). Although comparative data with other fine DNA fingerprinting methods such as RAPD (randomly amplified polymorphic DNA) are scarce for representatives of *Bacillus*, at least one study has revealed that the latter method is only slightly more discriminative than the automated ribotyping (riboprinting) for *B. cereus* isolates (Andersson *et al.* 1999). One can conclude with certainty that the ribotyping method is valuable for intraspecific studies in *Bacillus s.l.* The value of the method for taxonomic applications cannot yet be fully estimated until databases of the automated version of the technique are available.

Denaturing gradient gel electrophoresis (DGGE) and temperature gradient gel ectrophoresis (TGGE)

These methods are discussed together because they are based on the same principle.

PRINCIPLE OF THE METHODS

A mixture of small amplicons obtained via the PCR are separated via polyacrylamide gel electrophoresis. Establishing either a chemical (in the case of DGGE, e.g. Rosado *et al.* 1998) or thermal (in the case of TGGE, e.g. Nübel *et al.* 1996) denaturing gradient in the direction of the electrophoretic migration, results in a gradual opening of the double-stranded DNA amplicons during their migration in the gel. A G–C clamp of about 20 base pairs prevents complete denaturation. Each denatured amplicon in the pool of amplicons migrates towards a defined position corresponding to its base composition and general configuration (the exact denaturing conditions need to be determined beforehand). As a result, the method allows the separation of oligonucleotides with only a few differences in their sequences. The selected targets for PCR are usually small, hypervariable regions such as V6 or V8 of the 16S rRNA operon, but other regions (e.g. *Nif* genes; Rosado *et al.* 1998) of the bacterial genome may be chosen also.

TAXONOMIC EVALUATION AND APPLICATION TO *BACILLUS*

In general, DGGE and TGGE may allow discrimination at the strain level, and certainly at the subspecies level. Furthermore, sequence differences of multicopy genes such as rRNA operons and the above mentioned *Nif* genes as observed via these techniques do affect our view on the taxonomic resolution of rDNA sequences. Indeed, Nübel *et al.* (1996) demonstrated that the hypervariable parts of different copies of the rRNA operon within a single *Paenibacillus polymyxa* strain might exhibit up to 2.3% difference in partial sequence. This led to the conclusion that genetic variation of rRNA sequences in a microbial population of an environmental sample may exceed its organismic diversity. Random cloning of a single 16S rDNA copy masks this heterogeneity and could result in slightly different phylogenetic information on repeated determination of the 16S rDNA sequence, hence affecting (slightly) the taxonomic interpretation. However, it should be appreciated that these intrastrain variations do not interfere (Nübel *et al.* 1996) with the rule of 97% sequence similarity to delineate bacterial species (Stackebrandt & Goebel 1994).

Nübel *et al.* (1996) also revealed that a particular band position in the TGGE (but obviously also in DDGE) pattern may be occupied by oligonucleotides exhibiting a complete different sequence. Hence, the nature of the band has to be confirmed by sequencing (after excision of the band) or by hybridization with a specific probe on a blotted profile.

As patterns from both techniques can be obtained after amplification of target DNA obtained from nonpurified biological material such as soil or water samples, the methods allow visualization of the genetic biodiversity, including the nonculturable bacterial components of biotopes. Comparison of the sequences of the dominant bands visualized by DGGE or TGGE, with databases such as EMBL, may indicate the dominant bacterial component of the biotope under study.

Using TGGE, for example, an unknown group of *Bacillus* members has been discovered as the main bacterial component in Drentse grassland in the Netherlands (Felske *et al.* 1998; Felske 1999).

Methods addressing the complete bacterial genome

Percentage G+C

The mol% G+C composition of the overall bacterial genome is a stable biological parameter of a particular organism. In well-defined bacterial species, the variation in % G+C composition is not more than 3% and in well-defined genera, not more than 10% (Stackebrandt & Liesack 1993). It is generally required that % G+C composition is included in new taxon descriptions at the species and genus levels.

METHODOLOGY

Previously, mol% G+C composition was measured by indirect methods such as the buoyant density method, or the liquid denaturation method. Nowadays, an HPLC-based method of direct determination of the chemical composition of the DNA is used (Mesbah *et al.* 1989). Values of mol% G+C were reported for *Bacillus* species for the first time in the early 1970s by Bonde and Jackson (1971) and for *Bacillus*-type strains in the 1980s by Fahmy *et al.* (1985). The % G+C values of at least the type strains were also reported, of course, in the many new descriptions of species and genera of *Bacillus s.l.* The % G+C range for the different genera that emerged from *Bacillus s.l.*, and for other genera belonging to this phylogenetic group, are shown in table 10.2.

DNA–DNA relatedness measured by hybridization of overall genomes

The phylogenetic species concept of Wayne *et al.* (1987) imposes a challenge for most bacterial taxonomists: in particular how to describe new species. Indeed,

Table 10.2 Percentage G+C distribution of *Bacillus* and relatives.

Taxon	% G+C range
Alicyclobacillus	51.6–62.3
Amphibacillus	36
Bacillus sensu stricto (rRNA groups 1 and 2)*	33.2–64.6
Brevibacillus	42.8–57.4
Gracilibacillus	38–39
Halobacillus	40–43
Paenibacillus	39.3–54
Virgibacillus	37
Thermophiles (*B. stearothermophilus* and relatives; rRNA group 5*)	36–53.7

* See Ash *et al.* (1991, 1993).

editors and referees of taxonomic journals refer to this species concept as being decisive for species delineation when comparative 16S rDNA sequence analysis between strains are matching at or above 97% similarity with existing taxa. This need for DNA–DNA hybridization for *Bacillus* is supported by the observation that even identical 16S rDNA sequences do not guarantee species identity (Fox *et al.* 1992). Different methods for DNA–DNA hybridizations (Brenner *et al.* 1969; De Ley *et al.* 1970; Crosa *et al.* 1973; Popoff & Coynault 1980; Tjernberg *et al.* 1989; Ezaki *et al.* 1989) have been published, but it is beyond the scope of this chapter to discuss these techniques in detail. Only two methods are currently used for species delineation in *Bacillus*: (i) the liquid renaturation method (De Ley *et al.* 1970), or a variant, and the relatively recent microplate method of Ezaki *et al.* (1989). Data obtained by both methods have been evaluated and compared (Goris *et al.* 1998). It has to be mentioned that neither of these methods allows the determination of Δthermostability (expressed as ΔTm) of the hybrid. However, the difference between the thermal stability of the hybrid and the homologous duplex is important and can be decisive for taxonomic conclusions (Grimont *et al.* 1982).

PRINCIPLE OF THE LIQUID RENATURATION METHOD

This method is based on the observation that absorption of light at 360 nm is enhanced by about 30% when DNA in solution is denatured by heat or alkali. An equimolecular mixture of two different denatured DNAs is allowed to renature under optimal salinity and temperature conditions. The renaturation can be followed spectrophotometrically and the initial renaturation rate is a measure of the overall sequence similarity of the DNAs (De Ley *et al.* 1970). The method is reproducible but requires large amounts of purified, high-molecular-weight DNA and obtaining this is very time-consuming and laborious.

PRINCIPLE OF THE MICROPLATE METHOD

This is a miniaturized method for which only small amounts of DNA are needed. Denatured high-molecular-weight DNA of the first organism is bound to chemically treated microplates (Nunc, Denmark). Photobiotin is linked with denatured and sheared DNA of a second organism under illumination at 400 nm. The biotinylated DNA is hybridized with the fixed DNA of the first organism under optimal salinity and temperature conditions. Streptavidin-β-D-galactosidase is added which attaches to the biotinyl group extending from the hybrid. The final enzymatic reaction of the galactosidase with the 4-methylumbeliferyl-β-D-galactoside results in a fluorescent signal that develops with time and can be measured via a microplate reader. The signal obtained is related directly to the amount of DNA hybridized and can be expressed, after subtraction of background, as % DNA relatedness.

APPLICATION TO BACILLUS

The number of DNA–DNA relatedness studies between members of the genus *Bacillus s.l.* is enormous and it is impossible to mention more than a small part of these data here. A wide variety of techniques has been used; for example the measurement of renaturation rates (e.g. Nakamura & Swezey 1983 and many other publications where Nakamura is author or co-author), the filter method (e.g. Seki *et al.* 1983) and the more recent microplate method (e.g. Kuroshima *et al.* 1996).

Genotypic typing methods

A number of typing methods based on indirect comparative analysis of nucleic acid characteristics has been developed in the last decade. These methods aim originally at the discrimination of species, subspecies and even strains in order to perform epidemiological studies. Broadly speaking, the typing methods can be divided into two categories: (i) techniques that visualize the whole bacterial genome and (ii) techniques that visualize only more or less randomly selected parts of the bacterial genomes.

Typing techniques visualizing the complete genome

RESTRICTION FRAGMENT LENGTH POLYMORPHISM (RFLP) OF THE
BACTERIAL GENOME AND PULSED FIELD GEL ELECTROPHORESIS (PFGE)

Initially, a simple RFLP analysis of the bacterial genomes is visualized. This results in very complex patterns of DNA fragments that are difficult to compare because they appear more as smear-like patterning. The use of rare cutting restriction enzymes drastically reduces the number of DNA fragments. However, the high molecular weight of these fragments has required the development of a specific technique known as PFGE for the separation on agarose gels (see, e.g., Tenover *et al.* 1995).

Application to Bacillus *taxonomy.* The method is not only applied to differentiate between strains of a particular *Bacillus* species such as *B. sphaericus* (Zahner *et al.* 1998) or *Brevibacillus laterosporus* (Zahner *et al.* 1999), but also to differentiate between very closely related species such as *Bacillus anthracis*, *B. cereus*, *B. mycoides* and *B. thuringiensis* (Carlson *et al.* 1994; Harrell *et al.* 1995; Liu *et al.* 1997; Helgason *et al.* 2000). The last two of these studies dealt with rarely reported clinical infections by *B. cereus*.

Genomic typing techniques that visualize randomly selected parts of bacterial genomes

The philosophy behind these techniques refers to a random reduction of the visualized banding of the complex genetic pattern in order to obtain a workable blueprint of the variation of the bacterial genome. One way of achieving this reduction has been mentioned above as ribotyping; in that method, the number of visualized bands is reduced by a hybridization procedure. For the other methods discussed here, the introduction of the PCR methodology has been fundamental. Because of the simplicity of the PCR procedure, these methods are attractive and generally applicable to Gram-negative and Gram-positive bacteria.

RANDOMLY AMPLIFIED POLYMORPHIC DNA (RAPD)

Principle. The first class of the PCR-based methods is known as RAPD. Oligonucleotides of about 10–20 bases long are used as arbitrary primers in a PCR (Williams *et al.* 1990). The banding pattern obtained is not always very reproducible (although the method has been commercialized, e.g. Amersham-Pharmacia–Biotech, Sweden) and, hence, is not suited for comparative analysis in the long term. This is because of the less stringent PCR conditions, and even the type of thermocycler used affects the banding pattern. As a consequence, databases are of limited use, and data exchanged between laboratories have to be interpreted with great care.

Application to identification and taxonomic studies of Bacillus *and relatives.* Despite its drawbacks, the RAPD method has been applied to the discrimination of *Bacillus thuringiensis* (e.g. Brousseau *et al.* 1993), *B. sphaericus* (Woodburn *et al.* 1995) and thermophilic *Bacillus* members (Ronimus *et al.* 1997).

REP-PCR

A second group of PCR-based typing methods – generally known as rep-PCR – is based upon the observation that repetitive elements are dispersed throughout genomes of Gram-negative as well as Gram-positive organisms. Consensus motifs deduced from the sequence of the repetitive elements can be used as primers for PCR.

Different, relatively short, repetitive sequences are commonly found in prokaryotic as well as eukaryotic genomes. These repetitive sequences may be classified in different families such as Repetitive Extragenic Palindromes (REP), and Enterobacterial Repetitive Intergenic Consensus (ERIC). While sequences of the first family are often present in 500–1000 copies containing 35–40 bp as inverted repeats, the second family of repetitive sequences is only represented in 30–50 copies of about 124 bp long which are spread independently over the genome. Both repetitive sequences were discovered among representatives of the *Enterobacteriaceae*, but their existence was also shown via probe hybridizations in many other Gram-negative and Gram-positive bacteria. Primers were developed for PCR-amplifying DNA regions in between these repetitive sequences. The electrophoretic patterns allow discrimination at the within-species level and sometimes at the strain level. It is clear that this technique offers a tool to unravel the very fine taxonomic structure of a species.

Principle. The method is simple and uses instrumentation commonly available in molecular biology laboratories. Different protocols for rep-PCR are established and can be found in the original literature or in laboratory handbooks for molecular methods (e.g. Rademaker & De Bruyn 1997). Because of the use of consensus primers in the PCR, stringency of PCR is not maximal, and this may have a negative effect on reproducibility. In consequence, comparative analyses between repeated PCR products (either from different or from the same PCR run) need to be treated with great care. Application of these methods in comparative studies, but spread over different PCR and electrophoretic experiments, may be meaningful if reference patterns are included, in combination with a molecular weight ladder for normalization by computer-assisted data-handling methods. These reference patterns can then be used for normalization. Another approach may be to use more than one set of rep-primers (e.g. REP, BOX and/or ERIC). This would offer the opportunity of integrating the different patterns into one, more complex, combined pattern. If it is assumed that an acceptable reproducibility can be obtained (see above), this combination REP-PCR technique should enhance the discriminatory possibilities. Such an approach has not yet been applied to *Bacillus*.

Application to Bacillus *and related organisms*. This typing method has been used to unravel the genetic diversity of *B. sphaericus* (Da Silva *et al.* 1999; Miteva *et al.* 1999) and *B. larvae* ssp. *larvae* (Alippi & Aguilar 1998). In the latter case, BOX-PCR revealed a unique pattern that could identify the American foulbrood agent that is of economical importance to beekeepers. Furthermore, REP-PCR has been used to demonstrate the presence of the thermoresistant organism *B. sporothermodurans* in UHT treated milk (Klijn *et al.* 1997), while BOX-PCR in combination (but not in combining the patterns of the individual techniques into one single extended pattern) with RAPD was applied to subtype *Paenibacillus azotofixans* strains from the rhizoplane and rhizosphere soils of grasses (Rosado *et al.* 1998).

AMPLIFIED FRAGMENT LENGTH POLYMORPHISM (AFLP)

The method is based upon a specific combination of PCR and restriction methodologies (Zabeau & Vos 1993). Compared to the RAPD and the rep-PCR methods mentioned above, the AFLP technique is much more complex but also much more reproducible.

Principle. In the first step, a restriction of the bacterial genome is performed by two different restriction enzymes, randomly yielding three different kinds of DNA fragments in relation to their sticky ends. Short oligonucleotides (adaptors) are now ligated to the sticky ends (templates). Selective PCR amplification is performed by using primers with the same sequences as the adaptors, but extended with the sequence of the restriction site as well as an extra base. This selective amplification results in a strong reduction of the complexity of the banding pattern into a workable and reproducible pattern of about 40–100 bands that can be digitized and handled by suitable computer software for numerical analysis (Janssen *et al.* 1996).

Application to Bacillus. So far the AFLP method as described above has not often been included in studies on *Bacillus* taxonomy. It has been used to confirm the genotypic discrimination between *B. larvae* ssp. *larvae* and *B. larvae* ssp. *pulvifaciens* (Heyndrickx *et al.* 1996a) and more recently in an epidemiological study of *B. cereus* (Ripabelli *et al.* 2000). Furthermore, an AFLP approach has been used for the genetic comparison of *B. anthracis* and its closest relatives (Jackson *et al.* 1999; Keim *et al.* 1999; see also Chapter 3 of this book).

SDS-PAGE OF WHOLE-CELL PROTEINS

Introduction

Whole-cell protein patterning (SDS-PAGE) has proven in the past to be a rapid and cost-effective method for the comparison of large groups of bacteria. As an initial step in polyphasic characterization, the method allows the grouping of strains according to similarities in their protein patterns. This may be evaluated visually or computer-assisted.

The comparison of whole-cell protein patterns is relevant only when highly standardized conditions of growth are combined with a rigorously standardized procedure for analysis. The groupings obtained then need to be evaluated taxonomically, and the cut-off levels for species discrimination have to be determined for the set of strains under study. This cut-off level is not always straightforward to identify, and needs to be verified ultimately by DNA–DNA hybridization. However, generally speaking, an 80% protein pattern similarity in an unweighted pair group method analysis (UPGMA) can be considered as an acceptable basis for species delineation and usually corresponds to 70% or more DNA relatedness between the groups.

Methodology

General and specific aspects of the methodology of SDS-PAGE of whole-cell proteins in bacteria can easily be found in the literature (Vauterin *et al.* 1993; Pot *et al.* 1994). Normalization of the data is important for computer-assisted comparison of the results. Normalization between different electrophoretic runs can be achieved by the inclusion of a carefully chosen bacterial pattern, as applied by the group of the Ghent University. In this case, a protein extract of *Psychrobacter immobilis* LMG 1121 is loaded on both of the outer lanes and the central lane of the gel. If a molecular weight marker is loaded as well, the molecular weight of the protein bands can be estimated easily.

Computer-assisted methods (e.g. the GelCompar software package, Applied Maths, Kortrijk, Belgium) allow the construction of databases for identifica-

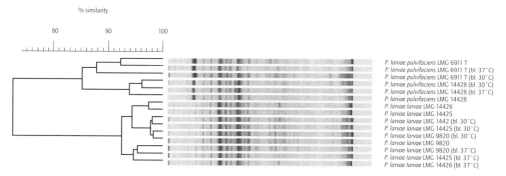

Figure 10.3 The effect of the medium composition on the protein-banding pattern (bl. means that blood has been added to the normal nutrient agar + 1% glucose medium). *Paenibacillus larvae* ssp. *larvae* strains show only weak growth on the medium without blood (= nutrient agar + 1% glucose) if they have been pre-cultured on blood medium.

tion. Sometimes the banding pattern of one-dimensional PAGE electrophore-grams of whole-cell proteins can be disturbed by the so-called 'smiling effect' (Vandamme *et al.* 1999). When this phenomenon occurs, it might be difficult or even impossible to obtain reproducible patterns that can be used for com-puter-assisted interpretation, and visual interpretation is then the only possible alternative.

The effect of medium composition on the protein pattern and, hence, on the clustering of the strains must also be stressed. SDS-PAGE patterns of bacteria that have been cultivated under different conditions cannot be compared reliably. As can be seen in figure 10.3, representatives of the two subspecies of *Paenibacillus larvae* show (slightly) different protein patterns when grown on different media and at different temperatures. For *P. larvae* ssp. *pulvifaciens* the medium affects the clustering to a greater extent than does the reproducibility (fixed at 93% sim-ilarity) of the method itself. Hence, the protein pattern needs to be obtained under rigorously standardized growth conditions.

Applications to *Bacillus*

At Ghent, the technique of whole-cell protein patterning has been used since the 1970s (with suitably regular updating of the methodology) as an initial approach to the taxonomic study of groups of strains of Gram-negative (e.g. *Xanthomonas*: Vauterin *et al.* 1991) and Gram-positive bacteria (e.g. *Lactobacillus*: Hertel *et al.* 1993; Gancheva *et al.* 1999).

In *Bacillus* and relatives, SDS-PAGE turned out to be very helpful in the polyphasic characterization of *Paenibacillus* (Heyndrickx *et al.* 1996a), *Aneurinibacillus* (Heyndrickx *et al.* 1997) and *Virgibacillus* (Heyndrickx *et al.* 1998). A special application of the method is demonstrated in figure 10.4 where it is shown that *Paenibacillus larvae* ssp. *larvae* can be differentiated at the sub-

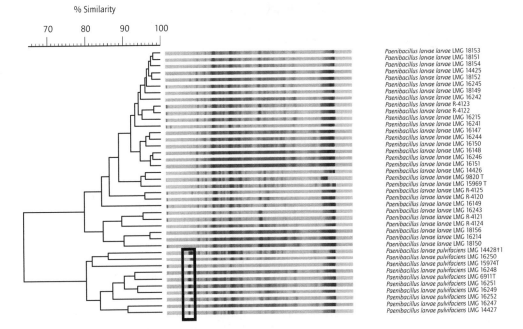

% Similarity

70   80   90   100

Paenibacillus larvae larvae LMG 18153
Paenibacillus larvae larvae LMG 18151
Paenibacillus larvae larvae LMG 18154
Paenibacillus larvae larvae LMG 14425
Paenibacillus larvae larvae LMG 18152
Paenibacillus larvae larvae LMG 16245
Paenibacillus larvae larvae LMG 18149
Paenibacillus larvae larvae LMG 16242
Paenibacillus larvae larvae R-4123
Paenibacillus larvae larvae R-4122
Paenibacillus larvae larvae LMG 16215
Paenibacillus larvae larvae LMG 16241
Paenibacillus larvae larvae LMG 16147
Paenibacillus larvae larvae LMG 16244
Paenibacillus larvae larvae LMG 16150
Paenibacillus larvae larvae LMG 16148
Paenibacillus larvae larvae LMG 16246
Paenibacillus larvae larvae LMG 16151
Paenibacillus larvae larvae LMG 14426
Paenibacillus larvae larvae LMG 9820 T
Paenibacillus larvae larvae LMG 15969 T
Paenibacillus larvae larvae LMG R-4125
Paenibacillus larvae larvae LMG R-4120
Paenibacillus larvae larvae LMG 16149
Paenibacillus larvae larvae LMG 16243
Paenibacillus larvae larvae LMG R-4121
Paenibacillus larvae larvae LMG R-4124
Paenibacillus larvae larvae LMG 18156
Paenibacillus larvae larvae LMG 16214
Paenibacillus larvae larvae LMG 18150
Paenibacillus larvae pulvifaciens LMG 14428†1
Paenibacillus larvae pulvifaciens LMG 16250
Paenibacillus larvae pulvifaciens LMG 15974T
Paenibacillus larvae pulvifaciens LMG 16248
Paenibacillus larvae pulvifaciens LMG 6911T
Paenibacillus larvae pulvifaciens LMG 16251
Paenibacillus larvae pulvifaciens LMG 16249
Paenibacillus larvae pulvifaciens LMG 16252
Paenibacillus larvae pulvifaciens LMG 16247
Paenibacillus larvae pulvifaciens LMG 14427

Figure 10.4 Discrimination of *Paenibacillus larvae* ssp. *larvae* from *P. larvae* ssp. *pulvifaciens* using SDS-PAGE of whole-cell proteins. The high-molecular-weight, dense protein band in the pattern of *P. larvae* ssp. *pulvifaciens* is marked by a rectangle.

species level from *P. larvae* ssp. *pulvifaciens*. Indeed, all strains of the latter sub-species show a a high-molecular-weight protein band (Heyndrickx *et al.* 1996a), which is absent in *P. larvae* ssp. *larvae*. The very stable pattern of *Bacillus larvae* ssp. *larvae* has also been demonstrated by Hornitzky and Djordjevic (1992), who investigated *P. larvae* ssp. *larvae* strains from geographically diverse regions in the southern hemisphere. The major disadvantage of techniques such as SDS-PAGE as diagnostic tools of course is the need for pure cultures. This constraint also holds for the BOX-PCR approach of Alippi and Aguilar (1998), and so in the case of American foulbrood, for example, diagnostic methods should be preferentially based on a specific PCR, or even DGGE or TGGE.

Conclusions

SDS-PAGE of whole-cell proteins is an important initial step in a polyphasic approach to the classification of larger groups of bacterial strains. It is clear that the obtained grouping needs to be supported by genetically based methods such as BOX-PCR and ultimately by DNA–DNA hybridizations. For some of these approaches, computer-assisted databases can be constructed for subsequent identification.

Acknowledgement

The author thanks the Fund for Scientific Research, Flanders, for his position as Research Director.

References

Akhurst, R.J., Lyness, E.W., Zhang, O.Y., Cooper, D.J. & Pinnock, D.E. (1997) A 16S rRNA gene oligonucleotide probe for identification of *Bacillus* isolates from sheep fleece. *Journal of Invertebrate Pathology* **69**, 24–30.

Alippi, A.M. & Aguilar, O.M. (1998) Characterization of isolates of *Paenibacillus larvae* ssp. *larvae* from diverse geographical origin by the polymerase chain reaction and BOX primers. *Journal of Invertebrate Pathology* **72**, 21–27.

Andersson, A., Svensson, B., Christiansson, A. & Ronner, U. (1999) Comparison between automatic ribotyping and random amplified polymorphic DNA analysis of *Bacillus cereus* isolates from the dairy industry. *International Journal of Food Microbiology* **47**, 147–151.

Ash, C., Farrow, J.A.E., Wallbanks, S. & Collins, M.D. (1991) Phylogenetic heterogeneity of the genus *Bacillus* revealed by comparative analyses of small subunit-ribosomal RNA sequences. *Letters in Applied Microbiology* **13**, 202–206.

Ash, C., Priest, F.G. & Collins, M.D. (1993) Molecular identifiaction of rRNA group 3 bacilli (Ash, Wallbanks and Collins) using a PCR probe tests. *Antoine van Leeuwenhoek* **64**, 253–260.

Bonde, G.J. & Jackson, D.K. (1971) DNA base ratios of *Bacillus* strains related to numerical and classical taxonomy. *Journal of General Microbiology* **69**, R7.

Brenner, D.J., Fanning, G.R., Rake, A. & Johnson, K.E. (1969) Batch procedure for thermal elution of DNA from hydroxyapatite. *Analytical Biochemistry* **28**, 447–459.

Brousseau, R., Saint-Onge, A., Prefontaine, G., Masson, L. & Cabana, J. (1993) Arbitrary primer polymerase chain reaction, a powerful method to identify *Bacillus thuringiensis* serovars and strains. *Applied and Environmental Microbiology* **59**, 114–119.

Carlson, C.R., Caugant, D.A. & Kolsto, A.B. (1994) Genotypic diversity of *Bacillus cereus* and *Bacillus thuringiensis* strains. *Applied and Environmental Microbiology* **60**, 1719–1725.

Claus, D. & Berkeley, R.C.W. (1986) Genus *Bacillus* Cohn 1872. In: *Bergey's Manual of Systematic Bacteriology* (eds P.H.A. Sneath *et al.*), Vol. 2, pp. 1105–1140. Williams & Wilkins, Baltimore, MD.

Crosa, J.H., Brenner, D.J. & Falkow, S. (1973) Use of a single-strand specific nuclease for analysis of bacterial and plasmid deoxyribonucleic acid homo- and heteroduplexes. *Journal of Bacteriology* **115**, 904–911.

Da Silva, K.R.A., Rabinovitch, L. & Seldin, L. (1999) Phenotypic and genotypic diversity among *Bacillus sphaericus* strains isolated in Brazil, potentially useful as biological control agents against mosquito larvae. *Research in Microbiology* **150**, 153–160.

De Ley, J., Cattoir, H. & Reynaerts, A. (1970) The quantitative measurement of DNA hybridization from renaturation rates. *European Journal of Biochemistry* **12**, 133–142.

De Muro, M.A., Mictchell, W.J. & Priest, F.G. (1992) Differentiation of mosquito-pathogenic strains of *Bacillus sphaericus* from non-toxic varieties by ribosomal RNA gene restriction patterns. *Journal of General Microbiology* **138**, 1159–1166.

Ezaki, T., Hashimoto, Y. & Yabuuchi, E. (1989) Fluorometric deoxyribonucleic acid-deoxyribonucleic acid hybridization in microdilution wells as an alternative to membrane filter hybridization in which radioisotopes are used to determine genetic relatedness among bacterial strains. *International Journal of Systematic Bacteriology* **39**, 224–229.

Fahmy, F., Fossdorf, J. & Claus, D. (1985) The DNA base composition of the type strains of the genus *Bacillus*. *Systematic and Applied Microbiology* **6**, 60–65.

Farrow, J.A.E., Ash, C., Wallbanks, S. & Collins, M.D. (1992) Phylogenetic analysis of the genera *Planococcus*, *Marinococcus* and *Sporosarcina* and their relationships to members of the genus *Bacillus*. *FEMS Microbiology Letters* **93**, 167–172.

Farrow, J.A.E., Wallbanks, S. & Collins, M.D. (1994) Phylogenetic interrelationships of round-spore-forming bacilli containing cell-walls based on lysine and the non-spore-forming genera *Caryophanon*, *Exiguobacterium*, *Kurthia* and *Plannococcus*. *International Journal of Systematic Bacteriology* **44**, 74–82.

Felske, A. (1999) Reviewing the DA001 files: a 16S rRNA chase on suspect X99967, a *Bacillus* and

Dutch underground activist. *Journal of Microbiological Methods* 36, 77–93.

Felske, A., Akkermans, A.D.L. & De Vos, W. (1998) *In situ* detection of an uncultured predominant *Bacillus* in Dutch grassland soils. *Applied and Environmental Microbiology* 64, 4588–4590.

Fox, G.E., Wisotzkey, J.D. & Jurtshuk, P., Jr (1992) How close is close: 16S rRNA sequence identity may not be sufficient to guarantee species identity. *International Journal of Systematic Bacteriology* 42, 166–170.

Gancheva A., Pot, B., Vanhonacker, K., Hoste, B. & Kersters, K. (1999) A polyphasic approach towards the identification of strains belonging to *Lactobacillus acidophilus* and related species. *Systematic and Applied Microbiology* 22, 573–585.

Goris, J., Suzuki, K-I., De Vos, P., Nakase, T. & Kersters, K. (1998) Evaluation of a microplate DNA-DNA hybridization method compared with the initial renaturation method. *Canadian Journal of Microbiology* 44, 1148–1153.

Grimont, F. & Grimont, P. (1986) Ribosomal ribonucleic acid gene restriction patterns as possible taxonomic tools. *Annales de l'Institut Pasteur/Microbiologie (Paris)* 137B, 165–175.

Grimont, P.A.D., Irino, K. & Grimont, F. (1982) The *Serratia liquefaciens* – *S. proteomaculans* – *S. grimesii* complex: DNA relatedness. *Current Microbiology* 7, 63–68.

Harrell, L.J., Andersen, G.L. & Wilson, K.H. (1995) Genetic variability of *Bacillus anthracis* and related species. *Journal of Clinical Microbiology* 33, 1847–1850.

Helgason, E., Caugant, D.A., Olsen, I. & Kosto, A.B. (2000) Genetic structure of population of *Bacillus cereus* and *B. thuringiensis* isolates associated with peridontitis and other human infections. *Journal of Clinical Microbiology* 38, 1615–1622.

Hertel, C., Ludwig, W., Pot, B., Kersters, K. & Schleifer, K.-H. (1993) Differentiation of lactobacilli occurring in fermented milk products by using oligonucleotide probes and electrophoretic protein profiles. *Systematic and Applied Microbiology* 16, 463–467.

Heyndrickx, M., Vandemeulebroecke, K., Hoste, B. et al. (1996a) Reclassification of *Paenibacillus* (formerly *Bacillus*) *pulvifaciens* (Nakamura 1984) Ash et al. 1994, a later subjective synonym of *Paenibacillus* (formerly *Bacillus*) *larvae* (White 1906) Ash et al. 1994, as a subspecies of *P. larvae*, with emended description of *P. larvae* as *P. larvae* ssp. *larvae* and *P. larvae* ssp. *pulvifaciens*. *International Journal of Systematic Bacteriology* 46, 270–279.

Heyndrickx, M., Vauterin, L., Vandamme, P., Kersters, K. & De Vos, P. (1996b) Applicability of combined amplified ribosomal DNA restriction analysis (ARDRA) patterns in bacterial phylogeny and taxonomy. *Journal of Microbiological Methods* 26, 247–259.

Heyndrickx, M., Lebbe, L., Vancanneyt, M. et al. (1997) A polyphasic reassessment of the genus *Aneurinibacillus*, reclassification of *Bacillus thermoaerophilus* (Meier-Stauffer et al. 1996) as *Aneurinibacillus thermoaerophilus* comb. nov., and emended descriptions of *A. aneurinilyticus* corrig., *A. migulanus*, and *A. thermoaerophilus*. *International Journal of Systematic Bacteriology* 47, 808–817.

Heyndrickx, M., Lebbe, L., Kersters, K., De Vos, P., Forsyth, G. & Logan, N.A. (1998) *Virgibacillus*: a new genus to accomodate *Bacillus pantothentitcus* (Proom and Knight 1950). Emended description of *Virgibacillus pantothenticus*. *International Journal of Systematic Bacteriology* 48, 99–106.

Hornitzky, M.A.Z. & Djordjevic, S.P. (1992) Sodium dodecyl sulphate-polyacrylamide profiles and western blots of *Bacillus larvae*. *Journal of Apicultural Research* 31, 47–49.

Jackson, P.J., Hill, K.K., Laker, M.T., Ticknor, L.O. & Keim, P. (1999) Genetic comparison of *Bacillus anthracis* and its close relatives using amplified fragment length polymorphism and polymerase chain reaction analysis. *Journal of Applied Microbiology* 87, 263–269.

Janssen, P., Coopman, R., Huys, G. et al. (1996) Evaluation of the DNA fingerprinting method AFLP as a new tool in bacterial taxonomy. *Microbiology* 142, 1881–1893.

Johansen, T., Carlson, C.R. & Kolstø, A.-B. (1996) Variable numbers of rRNA gene operons in *Bacillus cereus* strains. *FEMS Microbiology Letters* 136, 325–328.

Keim, P., Kalif, A., Schupp, J. et al. (1999) Molecular evolution and diversity in *Bacillus anthracis* as detected by amplified fragment length polymorphism markers. *Journal of Bacteriology* 179, 818–824.

Klijn, N., Herman, L., Langeveld, L. et al. (1997) Genotypical and phenotypical characterization of *Bacillus sporothermodurans* surviving UHT sterilisation. *International Dairy Journal* 7, 421–428.

Kunst, R., Ogasawara, N., Moszer, I. et al. (1997) The complete genome sequence of the Gram-positive bacterium *Bacillus subtilis*. *Nature* 390, 249–256.

Kuroshima, K., Sakane, T., Takata, R. & Yokota, A. (1996) *Bacillus ehimensis* sp. nov. and *Bacillus chitinolyticus* sp. nov., new chitinolytic members of the genus *Bacillus*. *International Journal of Systematic Bacteriology* 46, 76–80.

Liu, P.Y.F., Ke, S.C. & Chen, S.L. (1997) Use of pulsed field electrophoresis to investigate a pseudo-outbreak of *Bacillus cereus* in a pediatric unit. *Journal of Clinical Microbiology* 35, 1533–1535.

Logan, N.A., Lebbe, L., Hoste, B. *et al.* (2000) Aerobic endospore-forming bacteria from geothermal environments in northern Victoria Land, Antarctica, and Candlemas Island, South Sandwich archipelago, with the proposal of *Bacillus fumarioli* sp. nov. *International Journal of Systematic and Evolutionary Microbiology* 50, 1741–1753.

Lucchini, G.M. & Altwegg, M. (1992) rRNA gene restriction patterns as taxonomic tools for the genus *Aeromonas*. *International Journal of Systematic Bacteriology* 42, 384–389.

Manachini, P.L., Fortina, M.G., Levati, L. & Parini, C. (1998) Contribution to phenotypic and genotypic characterization of *Bacillus licheniformis* and description of new genomovars. *Systematic and Applied Microbiology* 21, 520–529.

Manachini, P.L., Mora, D., Nicastro, G. *et al.* (2000) *Bacillus thermodenitrificans* sp. nov., nom. rev. *International Journal of Systematic and Evolutionary Microbiology* 50, 1331–1337.

Mesbah, M., Premachandran, U. & Whitman, W.B. (1989) Precise measurement of the G+C content of deoxyribonucleic acid by high-performance liquid chromatography. *International Journal of Systematic Bacteriology* 39, 159–167.

Miteva, V., Selensk-Pobell, S. & Mitev, V. (1999) Random and repetitive primer amplified polymorphic DNA analysis of *Bacillus sphaericus*. *Journal of Applied Microbiology* 86, 928–936.

Nakamura, L.K. & Swezey, J. (1983) Deoxyribonucleic acid relatedness of *Bacillus circulans* Jordan 1890 strains. *International Journal of Systematic Bacteriology* 33, 703–708.

Niimura, Y., Koh E., Yanagida, F., Suzuki, K-I., Komagata, K. & Kozaki, M. (1990) *Amphibacillus xylanus* gen. nov., sp. nov., a facultatively anaerobic sporeforming xylan-digesting bacterium which lacks cytochrome, quinone and catalase. *International Journal of Systematic Bacteriology* 40, 297–301.

Nübel, U., Engelen, B., Felske, A. *et al.* (1996) Sequence heterogeneities of genes encoding 16S rRNAs in *Paenibacillus polymyxa* detected by temperature gradient gel electrophoresis. *Journal of Bacteriology* 178, 5636–5643.

Okamoto, K., Serror, P., Azevedo, V. & Vold, B. (1993) Physical mapping for stable RNA genes in *Bacillus subtilis* using polymerase chain-reaction amplification from a yeast artificial chromosome library. *Journal of Bacteriology* 175, 4290–4297.

Pittijarvi, T.S.M., Andersson, M.A., Scoging, M.A. & Salkino-Salonen, M.S. (1999) Evaluation of methods for recognising strains of the *Bacillus cereus* group with food poisoning potential among industrial and environmental contaminants. *Systematic and Applied Microbiology* 22, 133–144.

Popoff, M. & Coynault, C. (1980) Use of DEAE-cellulose filters in the S1 nuclease method for bacterial deoxyribonucleic acid hybridisation. *Annales de Microbiologie* 131A, 151–155.

Pot, B., Vandamme, P. & Kersters, K. (1994) Analysis of electrophoretic whole organism protein fingerprints. In: *Chemical Methods in Prokaryotic Systematics* (eds M. Goodfellow & A.G. O'Donnell), pp. 493–521. Wiley, Chichester.

Pukall, R., Brambilla, E. & Stackebrandt, E. (1998) Automated fragment length analysis of fluorescently labeled 16S rDNA after digestion with 4-base cutting restriction enzymes. *Journal of Microbiological Methods* 32, 55–64.

Rademaker, J.L.W. & De Bruyn, F. (1997) Characterization and classification of microbes by rep-PCR genomic fingerprinting and computer assisted pattern analysis. In: *DNA Markers: Protocols, Applications and Overviews* (eds G. Caetano-Anollés & M.P. Gresshoff), pp. 151–171. Wiley, New York.

Ripabelli, G., McLauchlin, J., Mithani, V. & Threlfall, E.J. (2000) Epidemiological typing of *Bacillus cereus* by amplified fragment length polymorphism. *Letters in Applied Microbiology* 30, 358–363.

Ronimus, R.S., Parker, L.E. & Morgan, H.W. (1997) The utilization of RAPD-PCR for identifying thermophilic and mesophilic *Bacillus* species. *FEMS Microbiology Letters* 147, 75–79.

Rosado, A.S., Duarte, G.F, Seldin, L. & Van Elsas, J.D. (1998) Genetic diversity of *nif*H gene sequences in *Paenibacillus azotofixans* strains and soil samples analyzed by denaturing gradient gel electrophoresis of PCR-amplified gene fragments. *Applied and Environmental Microbiology* 64, 2770–2779.

Salkinoja-Salonen M.S., Vuorio, R., Andersson, M.A. *et al.* (1999) Toxigenic strains of *Bacillus licheniformis* related to food poisoning. *Applied and Environmental Mcirobiology* 65, 4637–4645.

Seki, T., Minoda, M., Yagi, J-I. & Oshima, Y. (1983) Deoxyribonucleic acid reassociation between strains of *Bacillus firmus*, *Bacillus lentus*, and intermediate strains. *International Journal of Systematic Bacteriology* 33, 401–403.

Shida, O., Takagi, H., Kadowaki, K. & Komagata, K. (1996) Proposal for two new genera,

Brevibacillus gen. nov. and *Aneurinibacillus* gen. nov. *International Journal of Systematic Bacteriology* 46, 939–946.

Spring, S., Ludwig, W., Marquez, M.C., Ventosa, A. & Schleifer, K.-H. (1996) *Halobacillus* gen. nov., with descriptions of *Halobacillus litoralis* sp. nov. and *Halobacillus treuperi* sp. nov., and transfer of *Sporosarcina halophila* to *Halobacillus halophilus* comb. nov. *International Journal of Systematic Bacteriology* 46, 492–496.

Stackebrandt, E. & Goebel, B.M. (1994) Taxonomic note, a place for DNA-DNA reassociation and 16S rDNA sequence analysis in the present species definition in bacteriology. *International Journal of Systematic Bacteriology* 44, 846–849.

Stackebrandt, E. & Liesack, W. (1993) Nucleic acids and classification. In: *Handbook of New Bacterial Systematics* (eds M. Goodfellow & A.G. O'Donnell), pp. 151–194. Academic Press, London.

Tenover, F.C., Arbeit, R.D., Goering, R.V. *et al.* (1995) Interpreting chromosomal DNA restriction patterns produced by pulsed-field gel electrophoresis: criteria for bacterial strain typing. *Journal of Clinical Microbiology* 33, 2233–2239.

Tjernberg, I., Lindh, E. & Ursing, J. (1989) A quantitative bacterial dot blot method for DNA-DNA hybridization and its correlation to the hydroxyapatite method. *Current Microbiology* 18, 77–81.

Van De Peer, Y. & De Wachter, R. (1997) Construction of evolutionary distance trees with TREECON for Windows: accounting for variation in nucleotide substitution rate among sites. *Computer Applications in Biosciences* 13, 227–230.

Vandamme, P., Pot, B., Gillis, M., De Vos, P., Kersters, K. & Swings, J. (1996) Polyphasic taxonomy, a consensus approach to bacterial systematics. *Microbiological Reviews* 60, 407–438.

Vandamme, P., Goris, J., Coenye, T. *et al.* (1999) Assigment of centers for disease control group Ivc-2 to the genus *Ralstonia* as *Ralstonia paucula* sp. nov. *International Journal of Systematic Bacteriology* 49, 663–669.

Vauterin, L., Swings, J. & Kersters, K. (1991) Grouping of *Xanthomonas campestris* pathovars by SDS-PAGE of proteins. *Journal of General Microbiology* 137, 1677–1687.

Vauterin, L., Swings, J. & Kersters, K. (1993) Protein electrophoresis and classification. In: *Handbook of New Bacterial Systematics* (eds M. Goodfellow &

A.G. O'Donnell), pp. 251–280. Academic Press, London.

Wainö, M., Tindall, B.J., Schuman, P. & Ingvorsen, K. (1999) *Gracilibacillus* gen. nov., with description of *Gracilibacillus halotolerans* gen. nov., sp. nov.; transfer of *Bacillus dipsosauri* to *Gracilibacillus dipsosauri* comb. nov., and *Bacillus salexigens* to the genus *Salibacillus* gen. nov.; as *Salibacillus salexigens* comb. nov. *International Journal of Systematic Bacteriology* 49, 821–831.

Wayne, L.G., Brenner, D.J., Colwell, R.R. *et al.* (1987) Report of the ad hoc committee on reconciliation of approaches to bacterial systematics. *International Journal of Systematic Bacteriology* 37, 463–464.

Williams, J.G.K., Kubelic, A.R., Livak, K.J., Rafalski, J.A. & Tingey, S.V. (1990) DNA polymorphisms amplified by arbitrary primers are useful as genetic markers. *Nucleic Acids Research* 18, 6531–6535.

Wipat, A. & Harwood, C.R.C. (1999) The *Bacillus subtilis* genome sequence: the molecular blueprint of a soil bacterium. *FEMS Microbiology Ecology* 28, 1–9.

Wisotzkey, J.D., Jurtshuk, P., Jr, Fox, G.E., Deinhard, G. & Poralla, K. (1992) Comparative sequence analyses on the 16S rRNA (rDNA) of *Bacillus acidocaldarius*, *Bacillus acidoterrestris*, and *Bacillus cycloheptanicus* and proposal for the creation of a new genus, *Alicyclobacillus* gen. nov. *International Journal of Systematic Bacteriology* 42, 263–269.

Woese, C.R. (1987) Bacterial evolution. *Microbiological Reviews* 51, 221–271.

Woodburn, M.A., Younsten, A.A. & Hilu, K.H. (1995) Random amplified polymorphic DNA fingerprinting of mosquito-pathogenic and non-pathogenic strains of *Bacillus sphaericus*. *International Journal of Systematic Bacteriology* 45, 212–217.

Zabeau, M. & Vos, P. (1993) *Selective Restriction Fragment Amplification: a general method for DNA fingerprinting.* Publication no. 0534 858 A1. European Patent Office, The Hague.

Zahner, V., Momen, H. & Priest, F. (1998) Serotype H5aH5b is a major clone within mosquito-pathogenic strains of *Bacillus sphaericus*. *Systematic and Applied Microbiology* 21, 162–170.

Zahner, V., Rabdinovitch, L., Suffys, P. & Momen, H. (1999) Genotypic diversity among *Brevibacillus laterosporus* strains. *Applied and Environmental Microbiology* 65, 5182–5185.

Chapter 11

Bacillus thuringiensis Insecticides

Alistair Bishop

Introduction

Bacillus thuringiensis (*Bt*) would be a rather undistinguished Gram-positive bacterium were it not for a remarkable and defining characteristic. This feature has made it one of the most actively produced and researched bacteria of biotechnological value. Microscopic examination of the spore mother cell of *Bt* reveals large, distinctive, parasporal bodies adjacent to the spore (Agaisse & Lereclus 1995). These inclusions are composed of crystalline protein and are variously termed the δ-endotoxins or insecticidal crystal proteins (ICPs). The latter name reveals the reason for the interest in *Bt*: it is a highly amenable and adaptable biopesticide and the genes for δ-endotoxins production have been in the vanguard of the development of transgenic crop plants (see Chapter 12).

The δ-endotoxins of *Bt* are, with some exceptions (Sekar 1988), produced as the bacterium enters the sporulation phase. At the completion of sporulation the ICPs can represent up to 30% of the dry weight of the spore mother cell of natural isolates (Baum & Malvar 1995). Such a prodigious synthesis of crystalline proteins contributes significantly to the economic viability of *Bt* as a biopesticide; the regulatory mechanisms governing the synthesis of δ-endotoxins have been reviewed by Baum and Malvar (1995) and Agaisse and Lereclus (1995). The ICPs so produced may be composed of a single type of δ-endotoxin molecule or several different types, comprising one or more inclusion bodies. A large family of these proteins exists and has recently been reviewed (Schnepf *et al.* 1998) and a new system for their nomenclature has become adopted (Crickmore *et al.* 1998).

A family of crystal proteins

The term 'insecticidal crystal protein' is actually a misnomer, because several invertebrates other than insects are specifically attacked by the δ-endotoxins. The list of susceptible organisms includes protozoa, nematodes, flatworms and mites (Feitelson *et al.* 1992). This is in addition to the more established orders of target insects such as Lepidoptera, Coleoptera and Diptera.

The δ-endotoxin genes are designated by the '*cry*' prefix to indicate the crystalline nature of their protein product. Currently there are over 80 different classes and subclasses of Cry proteins (table 11.1), representing at least four distinct protein families (Crickmore *et al.* 1998). In addition to the Cry proteins

Table 11.1 Summary of known holotypes of crystal (Cry) and cytolytic (Cyt) proteins. The classification of δ-endotoxins may involve up to four ranks of delineation: those sharing a common primary rank (Arabic numeral) share approximately 95% sequence identity and often affect the same order of insect; the second and third ranks of similarity (upper and lower case letter, respectively) indicate decreasing sequence identity and increasing variation in potency and spectrum of activity within an insect order. A quaternary level of ranking of successive Arabic numerals, not shown here, recognizes different versions of effectively the same protein, exhibiting limited mutational changes.

Revised nomenclature of pesticidal crystal proteins				
Cry1Aa	Cry1Ha	Cry5Aa	Cry12Aa	Cry23Aa
Cry1Ab	Cry1Hb	Cry5Ab		
Cry1Ac		Cry5Ac	Cry13Aa	Cry24Aa
Cry1Ad	Cry1Ia			
Cry1Ae	Cry1Ib	Cry5Ba	Cry14Aa	Cry25A
sCry1Af	Cry1Ic			
Cry1Ag	Cry1Id	Cry6Aa	Cry15Aa	Cry26Aa
Cry1Ah	Cry1Ie	Cry6Ba		
			Cry16Aa	Cry27Aa
Cry1Ba	Cry1Ja	Cry7Aa		
Cry1Bb	Cry1Jb	Cry7Ab	Cry17Aa	Cry28Aa
Cry1Bc	Cry1Jc			
Cry1Bd		Cry8Aa	Cry18Aa	Cry29Aa
Cry1Be	Cry1Ka	Cry8Ba		
		Cry8Ca	Cry18Ba	Cry30Aa
Cry1Ca	Cry2Aa			
Cry1Cb	Cry2Ab	Cry9Aa	Cry18Ca	Cry31Aa
	Cry2Ac	Cry9Ba		
Cry1Da	Cry2Ad	Cry9Ca	Cry19Aa	Cry32Aa
Cry1Db		Cry9Da		
	Cry3Aa	Cry9Ea	Cry19Ba	
Cry1Ea	Cry3Ba			Cyt1Aa
Cry1Eb	Cry3Bb	Cry10Aa	Cry20Aa	Cyt1Ab
	Cry3Ca			Cyt1Ba
Cry1Fa		Cry11Aa	Cry21Aa	
Cry1Fb	Cry4Aa			Cyt2Aa
		Cry11Ba	Cry22Aa	Cyt2Ba
Cry1Ga	Cry4Ba			Cyt2Bb
Cry1Gb		Cry11Bb		

produced by *Bt*, related proteins have been found to be produced by *Clostridium bifermentans* (Barloy *et al.* 1996); these are classified as Cry 16Aa and Cry 17Aa (table 11.1). Similarly, *Bacillus popilliae* produces Cry 18Aa (Zhang *et al.* 1997). A structurally unrelated group of crystalline proteins termed Cyt toxins are synthesized by some dipteran-active strains of *Bt* (Ellar 1997). The Cyt toxins have general cytolytic properties at neutral pH values but have a synergistic, mosquitocidal effect with certain Cry toxins *in vivo* (Ellar 1997). A web page (http://www.biols.susx.ac.uk/Home/Neil_Crickmore/*Bt*/) is maintained detailing the full list of Cry and Cyt toxins.

Extensive reviews exist regarding the protein structure and mode of action of the δ-endotoxins (Ellar 1997; Schnepf *et al.* 1998). Briefly, the δ-endotoxins

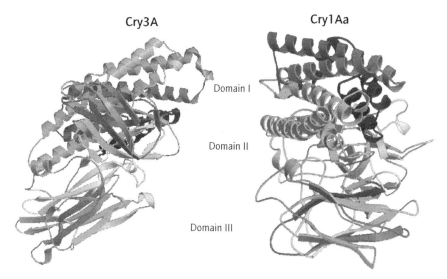

Figure 11.1 Three-dimensional structures of Cry3A and Cry1Aa, two insecticidal protoxins produced by *Bacillus thuringiensis*.

generally have a molecular mass of 130–140 kDa (e.g. Cry1Aa) or about 70 kDa (e.g. Cry3A). Regardless of size, the proteins conform to a three-domain structure (Li *et al.* 1991; Groschulski *et al.* 1995), as illustrated in figure 11.1.

In general terms, Domain I is recognized as a membrane-inserting region which forms pores in the gut epithelia of susceptible species; Domain II exhibits receptor-binding properties, thus having a role in determining the spectrum of activity of the δ-endotoxin, while a number of roles have been proposed for Domain III, including receptor binding. Much detail has been discovered beyond this overview and is reviewed by Dean *et al.* (1996). It has been suggested, for example, that Domain III is involved in the initial, reversible, binding to the receptor while Domain II mediates the second, irreversible, interaction (Lee *et al.* 1999). Two different δ-endotoxin-binding proteins have now been identified in the gut epithelia of susceptible insect larvae: an aminopeptidase N (Knight *et al.* 1994) and a cadherin-like glycoprotein (Vladmudi *et al.* 1995).

The δ-endotoxins are, in fact, protoxins that require proteolytic activation after ingestion in order to reveal their toxic properties. Typically, before proteolysis, a δ-endotoxin becomes solubilized by the highly alkaline conditions within the midgut of a susceptible insect. This occurs by proteolytic processing of either the N-terminus of the protein, the C-terminus, or both, by proteases within the gut. Part of the activated toxin inserts into the membrane after binding to a specific receptor on the brush border membrane of the midgut epithelial cells. The channels so formed allow the passage of ions and small molecules. The cells swell and eventually lyse owing to an osmotically driven influx of water (Ellar 1997). The highly alkaline environment found, for example, in the midguts of lepidopteran larvae cannot, then, be maintained. As a neutral pH value is approached it is hypothesized that (*Bt*) spore germination ensues and the resulting septicaemia brings about the eventual death of the insect (Ellar 1990). Although it is more common for susceptible insects to have highly alkaline midguts, examples exist of gut pH values in the neutral to acidic range. Solubilization of δ-endotoxins under such conditions does not occur to a significant extent *in vitro*. It has been suggested that proteolytic nicking of the δ-endotoxins *in vivo* increases their solubility (Carroll *et al*. 1997). It has also been shown experimentally (Schwarz *et al*. 1993) that δ-endotoxins form pores with different characteristics at different pH values. Two models by which pore formation may take place have been proposed and have been reviewed by Knowles (1994).

The phylogenetic relationship between the δ-endotoxins has been analysed by Bravo (1997). The analytical methods used indicated that there was a correlation between the phylogenetic relatedness of Domain I segments and the invertebrate groups that the whole toxin was active against. Furthermore, it was indicated that the different characteristics of Domain I were necessary for insertion into membranes with potentially different protein and lipid compositions. A potential problem with this theory occurs where one δ-endotoxin has activity against larvae with widely different gut pH values. Prime examples of multiple toxicity exist in the Cry1B group of toxins that were formerly recognized as solely active against lepidopteran larvae (alkaline guts). Johnson *et al*. (1998) reported Cry1B toxins which were also active against larvae of houseflies and blowflies (neutral or acidic guts); Zhong *et al*. (2000) reported a toxin which, in addition to lepidopteran and dipteran (housefly) activity, also killed coleopterans (neutral to acidic guts).

With respect to Domain II, evolution seems to have arisen from three different origins (Bravo 1997). It is of interest that toxins with closely related Domain IIs may bind to the two different types of receptor so far identified (Knight *et al*. 1994; Vladmudi *et al*. 1995). Explanations for this apparent contradiction include the likelihood that small differences in the amino acid sequence in Domain II could account for binding to different receptors, that other domains influence binding specificity, or that different carbohydrates on the receptors determine binding specificity. Domain III sequences apparently have evolved from two different origins (Bravo 1997). Shuffling of this domain may have produced δ-endotoxins with new invertebrate specificities.

The evolutionary origin of the δ-endotoxins is unclear but several factors make their subsequent diversification and dissemination very logical: (i) the genes encoding them are almost exclusively located on conjugative plasmids (Aronson 1993), albeit of large size (40–150 MDa); (ii) transposons and insertion elements have been found associated with δ-endotoxin genes; (iii) conjugal transfer between strains of *Bt* has been demonstrated during pathogenesis (Jarrett & Stephenson 1990; Thomas *et al*. 2000); and (iv) the possession of δ-endotoxin genes offers a selective advantage by potentially giving the producer organism access to the considerable food source of an insect cadaver (Ellar 1990).

Current products

There are currently 26 products based on *Bt* that are registered with the Environmental Protection Agency (EPA) of America and marketed worldwide.

Of these, 15 contain naturally occurring strains (Copping & Menn 2000). Three products are derived from recombination of conjugative plasmids resulting in new combinations of δ-endotoxin genes. Estimates vary of the global market for *Bt* but all agree that the market share is very small, representing about US $100 million per year (Copping & Menn 2000). Such estimates may not include the *Bt* pesticides produced at a local level or under the auspices of governments or aid organizations. The worldwide usage of *Bt* is, nevertheless, very small at less than 2% of total insecticide sales. In recent years the use of sprayed *Bt* products has declined in favour of transgenic *Bt* crops containing δ-endotoxin genes (see Chapter 12). An attractive aspect of *Bt* insecticides is that they have lower costs for development and registration compared to chemical products. It was estimated that these costs for the mosquitocidal strain, *Bacillus thuringiensis* ssp. *israelensis*, were 1/40 of what a comparable, synthetic product would have been (Becker & Margalit 1993). A registration period of 12 months is typical for a biopesticide compared to 36–45 months for a conventional chemical pesticide (Menn & Hall 1999).

Of the thousands of isolates of *Bt* that have been gathered over the years many have shown negligible toxicity towards the invertebrates against which they were screened. Either they have not been screened against the correct, susceptible species or they just produce nontoxic molecules. Strains with important, novel activities have been found in the past by screening and this success may be repeated. An alternative to this very laborious task of bioassaying numerous isolates for activity against a chosen pest is to try to improve existing properties by genetic modifications. This could either be in terms of widening the spectrum of activity or improving qualities such as field persistence. Particularly from an agricultural point of view *Bt* has several drawbacks: (i) the toxicity of most strains is limited to one order of insects, and even within that order there will be significant variation in activity between species; (ii) insecticide sprays become rapidly inactivated by UV radiation from sunlight (Leong *et al.* 1980), the δ-endotoxin component being particularly sensitive to such inactivation; and (iii) early instar larvae are the most susceptible to the toxin. This necessitates application at the correct time for optimal presentation of the toxin. In recent years, genetic manipulation has been employed in an effort to ameliorate some of these drawbacks (reviewed in Baum *et al.* 1999). Obviously, with a final goal of widespread application, the types of genetic procedure available have regulatory limitations.

A number of commercial products have been obtained by conjugal transfer of toxin genes. Because this is a naturally occurring process, strains so produced are not regarded as genetically manipulated and are thus deemed acceptable for commercial application. Examples include: (i) the transfer of a self-transmissible *cry*1A-containing plasmid from a *B. thuringiensis* ssp. *aizawi* strain to a *B. thuringiensis* ssp. *kurstaki* recipient, resulting in improved lepidopteran activity, and (ii) the production of a dual lepidopteran- and coleopteran-specific strain by the transfer of an approximately 88 MDa Cry3A-encoding plasmid from a *B. thuringiensis* ssp. *morrisoni* strain to a lepidopteran-specific *B. thuringiensis* ssp. *kurstaki* strain. Increased production of the coleopteran-specific Cry3A protein has been obtained by mutagenesis with gamma radiation of a *B. thuringiensis* ssp. *tenebrionis* strain to yield an oligosporogenous variant synthesizing abnormally large crystals (Baum *et al.* 1999). The sporulation-independent nature of *cry*3A gene expression in this mutant is thought to contribute to protein over-production through a prolongation of synthesis time from the stationary phase onwards and an absence of degradative, sporulation-related proteases. An additional mechanism of over-production in this strain was that gene duplication of *cry*3A had occurred on its native plasmid (Adams *et al.* 1994).

Recombinant strains of *Bt* have been produced by Ecogen (Pennsylvania, USA) and gained EPA approval; registration has usually been given within a year of application (Baum *et al.* 1999), indicating the environmentally benign nature that these strains are perceived to possess. The success of these ventures has been facilitated by using a homologous recombination system to allow the deletion of foreign DNA, such as antibiotic resistance genes, after the necessary selection has taken place. These recombinant strains produce more ICP than their parents and/or have wider spectra of insecticidal activity resulting from the introduction of new *cry* genes (reviewed by Baum *et al.* 1999). Alternative approaches to the problems of delivery and persistence have been taken to produce transgenic microorganisms. One example is the expression of δ-endotoxin genes in *Pseudomonas fluorescens*. In order to gain regulatory approval, the cells have to be killed at the end of fermentation. As a result of this fixation step the cell wall becomes thickened, offering a two- to threefold increase in foliar persistence for the δ-endotoxins compared to the original *Bt* insecticide. These and other transgenic delivery systems are reviewed by Gelertner and Schwab (1993). The climate of public opinion, certainly in Europe, does not seem to be favourable for the acceptance of the commercial release of live, transgenic microorganisms.

One of the most appealing aspects of *Bt* insecticides is the fact that they lend themselves to 'village level' or 'cottage industry' production. This is particularly attractive in less-developed countries. The need to use hard currency to buy relatively costly chemical pesticides is thus avoided and by-products of agriculture can be used to devise effective growth media. Such feedstocks as pulse and grain flours, paper residues and fish meal (Salama & Morris, 1993) attest to the wide potential diversity of growth and sporulation media suitable for *Bt*. China has a long history of *Bt* production both by solid and liquid fermentation with over one million hectares of crops being treated with it (Smits 1997). In 1996, the output of liquid *Bt* formulation reached 5500 Mt, most of which was exported (Suguvanam & Xie 1999). Batch cultures using solid media, such as steamed wheat or peanut bran, may take a week or more to complete and produce $1–5 \times 10^9$ spores/mL. The normal application rate equates to $10^{13}–10^{14}$ spores/ha (Smits 1997). Over 1000 Mt of *Bt* insecticide were produced in Cuba in 1997: 24% by industrial fermentation and 76% in solid substrate cultures (Fernández-Larrea Vega 1999). The cost of treating the 200 000 ha of crops protected in this way amounted to between two and ten Cuban pesos per hectare.

Genomic content and classification

The genome size of different strains of *Bt* seems to show a wide range of variation (Kolstø 1998). In one study the chromosome of the very closely related *Bacillus cereus* (see below) was estimated at only 2.4 Mb (Carlson & Kolstø 1994). This is very small considering the metabolic activities required for a free-living, sporulating organism: the mean genome size for spore-forming, 'generalist' bacteria has been estimated at 4.9 Mb (Shimketts 1998).

The chromosome of *B. thuringiensis* ssp. *berliner* 1715 is reported at 5.7 Mb (Carlson *et al.* 1996). An additional complement of over 1.2 Mb is located extra-chromosomally. Plasmids, both circular and linear (Carlson *et al.* 1994), are very important components of the *Bt* genome, being generally where the *cry* genes reside (González *et al.* 1981). *Bt* harbours a wide range of transposable elements (reviewed by Mahillon *et al.* 1994). Transposons and insertion sequences are frequently associated with *cry* genes. These transposable elements may play a role in multiplying the copies of *cry* genes within the cell or facilitating their conjugal transfer between cells.

Serotyping isolates of *Bt* (de Barjac & Frachon 1990) from the reaction of antisera against the different types of (chromosomally encoded) flagella (H-antigen) has provided one method of classification which has recently been updated (Lecadet *et al.* 1999). New strains of *Bt* are screened by reference H-antisera. Antiserum is prepared against any *Bt* strain that is not agglutinated with the reference antisera tested. The H-antigens of all known serotypes are then screened with the new antiserum. If cross-reaction does not occur or, if additional sub-factors are demonstrated by the cross-saturation technique, the new strain would represent a new serovar. Biochemical characterization may sometimes offer additional, discriminatory information in the face of conflicting serovar identification. Distinct serovars of *Bt* are given subspecies names such as *kurstaki* and *israelensis*, the two terms being synonymous. A new H-serotype of *Bt* is numbered and registered at the International Entomopathogenic *Bacillus* Centre (IEBC) Collection at the Institut Pasteur, Paris, France. Unfortunately, there is little correlation between serotype and insecticidal toxicity, which, as stated above, is largely or exclusively encoded on plasmids. For example, different strains within the *Bt* subspecies *morrisoni* possess dipteran, lepidopteran or coleopteran activity. Furthermore, most reference strains of *Bt* collected by the IEBC, particularly those in the last decade, produce δ-endotoxins with no known insecticidal activity (Lecadet *et al.* 1999).

Ribotyping data have given good correlation with serotype, however (Priest *et al.* 1994), indicating that little chromosomal exchange between strains of *Bt* has occurred. This is in marked contrast to the generally haphazard collection of δ-endotoxin genes that any particular member of a serotype may possess. Priest *et al.* (1994) suggested that plasmids drive (short-term) evolutionary change because, upon transfer to a new cell, novel combinations or recombinations of δ-endotoxin genes might be made. New pathogenic potential may thereby be produced allowing that clone of cells to establish a selective advantage by association with a particular invertebrate target. Such a clone would represent one of the plethora of strains which abound, but which have clearly defined spectra of toxicity; the chromosomal content of these strains remaining relatively static, as indicated by ribotyping (te Giffel *et al.* 1997).

Unwelcome relatives

For many years *Bacillus anthracis*, *B. cereus*, *B. mycoides* and *Bt* have been considered members of the same phylogenetic group (Somerville & Jones 1972; Kaneko *et al.* 1978; Priest *et al.* 1988). More recently, however, powerful molecular genetic techniques have begun to throw more light on the relationships within this group (see Chapter 3).

It has been suggested (Helgason *et al.* 2000) that *Bt* and *B. anthacis* are relatively recent variants from a *B. cereus* ancestor, derived from extrachromosomal transfer: the horizontal transfer of one of the two virulence plasmids, pXO1 and pXO2, into a mildly virulent strain of *B. cereus*, already possessing the other plasmid, giving rise to the highly pathogenic bacterium known as *B. anthracis* (Keim *et al.* 1997). The feasibility of such a transfer has been demonstrated clearly by Reddy *et al.* (1987). These authors demonstrated conjugal transfer of large plasmids encoding the capability for plasmid transfer from *Bt* to *B. anthracis* and *B. cereus*. The recipient strains then themselves became fertile. Such a strain of *B. anthracis* was capable of transferring the pathogenicity plasmids pXO1 and pXO2 to plasmid-cured strains of *B. anthracis* and *B. cereus*. It seems likely, though, that chromosomally located determinants are required for the full phenotype of *B. anthracis* to be expressed. Bacteria formally identified as *B. cereus* have been shown by Southern blot analysis to possess remnants or complete, but unexpressed, δ-endotoxin genes (Carlson *et al.* 1994).

A recent analysis has tried to overcome the apparent taxonomical uniformity presented by studies of ribosomal RNA genes (Yamada *et al.* 1999). These authors cloned and sequenced *gyrB* genes from *B. anthracis*, *B. cereus* and *Bt*. In spite of minor differences in DNA sequence they designed PCR primers which differentiated between strains of these organisms, obtained from culture collections. The correct designation was given with most strains although some strains of *B. cereus* did not amplify with the *B. cereus*-specific primer pair whereas, for example, two strains of *Bt* did so. With respect to the 104 environmental isolates that were analysed, the correct reaction was also generally given. Anomalies did arise, however, such as 4% and 2% of the *B. cereus* isolates reacting in a manner characteristic of *Bt* and *B. anthracis*, respectively. Similarly, using primers specific for insertion sequences, Henderson *et al.* (1995) found uniformity amongst *B. anthracis* strains but some strains designated as *B. cereus* gave reactions characteristic of *Bt* or *B. anthracis*.

One may summarize these genetic investigations by saying that *B. cereus* is the progenitor of a group of bacteria, including *B. anthracis* and *Bt*, which may come to be regarded as members of the same species (Keim *et al.* 1997; Helgason *et al.* 2000), *B. mycoides* being considered somewhat more distantly related (Keim *et al.* 1997). The relatively small contribution to the genome made by the extrachromosomal genes of *B. anthracis* and *Bt* has had an inordinate, but understandable, impact on our perception of the separate identities of these bacteria. What, however, are the ramifications of these taxonomic considerations? For decades little emphasis has been put upon the recognized close-relatedness of *Bt* to the potential opportunistic pathogen *B. cereus*.

The safety of *Bt*

As a result of the extensive exposure of humans to *Bt* during production, application and usage of the treated product there have been several assessments of the effect of *Bt* on experimental animals, usually rodents, by various routes of inoculation. These have been reviewed by several authors (Burges 1981; Saiki *et al.* 1990; Siegel & Shadduck 1990; Ray 1991; Drobniewski 1994; Bishop *et al.* 1999). Concerns have recently been voiced (Bishop *et al.* 1999; Rivera *et al.* 2000) about the pathogenic potential of *Bt* and the methodology by which safety testing of new strains is carried out: the underlying imperative being to maintain the excellent safety record that biopesticides, including *Bt*, have enjoyed to date. A paradox certainly exists between recent laboratory observations and years of practical experience.

A survey of natural isolates of *Bt* (Perani *et al.* 1998) revealed that 83% of natural isolates of *Bt* tested positive for *B. cereus*-type enterotoxin. Rusul and Yaacob (1995) assessed the prevalence of *B. cereus* in various foods. Most of the cooked and all of the dried foods that they tested contained *B. cereus*, *Bt* or *B. mycoides*. Over 80% of the isolates tested were deemed to be enterotoxigenic. Rivera *et al.* (2000) screened 74 strains of *Bt* for their genetic capability to synthesize the three types of enterotoxin that *B. cereus* is known to be capable of making (Granum & Lund 1997). Positive results for all three factors were obtained in the great majority of strains. In addition, the supernatant culture fluid in all but one strain exhibited *in vitro* cytotoxicity equal to that shown by food poisoning isolates of *B. cereus*. Bishop *et al.* (1999) tested the capability of several strains of commercially employed *Bt* to produce enterotoxin. All of the strains tested reacted positively, to a greater or lesser extent, to the assays used. These authors showed that high levels of spores could remain after normal food processing of vegetables that had been recently sprayed with a commercial preparation of *Bt* insecticide. One of the perceived advantages of bioinsectides, compared to chemical products, is that there is no recommended post-application period during which the crop cannot be harvested and, during which, the pesticide levels would diminish. Improper food handling of such spore-laden vegetables, should they be strongly enterotoxigenic, could lead to food poisoning. Only one authenticated case of food poisoning attributed to *Bt* has been reported (Jackson *et al.* 1995) in spite of the potential for food poisoning that strains of *Bt* have. Rivera *et al.* (2000) recommended that, given their widespread occurrence in *Bt*, enterotoxin genes should be deleted from commercial strains.

In view of the lack of reported food poisoning cases with these strains and given the opportunity for food mishandling to occur, it might be assumed that the established commercial strains have been proven to be safe. The requirements for mammalian toxicity testing of microbial pesticides is dependent on the responses of rodents (McClintock 1999). It has been shown that rats given high oral doses of spores of a *Bt* strain capable of producing enterotoxin failed to show any signs of illness (Bishop *et al.* 1999). It was suggested that, because *Bt* only produces these toxins during vegetative growth, the gut physiology may differ from that in humans in that germination does not occur in rats. Should this be the case it calls into question the validity of the established methodology of toxicity testing. This may be of particular relevance in less developed countries where new strains of *Bt* are being sought to counter local pest problems.

Apart from microorganisms associated with food, humans have not been exposed to any other biotechnological bacterium by so many routes of entry and for so many decades as they have to *Bt*. Countless tonnes of *Bt* have been produced by technologies of widely different levels of sophistication; numerous workers have applied dusts, wettable powders and ultra-low volume sprays containing *Bt* and millions of people have been exposed to aerosols containing this bacterium and have consumed agricultural crops and water treated with strains of it. For example, an area of Oregon (USA) where 120 000 people resided was sprayed with *Bt* in order to control the gypsy moth (Green *et al.* 1990) yet no signs of illnesses attributable to *Bt* were reported by the inhabitants. Similarly, for several years, 18 000 km of rivers in West Africa have been treated with *Bt*, again with no reports of adverse effects on the two or three million people who live in the environs (Becker & Margalit 1993).

How can such a ubiquitous and biotechnologically exploited bacterium, with the possible capability of some strains to be enterotoxigenic, only have caused one authenticated outbreak of food poisoning? It has been suggested that more than 10^5 cfu/g food are required to cause *B. cereus*-type food poisoning (Rusul & Yaacob 1995). A similar figure was indicated by Tan *et al.* (1997). Granum and Lund (1997) suggested that levels of enterotoxigenic bacilli in foods exceeding 10^3 cfu/g could pose a health risk. It might be that one of the major deficiencies of *Bt* as a biocontrol agent, the relatively poor survival of the spores after application, prevents large inocula from frequently entering the food chain. It is also likely that *Bt* is responsible for more cases of food poisoning but is misidentified as *B. cereus*. The major biotechnological strains assayed by Bishop *et al.* (1999) produced rather low titres of enterotoxin production. Such weak expression may significantly diminish the risk of food poisoning.

Health risks from exposure to *Bt* other than by the oral route have also been tested (see reviews, above). In summary, there are no grounds for concern with respect to the inocula to which people might realistically be exposed. Only one report of human infection has been reliably attributed to *Bt*. Damgaard *et al.* (1997) authenticated four isolates of *Bt* from skin infections of two badly burned hospital patients. These isolates probably originated from hot water used to treat them and were not from any pesticidal applications. This case of *Bt* causing skin infections must be viewed as an anomaly as patients with deep burns are regarded as highly immunosuppressed and open to any opportunistic infection.

Future prospects

Food crops

The market for biopesticides is predicted to grow by 10–15% during the current decade, compared to 2% for chemical pesticides (Menn & Hall 1999). Even allowing for this predicted growth in the share of the pesticide market, *Bt* pesticides will represent only a small proportion of the total. The small but burgeoning consumer interest in organic food is a promising niche market: *Bt* represents the only acceptable, effective, biological intervention against many pest species. The demand for organic food is growing faster than any other food commodity in the developed world (Copping & Menn 2000). Consumers of organic food may also be tolerant of limited pest damage to fruit and vegetables. As a result of the somewhat longer 'knock down' time compared to chemical pesticides, *Bt* may not be a suitable choice where nothing less than totally blemish-free produce is acceptable. Any small, high-value areas of agriculture where existing chemical pesticides are unavailable are unlikely to be challenged by new synthetic products because the return on development costs for agrochemical companies would be uneconomic (Copping & Menn 2000). *Bt* is also favourably regarded as a component of integrated pest management schemes where it is applied in conjunction with synthetic products.

Forestry

Forestry has been a successful example of market penetration (van Frankenhuysen 1993). Over half the forests in eastern Europe and in Canada, for example, are treated with bioinsecticides of which *Bt* is the dominant agent (Evans 1997). In Poland over 600 000 ha of softwood forest have recently been treated with *Bt* giving 95% control against a serious infestation of nun moth caterpillars (Butt *et al.* 1999). Limited crop damage in this industry is less critical than in horticulture and the environmental benefits, particularly in more populated areas and in recreational woodland, are important considerations. Such factors suggest that *Bt* has a bright future in this sector of agriculture.

Disease vector control

The discovery of *B. thuringiensis* ssp. *israelensis* (*Bti*) in 1976 (Goldberg & Margalit 1977) added to the panoply of bioinsecticides with activity against mosquito and simulid larvae such as *B. sphaericus* (see Chapter 13). This occurred at a time when resistance to chemical agents was becoming a serious problem. Diseases transmitted by dipteran vectors that are susceptible to *Bti* threaten a large proportion of the world's population. The geographical areas concerned may also be enlarging as a consequence of the apparent climate change in different parts of the world. Up to 800 000 L of *Bti* insecticide a year have been applied in the Onchocerciasis Control Programme in West Africa (Becker & Margalit 1993). Highly successful campaigns against nuisance insects in Germany (Becker & Margalit 1993) will promulgate the use of *Bti*, particularly in such areas where public safety and minimal environmental disruption are prime considerations. Although *Bti* has drawbacks, in terms of persistence in water for example, its safe and successful usage to date – currently amounting to several hundred tonnes per annum (Smits 1997) – augurs well for its continued future use. This would also include being part of an integrated control programme with chemical insecticides.

Genetic manipulation

It seems likely that such live recombinant strains that do gain acceptance will
have to remain *Bt*-based rather than transgenic. As the understanding of the functional structure of δ-endotoxins has improved, the ability to produce chimaeric
toxins has been gained. *Spodoptera littoralis* and *S. exigua* are two lepidopteran
species that are notoriously insensitive to many δ-endotoxins. The production of
a *cry*1C/ *cry*1Ab chimaeric gene and its expression in an asporogenic strain of *Bt*
has led to an insecticide with improved toxicity against these target insects
(Sanchis *et al.* 1999). Increased environmental safety is claimed for this strain
owing to its inability to produce spores and, hence, its much diminished persistence in the environment. It should be borne in mind, however, that the presence
of a viable spore is an essential component of an effective *Bt* insecticide against
some insects (Heimpel & Angus 1959; Miyasono *et al.* 1994). After germination,
different strains of *Bt* are capable of producing various other virulence factors
such as: α- and β-exotoxins; a 'louse factor'(Hill & Pinnock 1998); exoenzymes
such as chitinases, phospholipase C and haemolysin; immune inhibitors (InA and
InB); and vegetative insecticidal proteins (Ellar 1997). The β-exotoxins (Levinson
et al. 1990) have nonspecific toxicities that include vertebrates, and are unacceptable in commercial strains. The chitinases, however, seem to have potentiating
effects on the activities of the δ-endotoxins (Sampson & Gooday 1998). These
other potency factors have evolved with the *cry* genes to make *Bt* an organism
that can be exploited as an effective biopesticide. As the roles of some of these factors become better known, biotechnologists may be persuaded to view *Bt* in its
entirety as an integrated pesticidal organism and optimize it as such, rather than
expressing the δ-endotoxins transgenically.

Vaccine production

It has been well proven experimentally that a vital factor in determining the
efficacy of a vaccine is the use of an adjuvant to enhance its immunogenic properties. Currently only alum and an oil-in-water emulsion, MF59, are licensed for
this purpose for human use (Singh & O'Hagan 1999). Because many infectious
diseases gain entry to the body across mucosal membranes, the need for vaccines
that elicit mucosal immunity is increasingly being recognized (Mahon *et al.*
1998). Ideally, vaccines of the future will not require parenteral administration. It
has been shown recently that the Cry1Ac protoxin is capable of producing as
powerful an immune response in mice as cholera toxin (Vázquez-Padrón *et al.*
1999). Cholera toxin has previously been proposed as a possibly useful adjuvant
but its potent toxicity currently precludes this. The Cry1Ac protoxin, however,
has been shown to induce IgM, IgG and IgA antibodies in all of the mucosal surfaces tested after intraperitoneal, intranasal and rectal dosage (Moreno-Fierrros
et al. 2000). The ease with which pure *Bt* proteins may be produced and their stability and safety are attractive attributes. Although these reports are very recent
and need much further verification, a new chapter may be opening in the history
of the usefulness of *Bt* to mankind.

References

Adams, L.F., Mathewes, S., O'Hara, P., Petersen, A. & Gürtler, H. (1994) Elucidation of the mechanisms of CryIIIA overproduction in a mutagenized strain of *Bacillus thuringiensis* var. *tenebrionis*. *Molecular Microbiology* **14**, 381–389.

Agaisse, H. & Lereclus, D. (1995) How does *Bacillus thuringiensis* produce so much insecticidal crystal protein? *Journal of Bacteriology* **17**, 6027–6032.

Aronson, A.I. (1993) The two faces of *Bacillus thuringiensis*: insecticidal proteins and post-exponential survival. *Molecular Biology* **7**, 489–496.

Barloy, F., Delecluse, A., Nicolas L. & Lecadet, M.-M. (1996) Cloning and expression of the first anaerobic toxin gene from *Clostridium bifermentans* subsp. *malaysia*, encoding a new mosquitocidal protein with homologies to *Bacillus thuringiensis* delta-endotoxins. *Journal of Bacteriology* **178**, 3099–3105.

Baum, J.A. & Malvar, T. (1995) Regulation of insecticidal crystal protein production in *Bacillus thuringiensis*. *Molecular Microbiology* **18**, 1–12.

Baum, J.A., Johnson, T.B. & Carlton, B.C. (1999) *Bacillus thuringiensis*: natural and recombinant products. In: *Biopesticides Use and Delivery* (eds F.R. Hall & J.J. Menn), pp. 189–210. Humana Press, Totowa, NJ.

Becker, N. & Margalit, J. (1993) Use of *Bacillus thuringiensis israelensis* against mosquitoes and blackflies. In: *Bacillus thuringiensis, an Environmental Biopesticide: Theory and Practice.* (eds P.F. Entwhistle *et al.*), pp. 147–170. Wiley, Chichester.

Bishop, A.H., Johnson, J. & Perani, M. (1999) The safety of *Bacillus thuringiensis* to mammals investigated by oral and subcutaneous dosage. *World Journal of Biotechnology* **15**, 375–380.

Bravo, A. (1997) Phylogenetic relationships of *Bacillus thuringiensis* δ-endotoxin family proteins and their functional domains. *Journal of Bacteriology* **179**, 2793–2801.

Burges, H.D. (1981) Safety, safety testing and quality control of microbial pesticides. In: *Microbial Control of Pests and Plant Diseases 1970–1980* (ed. H.D. Burges), pp. 737–767. Academic Press, London.

Butt, T.M., Harris, J.G. & Powell, K.A. (1999) Microbial pesticides: the European scene. In: *Biopesticides Use and Delivery* (eds F.R. Hall & J.J. Menn), pp. 23–44. Humana Press, Totowa, NJ.

Carlson, C.R. & Kolstø, A.-B. (1994) A small (2.4Mb) *Bacillus cereus* chromosome corresponds to a conserved region of a larger (5.3 Mb) *Bacillus cereus* chromosome. *Molecular Microbiology* **13**, 161–169.

Carlson, C.R., Caugant, D.A. & Kolstø, A.-B. (1994) Genotypic diversity among *Bacillus cereus* and *Bacillus thuringiensis* strains. *Applied and Environmental Microbiology* **60**, 1719–1725.

Carlson, C.R., Johansen, T., Lecadet, M.-M. & Kolstø, A.-B. (1996) Genomic organisation of the entomopathogenic bacterium *Bacillus thuringiensis* subsp. *berliner* 1715. *Microbiology* **142**, 1625–1634.

Carroll, J., Convents, D., Van Damme, J. *et al.* (1997) Intramolecular proteolytic cleavage of *Bacillus thuringiensis* Cry3A delta-endotoxin may facilitate its coleopteran toxicity. *Journal of Invertebrate Pathology* **70**, 41–49.

Copping, L.G. & Menn, J.J. (2000) Biopesticides: a review of their action, applications and efficacy. *Pest Management Science* **56**, 651–676.

Crickmore, N., Zeigler, D.R., Feitelson, J. *et al.* (1998) Revision of the nomenclature for the *Bacillus thuringiensis* pesticidal crystal proteins. *Microbiology and Molecular Biology Reviews* **62**, 807–813.

Damgaard, P.H., Granum, P.E., Bresciani, J. *et al.* (1997) Characterization of *Bacillus thuringiensis* isolated from infections in burn wounds. *FEMS Immunology and Medical Microbiology* **18**, 47–53.

de Barjac, H. & Frachon, E. (1990) Classification of *Bacillus thuringiensis* strains. *Entomophaga* **35**, 233–240.

Dean, D.H., Rajamohan, F., Lee, M.K. *et al.* (1996) Probing the mechanism of action of *Bacillus thuringiensis* insecticidal proteins by site-directed mutagenesis – a minireview. *Gene* **179**, 111–117.

Drobniewski, F.A. (1994) The safety of *Bacillus* species as insect vector control agents. *Journal of Applied Bacteriology* **76**, 101–109.

Ellar, D.J. (1990) Pathogenic determinants of entomopathogenic bacteria. In: *Fifth International Colloquium on Invertebrate Pathology and Microbial Control: 1990* (ed. D. Pinnock), pp. 298–302. Society for Invertebrate Pathology, San Diego, CA.

Ellar, D.J. (1997) The structure and function of *Bacillus thuringiensis* δ-endotoxins and prospects for biopesticide improvement. In: *Microbial Insecticides: Novelty or Necessity?* (ed. British Crop Protection Council), Symposium Proceedings 68, pp. 83–100. British Crop Protection Council, Farnham.

Evans, H.F. (1997) The role of microbial insecticides in forest pest management. In: *Microbial*

Insecticides: Novelty or Necessity? (ed. British Crop Protection Council), Symposium Proceedings 68, pp. 29–40. British Crop Protection Council, Farnham.

Feitelson, J.S., Payne, J. & Kim, L. (1992) *Bacillus thuringiensis*: insects and beyond. *Bio/Technology* 10, 271–275.

Fernández-Larrea Vega, O. (1999) A review of *Bacillus thuringiensis (Bt)* production and use in Cuba. *Biocontrol News and Information* 20, 47N–48N.

Gelertner, W. & Schwab, G.E. (1993) Transgenic bacteria, viruses, algae and other microorganisms as *Bacillus thuringiensis* toxin delivery systems. In: Bacillus thuringiensis, *an Environmental Biopesticide: Theory and Practice* (eds P.F. Entwhistle *et al.*), pp. 89–104. Wiley, Chichester.

Goldberg, L.H. & Margalit, J. (1977) A bacterial spore demonstrating rapid larvicidal activity against *Anopheles sergentii, Uranotaenia unguiculata, Culex univitattus, Aedes aegypti* and *Culex pipiens. Mosquito News* 37, 355–358.

González, J.M., Jr, Dulmage, H.T. & Carlton, B.C. (1981) Correlation between specific plasmids and δ-endotoxin production in *Bacillus thuringinesis. Plasmid* 5, 351–365.

Granum, P.E. & Lund, T. (1997) *Bacillus cereus* and its food poisoning toxins. *FEMS Microbiology Letters* 157, 223–228.

Green, M., Heumann, M., Sokolow, R. *et al.* (1990) Public health implications of the microbial pesticide *Bacillus thuringiensis*: an epidemiological study, Oregon, 1985–6. *American Journal of Public Health* 80, 848–852.

Groschulski, P., Masson, L., Borisova, M. *et al.* (1995) *Bacillus thuringiensis* CryIA(a) insecticidal toxin: crystal structure and channel formation. *Journal of Molecular Biology* 254, 447–464.

Heimpel, A.M. & Angus, T.A. (1959) The site of action of the crystalliferous bacteria in lepidopteran larvae. *Journal of Invertebrate Pathology* 1, 152–170.

Helgason, E., Økstad, O.A., Caugant, D.A. *et al.* (2000) *Bacillus anthracis, Bacillus cereus* and *Bacillus thuringiensis* – one species on the basis of genetic evidence. *Applied and Environmental Microbiology* 66, 2627–2630.

Henderson, I., Dongzheng, Y. & Turnbull, P.C.B. (1995) Differentiation of *Bacillus anthracis* and other '*Bacillus cereus* group' bacteria using IS231-derived sequences. *FEMS Microbiology Letters* 128, 113–118.

Hill, C.A. & Pinnock, D.E. (1998) Histopathological effects of *Bacillus thuringiensis* on the alimentary canal of the sheep louse, *Bovicola ovis. Journal of Applied Microbiology* 72, 9–20.

Jackson, S.G., Goodbrand, R.B., Ahmed, R. & Kasatiya, S. (1995). *Bacillus cereus* and *Bacillus thuringiensis* isolated in a gastroenteritis outbreak investigation. *Letters in Applied Microbiology* 21, 103–105.

Jarrett, P. & Stephenson, M. (1990) Plasmid transfer between strains of *Bacillus thuringiensis* infecting *Galleria mellonella* and *Spodoptera litoralis. Applied and Environmental Microbiology* 56, 1608–1614.

Johnson, C., Bishop, A.H. & Turner, C.L. (1998) Isolation and activity of strains of *Bacillus thuringiensis* toxic to larvae of the housefly (Diptera: Muscidae) and tropical blowflies (Diptera: Calliphoridae). *Journal of Invertebrate Pathology* 71, 138–144.

Kaneko, T., Nozaki, R. & Aizawi, K. (1978) Deoxyribonucleic acid relatedness between *Bacillus anthracis, Bacillus cereus* and *Bacillus thuringiensis. Microbiology and Immunology* 22, 639–641.

Keim, P., Kalif, A., Schupp, J. *et al.* (1997) Molecular evolution and diversity in *Bacillus anthracis* as detected by amplified fragment length polymorphism markers. *Journal of Bacteriology* 179, 818–824.

Knight, P.J.K., Crickmore, N. & Ellar, D.J. (1994) The receptor for *Bacillus thuringiensis* Cry1A(c) delta-endotoxin in the brush border membrane of the lepidopteran *Manduca sexta* is aminopeptidase N. *Molecular Microbiology* 11, 429–436.

Knowles, B.H. (1994) Mechanism of action of *Bacillus thuringiensis* insecticidal δ-endotoxin. *Advances in Insect Physiology* 24, 275–308.

Kolstø, A.-B. (1998) *Bacillus cereus/ Bacillus thuringiensis.* In: *Bacterial Genomes. Physical Structure and Analysis* (eds F.J. de Bruijn *et al.*), pp. 609–612. Chapman & Hall, New York.

Lecadet, M.-M., Frachon, E., Dumanoir, V.C. *et al.* (1999) Updating the H-antigen classification of *Bacillus thuringiensis. Journal of Applied Microbiology* 86, 660–672.

Lee, M.K., You, T.H., Gould, F.L. *et al.* (1999) Identification of residues in Domain III of *Bacillus thuringiensis* CryIAc toxin that affect binding and toxicity. *Applied and Environmental Microbiology* 65, 4513–4520.

Leong, K.L.H., Cano, R.J. & Kubinski, A.M. (1980) Factors affecting *Bacillus thuringiensis* total field persistence. *Environmental Entomology* 9, 593–599.

Levinson, B.L., Kasyan, K.L., Chiu, S.S., Currier, T.C. & Gonzalez, J.M., Jr (1990) Identification of β-exotoxin production, plasmids encoding β-exotoxin, and a new exotoxin in *Bacillus thuringiensis* by using high-performance liquid chromatography. *Journal of Bacteriology* **172**, 3172–3177.

Li, J., Carroll, J. & Ellar, D.J. (1991) Crystal structure of δ-endotoxin from *Bacillus thuringiensis* at 2.5Å resolution. *Nature* **353**, 815–821.

Mahillon, J., Rezsöhazy, R., Hallet, B. & Delcour, J. (1994) IS*231* and other *Bacillus thuringiensis* transposable elements: a review. *Genetica* **93**, 13–26.

Mahon, B.P., Moore, A., Johnson, P.A. & Mills, K.H.G. (1998) Approaches to new vaccines. *Critical Reviews in Biotechnology* **18**, 257–282.

McClintock, J.T. (1999) The Federal registration process and requirements for the United States. In: *Biopesticides Use and Delivery* (eds F.R. Hall & J.J. Menn), pp. 415–441. Humana Press, Totowa, NJ.

Menn, J.J. & Hall, F.R. (1999) Biopesticides: present status and future prospects. In: *Biopesticides Use and Delivery* (eds F.R. Hall & J.J. Menn), pp. 1–10. Humana Press, Totowa, NJ.

Miyasono, M., Inagaki, S., Yamamoto, M. *et al.* (1994) Enhancement of δ-endotoxin activity by toxin-free spore of *Bacillus thuringiensis* against the diamondback moth *Plutella xylostella*. *Journal of Invertebrate Pathology* **63**, 111–112.

Moreno-Fierros, L., García, N., Gutiérrez, R., López-Revilla & Vázquez-Padrón, R.I. (2000) Intranasal, rectal and intraperitoneal immunisation with protoxin Cry1Ac from *Bacillus thuringiensis* induces compartmentalised serum, intestinal, vaginal and pulmonary immune responses in Balb/c mice. *Microbes and Infection* **2**, 885–890.

Perani, M., Bishop, A.H. & Vaid, A. (1998) Prevalence of β-exotoxin, diarrhoeal toxin and specific δ-endotoxin in natural isolates of *Bacillus thuringiensis*. *FEMS Microbiology Letters* **160**, 55–60.

Priest, F.G., Kaji, D.A., Rosato, Y.B. & Canhos, V.P. (1994) Characterization of *Bacillus thuringiensis* and related bacteria by ribosomal RNA gene restriction fragment length polymorphisms. *Microbiology* **140**, 1015–1022.

Ray, D.E. (1991) Pesticides derived from plants and other organisms. In: *Handbook of Pesticide Toxicology* (eds W.J. Hayes Jr & E.R. Laws Jr), pp. 585–636. Academic Press, San Diego, CA.

Reddy, A., Battisti, L. & Thorne, C.B. (1987) Identification of self-transmissable plasmids in four *Bacillus thuringiensis* subspecies. *Journal of Bacteriology* **169**, 5263–5270.

Rivera, A.M.G., Granum, P.E. & Priest, F.G. (2000) Common occurrence of enterotoxin genes and enterotoxicity in *Bacillus thuringiensis*. *FEMS Microbiology Letters* **190**, 151–155.

Rusul, G. & Yaacob, N.H. (1995) Prevalence of *Bacillus cereus* in selected foods and detection of enterotoxin using TECRA-VIA and BCET-RPLA. *International Journal of Food Microbiology* **25**, 131–139.

Saiki, J.E., Lacey, L.A. & Lacey, C.M. (1990) Safety of microbial insecticides to vertebrates – domestic animals and wildlife. In: *Safety of Microbial Insecticides* (eds M. Laird *et al.*), pp. 115–134. CRC Press, Boca Raton, FL.

Salama, H.S. & Morris, O.N. (1993) The use of *Bacillus thuringiensis* in developing countries. In: Bacillus thuringiensis, *an Environmental Biopesticide: Theory and Practice* (eds P.F. Entwhistle *et al.*), pp. 237–253. Wiley, Chichester.

Sampson, M.N. & Gooday, G.W. (1998) Involvement of chitinases of *Bacillus thuringiensis* during pathogenesis in insects. *Microbiology* **144**, 2189–2194.

Sanchis, V., Gohar, M., Chaufaux, J. *et al.* (1999) Development and field performance of a broad-spectrum nonviable asporogenic recombinant strain of *Bacillus thuringiensis* with greater potency and UV resistance. *Applied and Environmental Microbiology* **65**, 4032–4039.

Schnepf, E., Crickmore, N., van Rie, J. *et al.* (1998) *Bacillus thuringiensis* and its pesticidal crystal proteins. *Microbiology and Molecular Biology Reviews* **62**, 775–806.

Schwarz, J.L., Garneau, L., Savaria, D. *et al.* (1993) Lepidopteran-specific crystal toxins from *Bacillus thuringiensis* form cation- and anion-selective channels in planar lipid bilayers. *Journal of Membrane Biology* **132**, 53–62.

Sekar, V. (1988) The insecticidal crystal protein gene is expressed in vegetative cells of *Bacillus thuringiensis* var. *tenebrionis*. *Current Microbiology* **17**, 347–349.

Shimketts, L.J. (1998) Structure and sizes of genomes of the Archaea and bacteria. In: *Bacterial Genomes. Physical Structure and Analysis* (eds F.J. de Bruijn *et al.*), pp. 5–11. Chapman & Hall, New York.

Siegel, J.P. & Shadduck, J.A. (1990) Safety of microbial insecticides to vertebrates – humans in safety of microbial insecticides. In: *Safety of Microbial Insecticides* (eds M. Laird *et al.*), pp. 101–114. CRC Press, Boca Raton, FL.

Singh, M. & O'Hagan, D. (1999) Advances in vaccine adjuvants. *Nature Biotechnology* **17**, 1075–1081.

Smits, P.H. (1997) Insect pathogens: their suitability as biopesticides. In: *Microbial Insecticides: Novelty or Necessity?* (ed. British Crop Protection Council), Symposium Proceedings 68, pp. 21–28. British Crop Protection Council, Farnham.

Somerville, H.J. & Jones, M.L. (1972) DNA competition experiments within the *Bacillus cereus* group of bacilli. *Journal of General Microbiology* **73**, 257–261.

Sugavanam, B. & Xie, T. (1999) Developing countries. In: *Biopesticides Use and Delivery* (eds F.R. Hall & J.J. Menn), pp. 45–54. Humana Press, Totowa, NJ.

Tan, A., Heaton, S., Farr, L. & Bates, J. (1997) The use of *Bacillus* diarrhoeal enterotoxin (BDE) detection using an ELISA technique in the confirmation of the aetiology of *Bacillus*-mediated diarrhoea. *Journal of Applied Microbiology* **82**, 677–682.

te Giffel, M.C., Beumer, R.R., Klijn, N., Wagendorp, A. & Rombouts, F.M. (1997) Discrimination between *Bacillus cereus* and *Bacillus thuringiensis* using the specific DNA probes based on variable regions of 16S rRNA. *FEMS Microbiology Letters* **146**, 47–51.

Thomas, D.J.I., Morgan, J.A.W., Whipps, J.M. & Saunders, J.R. (2000) Plasmid transfer between the *Bacillus thuringiensis* subspecies *kurstaki* and *tenebrionis* in laboratory culture and soil and in lepidopteran and coleopteran larvae. *Applied and Environmental Microbiology* **66**, 118–124.

van Frankenhuysen, K. (1993) The challenge of *Bacillus thuringiensis*. In: Bacillus thuringiensis, *an Envir-*

onmental Biopesticide: Theory and Practice (eds P.F. Entwistle *et al.*), pp. 1–35. Wiley, Chichester.

Vázquez-Padrón, R.I., Moreno-Fierros, L., Neri-Bazán, L. *et al.* (1999) *Bacillus thuringiensis* Cry1Ac protoxin is a potent systemic mucosal adjuvant. *Scandinavian Journal of Immunology* **46**, 578–584.

Vladmudi, R.K., Weber, E., Ji, T. & Bulla, L.A., Jr (1995) Cloning and expression of a receptor for an insecticidal toxin of *Bacillus thuringiensis*. *Journal of Biological Chemistry* **270**, 5490–5494.

Yamada, S., Ohashi, E., Agata, N. & Venkateswaran, K. (1999) Cloning and nucleotide sequence analysis of *gyrB* of *Bacillus cereus*, *B. thuringiensis*, *B. mycoides*, and *B. anthracis* and their application to the detection of *B. cereus* in rice. *Applied and Environmental Microbiology* **65**, 1483–1490.

Zhang, J., Hodgman, T.C., Krieger, L., Schnetter, W. & Schairer, H.U. (1997) Cloning and analysis of the first *cry* gene from *Bacillus popilliae*. *Journal of Bacteriology* **179**, 4336–4341.

Zhong, C., Ellar, D.J., Bishop, A. *et al.* (2000) Characterization of a *Bacillus thuringiensis* δ-endotoxin which is toxic to insects in three orders. *Journal of Invertebrate Pathology* **76**, 131–139.

Chapter 12

Bt Crops: a novel insect control tool

Jeroen Van Rie

Introduction

Despite the use of pesticides, crop losses in the 1980s were estimated to be about as high as several of the previous decades (Pimentel 1991). Other estimates indicate that current crop losses are even higher than those 35 years ago (Oerke *et al.* 1994). In rice, for example, the growing of high-yielding 'Green Revolution' varieties has increased yields, but these varieties are more susceptible to pathogens and to planthopper and leafhopper attack. Another contributing factor is the increase in certain crops grown in tropical and subtropical regions, where crop losses are high because of the higher incidence of harmful organisms and sometimes poor crop protection measures. Worldwide pre-harvest crop losses have been estimated to be 13.8% because of insects and other arthropods, 11.6% because of disease (fungi, bacteria and viruses) and 9.5% because of weeds (Chrispeels & Sadava 1994). Total crop losses in Africa and Asia, the continents with the largest annual human population increase, reach almost 50% (Oerke *et al.* 1994). In order to control insects efficiently in a sustainable way, synthetic insecticides must be integrated with alternative pest control methods. One of these involves the use of resistant plant varieties obtained through 'classical breeding' and genetically engineered plant resistance to pests and pathogens.

Plant engineering

Significant progress has been made since the first successful transformations of plants. The capacity to introduce and express foreign genes in plants now extends to over 120 species, including some previously classified as recalcitrant (Birch 1997). *Agrobacterium*-mediated transformation has proven an efficient and reliable method with which to engineer different traits in a wide range of crop plants, both dicotyledonous and monocotyledonous plants (Hansen & Wright 1999). The main advantages of this DNA transfer method are the low level of rearrangements in the transforming DNA and the high number of plants with a single insertion of the transgene. In contrast, direct gene delivery systems such as particle bombardment or protoplast electroporation frequently result in a higher frequency of complex patterns of transgene integration. Equally important has been the development of tissue culture techniques that allow the production of highly regenerable tissues from immature and undifferentiated tissue and the development of tools to control the expression of a transgene in a plant. Plant transformation is still a random process with respect to the integration site of the transgene into the plant genome, sometimes resulting in suboptimal transgene expression or a negative impact on the expression of endogenous plant genes. Somaclonal variation is another aspect that can potentially lead to transgenic plants with suboptimal characteristics. Gene silencing, probably owing to the presence of multiple copies of foreign gene sequences, has also been observed in transformed plants. Together, these phenomena necessitate the generation of a large number of transgenic plant lines (events) from which those plants with the best performance (elite events) have to be selected through several rounds of laboratory and field evaluations – a process sometimes referred to as elite event selection.

Insecticidal crystal proteins from *Bacillus thuringiensis*

Different proteins, either from plant or bacterial sources, are available for genetic modification of crop plants for protection against insect pests. Plant-derived insecticidal proteins include protease inhibitors, lectins, lectin-like proteins and hydrolytic enzymes. Insecticidal proteins from bacterial sources include the insecticidal crystal proteins (ICPs) from *B. thuringiensis* (*Bt*), vegetative insecticidal proteins (VIPs) from *Bacillus* species, and certain enzymes such as cholesterol oxidase (Warren 1997; Corbin *et al.* 1998). Formulations of *Bt* spore–crystal mixtures have been used for more than 40 years and have demonstrated that *Bt* is a very specific, effective and safe bioinsectide. However, under field conditions, *Bt* is rapidly degraded, primarily by UV light. Since single gene products harbour the insecticidal activity, it is logical to use plant transformation to transfer the active ingredient into the plant itself. During sporulation, *Bt* produces crystalline inclusions containing one or more ICP types. When ingested by susceptible insects, the crystals dissolve in the insect gut and the protoxins are liberated and proteolytically activated to a toxic fragment. This fragment passes through the peritrophic membrane, binds to a specific receptor on the brush border membrane of gut epithelial cells and (partially) inserts into the membrane, generating pores. The change in membrane permeability leads to colloid osmotic lysis of gut epithelial cells and, ultimately, to death of the insect (Schneph *et al.* 1998).

More than 150 ICP sequences are currently known and classified solely on the basis of sequence homology of the full-length proteins into 28 Cry classes and 2 Cyt classes (Crickmore *et al.* 1998, 2000). There is no simple correlation between sequence and insecticidal spectrum, but some generalizations can be made. For example, ICPs belonging to the Cry1 and Cry9 classes are active against lepidopteran larvae, whereas Cry3, Cry7 and Cry8 proteins are active against coleopteran larvae. However, within a certain class, ICPs may have widely differing activity spectra. This specificity is still one of the most intriguing aspects of ICPs. Any of the different steps of the mode of action can influence the activity spectrum.

Cry1 and Cry9 ICPs are protoxins of about 120–140 kDa that are proteolytically processed to an active toxic fragment of about 60–70 kDa. Characterization of the proteolytic fragment and fragments generated by the expression of truncated ICP genes have indicated that, while only few amino acids can be removed from the N-terminus without interfering with biological activity, about half of the protoxin can be removed at the C-terminus. Upon alignment of ICP amino acid sequences, it is clear that sequence variation is not distributed in a random fashion. Five conserved sequence blocks can be distinguished in the activated fragment of most Cry proteins.

Today, the crystal structures of three activated ICPs have been solved (Li *et al.* 1991; Grochulski *et al.* 1995; Morse *et al.* 1998). These three proteins, Cry3Aa, Cry1Aa and Cry2Aa, have very similar architectures and are composed of three structural domains. The N-terminal domain (residues 58–290 in Cry3A) contains seven α-helices with the central more hydrophobic helix (α5) encircled by six outer amphipathic helices. The second domain (residues 291–500) consists of three β-sheets, packed as three sides of a prism. The third, C-terminal, domain (residues 501–644) is a β sandwich with the outer sheet facing the solvent and the inner sheet facing the other two domains. The conserved sequence blocks in the crystal structure are located either within a structural domain or at domain interfaces, suggesting that most other ICPs possess a similar global architecture. Based on these structures and the characterization of ICP mutants and hybrids, the following hypotheses have been put forward regarding the function of the three structural domains of ICPs: the long amphipathic helices of Domain I are responsible for pore formation; Domain II plays a major role in receptor binding; Domain III also plays a role in receptor binding and perhaps modulates pore formation, apart from its role in providing structural integrity to the protein (Schneph *et al.* 1998).

Bt plants

The first experiments to create plants expressing *cry* genes (*Bt* plants) used T-DNA vectors in *Agrobacterium* carrying the coding sequence for Cry1A protoxins. Only very low levels of Cry proteins and no significant insecticidal activity related to *Bt* was observed (Adang *et al.* 1987; Barton *et al.* 1987; Vaeck *et al.* 1987). The first successes were obtained by expressing gene fragments encoding the toxic part of the ICP only. Expression of truncated *cry1Aa* (Barton *et al.* 1987) and *cry1Ab* (Vaeck *et al.* 1987) genes in tobacco resulted in significant levels of protein and high insecticidal activity to *Manduca sexta* larvae feeding from the leaves. Also, tomato plants engineered with a truncated *cry1Ac* gene (Fischhoff *et al.* 1987) proved to be protected from feeding damage by *M. sexta* and resulted in mortality or growth inhibition of *Heliothis virescens* and *Helicoverpa zea* larvae. Tubers from different potato varieties engineered with a truncated *cry1Ab* gene under the control of a wound-stimulated promoter and infested with *Phthorimaea operculella* larvae did not show tunnelling or feeding damage following a 2-month storage period (Peferoen *et al.* 1990). Cotton plants expressing truncated *cry1A* genes controlled *H. virescens* larvae (Perlak *et al.* 1990). When tested under agronomic conditions in the field, transgenic tomato (Delannay *et al.* 1989) and tobacco plants (Warren *et al.* 1992) expressing truncated ICP genes showed substantial levels of insecticidal activity against their primary pest insects. However, it became apparent that for certain crops or insect pests, higher ICP expression levels were needed to achieve complete insect control in the field.

Typically, expression levels of native truncated ICP genes in plants – usually about 0.001% of total soluble protein – were lower than levels obtained with other transgenes. Plant genes generally have a high G+C content wheras ICP genes typically have a high A+T content. A+T-rich regions in native ICP genes contain cryptic intron splice sites (Brown & Simpson 1998) and potential polyadenylation sites (Gallie 1993) resulting in aberrant splicing (Cornelissen *et al.* 1993) or premature polyadenylation (Diehn *et al.* 1998), both leading to nonfunctional mRNA. Furthermore, wild type ICP genes contain ATTTA motifs, known as messenger-destabilizing elements in eukaryotes, including plants (Ohme-Takagi *et al.* 1993). The codon usage in ICP and plant genes is significantly different. The presence of rare plant codons in native ICP genes could result in ribosomal pausing (Gallie 1993) and, perhaps, to accelerated degradation of the *cry* gene messenger. However, experimental evidence suggests that the presence of rare codons *per se* does not interfere dramatically with mRNA accumulation (Van Hoof & Green 1997; De Rocher *et al.* 1998). On the other hand, comparison of mRNA and protein levels in plants transformed with truncated *cry1Ab* genes with different degrees of modification led the authors to suggest that the presence of predominantly plant-preferred codons improved the overall ICP gene translational efficiency (Perlak *et al.* 1991).

Several authors have demonstrated that modifications in a specific region could result in significant improvements in expression. For example, Perlak *et al.* (1991) found that changes in the 5′ half of *cry1Ab* were more efficient in achieving increased expression levels than changes in the 3′ half. Within the 5′ half, a particular 37-nucleotide stretch appeared to have a dramatic impact on expression. By analysing transcript levels of various *cry1Ab* 3′ deletions in transgenic tobacco, Murray *et al.* (1991) concluded that at least some RNA instability elements are located within the first 570 bases. Cornelissen *et al.* (1991) identified a region between nucleotides 785 to 1285 in *cry1Ab* where transcript elongation was retarded. Tobacco plants transformed with a *cry1Ab* gene with 63 translationally neutral substitutions in this region showed up to 20-fold higher levels of *cry1Ab* transcript (van Aarsen *et al.* 1995). Furthermore, these authors demonstrated that modifications that removed cryptic splice sites caused further increases in transcript levels. Another limited modification thought to improve ICP expression levels is the introduction of a consensus translation start context (van Aarsen *et al.* 1995; Joshi *et al.* 1997). A *cry1Ab* gene with limited modifications (altered translation start context and 785–1285 region) was introduced in potato and 90% of the transgenic potato plants proved highly resistant to potato tubermoth larvae. Transgenic tubers suffered no feeding damage or tunnelling in a 2-month storage test with potato tubers infested with *P. operculella* eggs (Jansens *et al.* 1995).

Although modifications in a specific region can result in significant improvements in expression, translationally neutral nucleotide changes throughout the ICP coding region are probably required to obtain the highest levels of expression of *cry* genes integrated in the nuclear genome. The truncated *cry1Ab* gene was rendered more 'plant like' by removing potential polyadenylation signals and ATTTA sequences by changing 62 of the 1743 nucleotides (Perlak *et al.* 1991). Transformation of tobacco and tomato with constructs containing this modified gene (96.4% nucleotide sequence homology to the native sequence) resulted in a higher number of insecticidal plants and higher expression levels (0.02% of total soluble protein) compared to constructs containing the wild type genes.

Another modified *cry1Ab* gene, containing additional changes to increase overall G+C content and to introduce plant-preferred codons while having a nucleotide sequence homology of only 78.9% to the native sequence, increased expression levels up to about 0.2% of the total soluble proteins. Similar results were obtained for a modified truncated *cry1Ac* gene. Cotton engineered with these modified genes showed protection from feeding damage by their main lepidopteran pests (Wilson *et al.* 1992).

Synthetic *cry1A* genes have now been used for transformation of a number of additional plant species including rice, corn, peanut, chinese flowering cabbage, canola, coffee, broccoli and soybean (Mazier *et al.* 1997). For example, corn transformed with a truncated modified *cry1Ab* gene (Koziel *et al.* 1993), driven by either a constitutive promotor or the combination of a green tissue- and pollen-specific promotor, resulted in both cases in plants with high levels of expression (up to 4 µg/mg total plant protein). Field trials confirmed that tunnelling of corn stalks by *Ostrinia nubilalis* (European corn borer) was reduced dramatically in such plants. Similarly, corn lines transformed with a modified *cry9C* gene showed nearly complete inhibition of stalk tunnelling by *O. nubilalis* in greenhouse and field trials, and reduced feeding damage by *Agrotis ipsilon* (Jansens *et al.* 1997). Using a *cry3A* gene, rendered more plant-like by wholesale modifications, high expression levels of Cry3Aa were obtained in potato and full control of *Leptinotarsa decemlineata* (Colorado potato beetle) larvae was observed (Perlak *et al.* 1993). Similarly, a *cry1C* gene redesigned for high level expression in plants, provided protection to *Spodoptera littoralis* and *S. exigua* in transgenic tobacco and alfalfa (Strizhov *et al.* 1996) and protection from *Plutella xylostella* (diamondback moth) in transgenic broccoli (Cao *et al.* 1999). Certain modifications outside the coding region may also contribute to ICP expression in plants. In the case of monocotyledonous plants, introns are frequently introduced to increase ICP expression levels (Fujimoto *et al.* 1993; Armstrong *et al.* 1995; Wünn *et al.* 1996; Nayak *et al.* 1997; Alam *et al.* 1999). Furthermore, 5' untranslated leader sequences are sometimes included (e.g. Jansens *et al.* 1997). Currently, transgenic cotton, corn and potato varieties are commercially grown to control the major lepidopteran (*Bt* cotton, *Bt* corn) or coleopteran (*Bt* potato) pests on these crops (tables 12.1 and 12.2).

Insect resistance to *Bt*

Over 500 insect species have become resistant to single or multiple insecticides, demonstrating their enormous genetic plasticity (Georghiou & Lagunes-Tejeda 1991). In 1985, the first report on resistance to *Bt* was published: a 250-fold level

Table 12.1 Adoption of *Bt* crops in USA (from James 1998).

Crop	1996 area		1997 area		1998 area	
	ha	%	ha	%	ha	%
Bt corn	300 000	1	2 800 000	9	6 500 000	22
Bt cotton	700 000	13	<1 000 000	17	>1 000 000	20
Bt potato	4 000	1	10 000	2.5	20 000	5

Table 12.2 Registered *Bt* plant pesticides (adapted from EPA 1999).

Event/product	Year registered	Expiration date	Toxin	Crop	Company
NewLeaf	May 1995	None	Cry3A	Potato	Monsanto/ NatureMark
NewLeaf Plus	Dec 1998	None	Cry3A + viral resistance gene	Potato	Monsanto/ NatureMark
Bollgard	Oct 1995	Jan 2001	Cry1Ac	Cotton	Monsanto
Event 176 (KnockOut)	Aug 1995/ March 1998	April 2001	Cry1Ab	Field corn/ popcorn	Novartis
Event 176 (NatureGard)	Aug 1995	April 2001	Cry1Ab	Field corn	Mycogen
Bt 11 (YieldGard)	Oct 1996	April 2001	Cry1Ab	Field corn	Novartis
Bt 11 (Attribute)	March 1998	April 2001	Cry1Ab	Sweet corn	Novartis
Mon810 (YieldGard)	Dec 1996	April 2001	Cry1Ab	Field corn	Monsanto
DBT-418 (Bt-Xtra)	March 1997	April 2001	Cry1Ac	Field corn	DeKalb (Monsanto)
CBH-351 (StarLink)	May 1998	April 2001	Cry9C	Field corn	PGS/AgrEvo (Aventis)

of resistance to *Bt* was observed in a *Plodia interpunctella* population from grain bins that were regularly treated with *Bt*. Resistance was caused by a 50-fold decrease in binding affinity of one of the receptors for the Cry1Ab protein, one of the ICPs in the *Bt* insecticidal formulation used (Van Rie *et al.* 1990). However, this Cry1Ab-resistant population had an increased susceptibility for Cry1Ca, a protein that the insects had not been exposed to. This increased susceptibility corresponded to an increase in the binding site concentration on the midgut for the Cry1Ca protein. Since then, a substantial number of strains of different insect species with various levels of resistance to *Bt* ICPs has been obtained by laboratory selection experiments (Van Rie & Ferré 2000). These experiments demonstrate that insects possess different mechanisms (mainly altered receptor binding and altered proteolytic processing) to overcome the toxicity of ICPs. Laboratory selection is, of course, quite different from field adaptation; indeed, it frequently leads to polygenic resistance, whereas field-evolved resistance is often associated with a single gene (Roush 1993). Nevertheless, resistant colonies obtained through laboratory selection are valuable for assessing which biochemical and genetic resistance mechanisms can be associated with insect adaptation. The first case of field resistance to *Bt* was reported from Hawaii where different *Plutella xylostella* populations showed different levels of susceptibility to a formulated *Bt* product (DiPel). Populations from heavily treated areas proved less susceptible than populations that had been treated at lower levels, with the highest level of resistance at 30-fold (Tabashnik *et al.* 1990). Further selection in the laboratory increased resistance levels significantly. The resulting strain had high levels of (cross)-resistance to Cry1Aa, Cry1Ab, Cry1Ac, Cry1Fa and Cry1Ja but no significant cross-resistance to Cry1Ba, Cry1Bb, Cry1Ca, Cry1Da, Cry1Ia or Cry2Aa (Tabashnik *et al.* 1996). Binding of Cry1Ab and Cry1Ac, but not Cry1Ca, was reduced strongly in the resistant strain. A very similar pattern of resistance and binding characteristics was observed in a strain from Pennsylvania (Tabashnik *et al.* 1997).

The cross-resistance and binding data from these resistant strains can be understood more easily using the model for the ICP binding sites in (susceptible) *P. xylostella*: according to this model, one site is recognized by Cry1Aa, Cry1Ab, Cry1Ac and Cry1F, another site by Cry1Ba and another by Cry1Ca; Cry1Aa recognizes an additional site (Van Rie & Ferré 2000). Therefore, an altered Cry1Ab binding site would result in cross-resistance to other Cry1 proteins binding to that same site. Similarly, resistance to Cry1Ab, but not to Cry1Ba and Cry1Ca, in a field population of *P. xylostella* from the Philippines and in another population from Florida was caused by a change in the target site for the Cry1Ab protein, while the receptors for the Cry1Ba and Cry1Ca were fully functional (Ferré *et al.* 1991; Tang *et al.* 1996). In all the *P. xylostella* populations that developed resistance in the field and for which resistance mechanisms were studied, resistance proved to be recessive and related to an alteration at an ICP binding site. In those cases where resistance results from altered binding to a receptor, no resistance to ICPs binding to another receptor has been observed. The lack of such cross-resistance could be exploited in resistance management strategies.

Resistance management with *Bt* plants

In insects, resistance is a pre-adaptive phenomenon that develops by selection of rare individuals that can survive a certain insecticidal treatment of a population. Resistance management strategies try to prevent or diminish the selection of the rare individuals carrying resistance genes and, therefore, to keep the frequency of resistance genes below levels which would result in inefficient insect control.

There are several different strategies that, at least in theory, should slow down the development of insect resistance (Roush 1997), including tissue-specific ICP expression (protecting only the critical tissues), wound-induced ICP expression (resulting in expression only at the threshold of insect damage), chemical-induced ICP expression, ultra high ICP expression (making the plant a nonhost for the pest), rotation of crops expressing different types of ICPs, and pyramiding (expression of different ICP types in the same crop plant). Based on a consensus among population geneticists and insect resistance experts, current insect resistance management (IRM) tactics for *Bt* crops implement the high-dose strategy combined with a refuge. The principle of the high-dose strategy is that the plant tissues express an ICP dose high enough to kill all of the most common carriers of resistance, i.e. the resistant heterozygotes. A refuge is an area free of toxin-expressing plants that allows susceptible insects to survive. Any rare resistant insects emerging from the *Bt* crop area will more likely mate with insects from the refuge than with other resistant insects. Such crosses will result in heterozygous resistant progeny, which will be killed by the transgenic crop plants, and hence cause a dilution of resistance. Refugia that are temporally and spatially contiguous with the transgenic crop would ensure random mating between resistant and susceptible adults and should produce at least 500 susceptible moths for each resistant moth emerging from the *Bt* crop. Data on insect adaptation to *Bt* indicate that in most cases resistance is (at least partially) recessive (Van Rie & Ferré 2000). Using modified ICP genes, a 'high dose' (defined as 25 times the dose needed to kill all homozygous susceptible larvae) can be achieved in plants. These observations indicate that high-dose *Bt* plants should kill not only susceptible insects, but also nearly all heterozygous susceptible larvae. Indeed, progeny from a cross between a Cry1A-resistant and a susceptible *P. xylostella* colony, could not survive on transgenic broccoli expressing a truncated modified *cry1Ac* gene (Metz *et al.* 1995).

Although the high-dose/refuge strategy may be difficult to realize for sprayable insecticides, it seems likely to be efficacious for ICP-expressing plants. Whereas the validity of the high-dose/refuge strategy was originally based only on projections from computer models simulating insect population growth under various conditions, recent studies have provided experimental support for this strategy. Indeed, selection of *P. xylostella* under laboratory conditions resulted in resistance in colonies without refuge more rapidly than in those provided with a refuge (Liu & Tabashnik 1997). Furthermore, controlled field trials involving Cry1Ac-expressing broccoli plants and artificial *P. xylostella* populations, with known Cry1A resistance allele frequencies, have demonstrated that a 20% refuge, separated from the *Bt* plants, was more effective in maintaining the population of susceptible insects than a 20% mixed refuge, created by planting a mixture of seeds of *Bt* and non-*Bt* plants (Shelton *et al.* 2000). This probably holds true for any *Bt* crop where pest larvae can move between plants to any extent. The authors also reiterated the notion that the size and potential treatment of the refuge should be such that enough susceptible adults emerge from it to 'overwhelm' any resistant adults emerging from the *Bt* crop plants. Currently, the insect resistance management plan for *Bt* cotton requires a 20% non-*Bt* cotton refuge that can be sprayed or a 4% unsprayed refuge. For corn, the US EPA (Environmental Protection Agency) requires a structured refuge of at least 20% non-*Bt* corn and 50% in cotton-growing areas owing to the extra potential selection pressure on *H. zea* from Cry1A-expressing cotton. If the refuge is unsprayed, it should be within half a mile of the *Bt* field; if sprayed it should be within quarter of a mile (EPA 2000). Experience with *Bt* crops grown under different agronomic conditions will allow further optimization of resistance management tactics. Furthermore, it is clear that *Bt* plants are not a standalone product, but rather an additional insect control tool that can be integrated with other pest management tools, such as crop rotation, ICP rotation, spray-on insecticides, destruction of larval overwintering sites, etc.

Environmental impact of *Bt* plants

One of the environmental benefits of the adoption of *Bt* crops is the potential to reduce insecticide sprays. In 1996, 70% of *Bt* cotton in the USA reportedly did not require insecticide sprays for targeted insect pests (James 1998). The EPA cites figures from Monsanto that cotton insecticide use dropped by 3.6 million litres per year after *Bt* cotton entered the market (Lehrman 1999). In Australia, Ingard cotton received 50% less insecticides than conventional cotton according to a 1996/1997 survey. For *Helicoverpa* sprays, the reduction was greater than 50% (Constable *et al.* 1998). Significant reductions in insecticide applications have also been reported for *Bt* cotton in China (James 1998). In 1997, *Bt* potatoes required up to 40% less insecticide than non-*Bt* potatoes (James 1998).

An important element to consider for any novel insect control tool is its potential impact on nontarget insects, predators and parasites present in the cropping system. In view of the specificity of ICPs, it can be expected that the effect of *Bt* crops on most nontarget insects should be minimal, especially when compared to the effects of broad-spectrum insecticides as used in many insect control regimes. In the case of *Bt* cotton and corn plants, both expressing lepidopteran-specific ICPs (Cry1A or Cry9C), it can be expected that some nontarget Lepidoptera may be negatively affected when challenged with tissues of such *Bt* plants. Indeed, Losey *et al.* (1999) found that, under laboratory conditions, *Bt* corn pollen dusted over milkweed plants decreased survival of larvae of the monarch butterfly, *Danaus plexippus*. However, the critical question is not whether some Lepidoptera are susceptible to the ICP expressed in tissue of *Bt* plants, but whether their larvae are exposed to the protein under field conditions. Preliminary data from subsequent research by Sears *et al.* (2000) demonstrated that pollen dusted on milkweed plants at 135 grains of pollen/cm^2 had no greater effect on survival of *D. plexippus* larvae than non-*Bt* pollen, although it resulted in a decreased weight gain. This pollen density should be compared to the average values found on milkweed stands in cornfields (about 80 pollen/cm^2), at cornfield edges (about 30 pollen/cm^2) or at 5 m from cornfield edges (1 pollen/cm^2). From these observations, it follows that a substantial fraction of milkweed plants will be out of range of significant levels of *Bt* pollen deposition. Another important factor to consider in this respect is the type and amount of ICP produced in pollen from the various *Bt* corn varieties. The studies by Sears *et al.* (2000) were performed on neonates (usually the most susceptible larval stage of Lepidoptera) using pollen from event 176, as present in the KnockOut and NatureGard varieties, which expresses *Bt* in pollen at a dose approximately 100 times that in pollen from YieldGard varieties. Also, Cry9C, expressed in StarLink varieties, is expressed only marginally in pollen. Furthermore, the monarch larval stage, which takes about 2 weeks, does not necessarily overlap with the pollen shed period (usually about 10–14 days). Taken together, it seems likely that the majority of the monarch larvae will not be exposed to (harmful amounts of) *Bt* corn pollen, and hence the impact on monarch population densities should be minimal.

Recent studies report on the effects of Cry1Ab protein on larvae of the green lacewing (*Chrysoperla carnea*), a predator that feeds on *Ostrinia nubilalis* eggs and young larvae. These larvae showed an increase in mortality when offered *O. nubilalis* or *Spodoptera littoralis* larvae that had fed for 1 day on Cry1Ab-expressing *Bt* corn compared to non-*Bt*-corn-fed larvae (Hilbeck *et al.* 1998a). In an effort to differentiate between a direct effect of the toxin and an indirect effect, reduced nutritional quality of the *Bt*-fed prey, these authors compared survival of *C. carnea* larvae developing on *Ephestia kuehniella* eggs or on artificial diet with or without Cry1Ab. The use of the artificial diet increased the mortality to 30%, compared to 8% when using eggs as the diet. Inclusion of Cry1Ab at 100 μg/mL in the diet increased the mortality to 57% (Hilbeck *et al.* 1998b). It should be noted that this Cry1Ab level is about 20-fold higher than the Cry1Ab expression level in leaves of the *Bt* corn used in the first study. It seems highly unlikely that *C. carnea* would complete its development solely on *Bt*-fed *O. nubilalis* larvae under field conditions. Predators such as green lacewings are probably attracted to corn by the presence of resources such as pollen, aphids and thrips, which are major components of their diet, and feed on *O. nubilalis* eggs and young larvae opportunistically (Orr & Landis 1997). Moreover, lacewings will have only a limited opportunity to prey on *O. nubilalis* larvae, since *Bt* corn controls these larvae effectively. Field studies comparing densities of beneficial insects, including *C. carnea*, on *Bt* and non-*Bt* corn did not show significant differences (Orr & Landis 1997; Pilcher *et al.* 1997). Similarly, no differences were found in beneficial insect densities on Cry1Ab-expressing and non-*Bt* cotton and sweetcorn (Wilson *et al.* 1992; Fitt *et al.* 1994; Wold *et al.* 2001).

The above data illustrate that considerable care should be taken in extrapolating laboratory findings to natural field conditions. Factors such as the significance of the crop as a food source and the degree of specialization of the predator or parasite species are likely to be important in estimating the impact under field conditions. Clearly, in evaluating the risk to nontarget organisms, both toxicity and exposure should be taken into account. Also, any impact of *Bt* crops should be judged alongside conventional insect control methods. For example, as Fitt *et al.* (1994) pointed out, even if some beneficials would be less abundant in *Bt* cotton than in an unsprayed non-*Bt* cotton crop, it must be remembered that beneficials are virtually absent from commercial cotton crops where pesticides are used extensively.

Biologically active Cry1Ab toxin was reported to be present in soil rhizosphere samples taken from *Bt* corn seedlings (Saxena *et al.* 1999). Cry proteins rapidly degrade in soil environments (Sims & Holden 1996). It remains to be seen what impact the presence of this protein may have on soil organisms under field conditions. Based on the specificity of Cry1Ab for certain lepidopteran insect species, the small number of lepidopteran species that dwell in the soil during their larval stage and the rapid degradation of Cry proteins in soil, such impact would be expected to be minimal.

Certain *Fusarium* species, particularly *F. moniliforme*, can cause ear rot in corn. Insect larvae such as *O. nubilalis* can act as vectors of *Fusarium* spores from the plant surface to damaged kernels or the interiors of stalks, or provide entry wounds for fungi (Sobek & Munkvold 1999). Studies by Munkvold *et al.* (1997) have shown that the severity of *Fusarium* ear rot and the incidence of symptomless kernel infection were reduced in Cry1Ab-expressing corn hybrids compared to near-isogenic nontransgenic corn lines. The concentration of fumonisins – mycotoxins that can cause fatal leucoencephalomalacia in horses, pulmonary oedema in swine and cancer in laboratory rats, and that have been implicated in oesophageal cancer in humans – were reportedly lower in field-grown *Bt* corn lines expressing either Cry1Ab or Cry9C in kernels than in non-*Bt* corn (Munkvold *et al.* 1999).

Conclusion

Bt crops have provided farmers with an additional tool to control insect pests on corn, cotton and potatoes. When growing such crops, farmers must agree to implement certain IRM tactics, aimed at preventing or delaying the development of resistance in insect populations. These tactics are being refined as our knowledge about insect pest biology and insect–crop plant interactions expands. Field evaluations conducted so far have not found a negative impact of *Bt* crops on nontarget and beneficial insects. Multiyear studies are in progress to monitor more precisely the densities of such insects in *Bt* crop fields. Especially in cotton, significant reductions in synthetic insecticide sprays have been realized upon adoption of *Bt* crops. Judicious use of this novel insect control tool should result in sustainable benefits to farmers and the environment, and as a consequence, also to the consumer.

References

Adang, M.J., Firoozabady, E., Klein, J. *et al.* (1987) In: *Molecular Strategies for Crop Protection* (eds C.J. Arntzen & C. Ryan), p. 345. Alan R. Liss, New York.

Alam, M.F., Datta, K., Abrigo, E. *et al.* (1999) Transgenic insect-resistant maintainer line (IR68899B) for improvement of hybrid rice. *Plant Cell Reports* 18, 572–575.

Armstrong, C.L., Parker, G.B., Pershing, J.C. *et al.* (1995) Field evaluation of European corn borer control in progeny of 173 transgenic corn events expressing an insecticidal protein from *Bacillus thuringiensis*. *Crop Science* 35, 550–557.

Barton, K., Whiteley, H. & Yang, N.-S. (1987) *Bacillus thuringiensis* δ-endotoxin expressed in transgenic *Nicotiana tabacum* provides resistance to lepidopteran insects. *Plant Physiology* 85, 1103–1109.

Birch, R.G. (1997) Plant transformation: problems and strategies for practical application. *Annual Reviews in Plant Physiology and Plant Molecular Biology* 48, 297–326.

Brown, J.W.S. & Simpson, C.G. (1998) Splice site selection in plant pre-mRNA splicing. *Annual Reviews in Plant Physiology and Plant Molecular Biology* 49, 77–95.

Cao, J., Tang, J.D., Strizhov, N., Shelton, A.M. & Earle, E.D. (1999) Transgenic broccoli with high levels of *Bacillus thuringiensis* Cry1C protein control diamondback moth larvae resistant to Cry1A or Cry1C. *Molecular Breeding* 5, 131–141.

Chrispeels, M.J. & Sadava, D.E. (eds) (1994) *Plants, Genes, and Agriculture*. Jones & Bartlett, Boston, MA.

Constable, G.A., Llewellyn, D.J. & Reid, P.E. (1998) Biotechnology risks and benefits: the Ingard cotton example. http://life.csu.edu.au/agronomy/papers/invite/const/agronsoc.html.

Corbin, D.R., Greenplate, J.T. & Purcell, J.P. (1998) The identification and development of proteins for control of insects in genetically modified crops. *HortScience* 33, 614–617.

Cornelissen, M., Soetaert, P., Stam, M. & Dockx, J. (1991) Modified *Bacillus thuringiensis* insecticidal-crystal protein genes and their expression in plant cells. International application published under the patent cooperation treaty (PCT). WO 91/16432. World Intellectual Property Organization, Geneva.

Cornelissen, M., Soetaert, P., Stam, M., Dockx, J. & van Aarsen, R. (1993) Modified genes and their expression in plant cells. International application published under the patent cooperation treaty (PCT). WO 93/09218. World Intellectual Property Organization, Geneva.

Crickmore, N., Zeigler, D.R., Feitelson, J. *et al.* (1998) Revision of the nomenclature for *Bacillus thuringiensis* cry genes. *Microbiology and Molecular Biology Reviews* 62, 807–813.

Crickmore, N., Zeigler, D.R., Feitelson, J. *et al.* (2000) *Bacillus thuringiensis* toxin nomenclature. http://www.biols.susx.ac.uk/ Home/ Neil_Crickmore/*Bt*/index.html.

De Rocher, E.J., Vargo-Gogola, T.C., Diehn, S.H. & Green, P. (1998) Direct evidence for rapid degradation of *Bacillus thuringiensis* toxin mRNA as a cause of poor expression in plants. *Plant Physiology* 117, 1445–1461.

Delannay, X., LaVallee, B.J., Proksch, R.K. *et al.* (1989) Field performance of transgenic tomato plants expressing the *Bacillus thuringiensis* var. *kurstaki* insect control protein. *Bio/Technology* 7, 1265–1269.

Diehn, S.H., Chiu, W.-L., De Rocher, E.J. & Green, P.J. (1998) Premature polyadenylation at multiple sites within a *Bacillus thuringiensis* toxin gene-coding region. *Plant Physiology* 117, 1433–1443.

EPA (1999) www.epa.gov/oppbppd1/biopesticides/ otherdocs/bt_position_paper_618.htm.

EPA (2000) www.epa.gov/oppbppd1/biopesticides/ otherdocs/bt_corn_ltr.htm.

Ferré, J., Real, D.M., Van Rie, J., Jansens, S. & Peferoen, M. (1991) Mechanism of resistance to *Bacillus thuringiensis* in a field population of *Plutella xylostella*. *Proceedings of the National Academy of Sciences of the USA* 88, 5119–5123.

Fischhoff, D.A., Bowdish, K.S., Perlak, F.J. *et al.* (1987) Insect tolerant transgenic tomato plants. *Bio/Technology* 5, 807–813.

Fitt, G.R., Mares, C.L. & Llewellyn, D.J. (1994) Field evaluation and potential ecological impact of transgenic cottons (*Gossypium hirsutum*) in Australia. *Biocontrol Science and Technology* 4, 535–548.

Fujimoto, H., Itoh, K., Yamamoto, M., Kyozuka, J. & Shimamoto, K. (1993) Insect resistant rice generated by introduction of a modified δ-endotoxin gene of *Bacillus thuringiensis*. *Bio/Technology* 11, 1151–1155.

Gallie, D.R. (1993) Posttranscriptional regulation of gene expression in plants. *Annual Reviews in Plant Physiology and Plant Molecular Biology* 44, 77–105.

Georghiou, G.P. & Lugunes-Tejeda, A. (1991) *The Occurrence of Resistance to Pesticides in Arthropods*. FAO/ UN, Rome.

Grochulski, P., Masson, L., Borisova, S. *et al.* (1995) *Bacillus thuringiensis* Cry1A(a) insecticidal toxin: crystal structure and channel formation. *Journal of Molecular Biology* 254, 447–464.

Hansen, G. & Wright, M.S. (1999) Recent advances in the transformation of plants. *Trends in Plant Science* 4, 226–231.

Hilbeck, A., Baumgartner, M., Fried, P.M. & Bigler, F. (1998a) Effects of transgenic *Bacillus thuringiensis* corn-fed prey on mortality and development time of immature *Chrysoperla carnea* (Neuroptera: Chrysopidae). *Environmental Entomology* 27, 480–487.

Hilbeck, A., Moar, W.J., Pusztai-Carey, M., Filippini, A. & Bigler, F. (1998b) Toxicity of *Bacillus thuringiensis* Cry1Ab toxin to the predator *Chrysoperla carnea* (Neuroptera: Chrysopidae). *Environmental Entomology* 27, 1255–1263.

James, C. (1998) *Global Review of Commercialized Transgenic Crops 1998*. ISAAA Briefs 8. ISAAA, Ithaca, NY.

Jansens, S., Cornelissen, M., De Clercq, R., Reynaerts, A. & Peferoen, M. (1995) *Phthorimaea operculella* (Lepidoptera: Gelechiidae) resistance in potatoes by expression of the *Bacillus thuringiensis* Cry1A(b) insecticidal crystal protein. *Journal of Economic Entomology* 88, 1469–1476.

Jansens, S., van Vliet, A., Dickburt, C. *et al.* (1997) Transgenic corn expressing a Cry9C insecticidal protein from *Bacillus thuringiensis* protected from European corn borer damage. *Crop Science* 37, 1616–1624.

Joshi, C.P., Zhou, H., Huang, X. & Chiang, V.L. (1997) Context sequences of translation initiation codon in plants. *Plant Molecular Biology* 35, 993–1001.

Koziel, M.G., Beland, G.L., Bowman, C. *et al.* (1993) Field performance of elite transgenic maize plants expressing an insecticidal protein derived from *Bacillus thuringiensis*. *Bio/Technology* 11, 194–200.

Lehrman, S. (1999) GM backlash leaves US farmers wondering how to sell their crops. *Nature* 401, 107.

Li, J., Carroll, J. & Ellar, D.J. (1991) Crystal structure of insecticidal δ-endotoxin from *Bacillus thuringiensis* at 2.5 Å resolution. *Nature* 353, 815–821.

Liu, Y.-B. & Tabashnik, B.E. (1997) Experimental evidence that refuges delay insect adaptation to *Bacillus thuringiensis*. *Proceedings of the Royal Society of London B* 264, 605–610.

Losey, J.E., Rayor, L.S. & Carter, M.E. (1999) Transgenic pollen harms monarch larvae. *Nature* 399, 214.

Mazier, M., Pannetier, C., Tourneur, J., Jouanin, L. & Giband, M. (1997) The expression of *Bacillus thuringiensis* toxin genes in plant cells. In:

Biotechnology Annual Reviews (ed. M.R. El-Gewely), Vol. 3, pp. 313–347. Elsevier, Amsterdam.

Metz, T.D., Roush, R.T., Tang, J.D., Shelton, A.M. & Earle, E.D. (1995) Transgenic broccoli expressing a *Bacillus thuringiensis* insecticidal crystal protein: implications for pest management strategies. *Molecular Breeding* 1, 309–317.

Morse, R.J., Powell, G., Ramalingam, V., Yamamoto, T. & Stroud, R.M. (1998) Crystal structure of Cry2Aa from *Bacillus thuringiensis* at 2.2 Angstroms: structural basis of dual specificity. In: *Proceedings of the VIIth International Colloquium on Invertebrate Pathology and Microbial Control, Sapporo, August 23–28, 1998*, pp. 1–2. Society of Invertebrate Pathology, Sapporo.

Munkvold, G.P., Hellmich, R.L. & Showers, W.B. (1997) Reduced *Fusarium* ear rot and symptomless infection in kernels of maize genetically engineered for European corn borer resistance. *Phytopathology* 87, 1071–1077.

Munkvold, G.P., Hellmich, R.L. & Rice, L.G. (1999) Comparison of fumonisin concentrations in kernels of transgenic *Bt* maize hybrids and nontransgenic hybrids. *Plant Disease* 83, 130–138.

Murray, E.E., Rocheleau, T., Eberle, M., Stock, C., Sekar, V. & Adang, M. (1991) Analysis of unstable RNA transcripts of insecticidal crystal protein genes of *Bacillus thuringiensis* in transgenic plants and electroporated protoplasts. *Plant Molecular Biology* 16, 1035–1050.

Nayak, P., Basu, D., Das, S. *et al.* (1997) Transgenic elite indica rice plants expressing Cry1Ac δ-endotoxin of *Bacillus thuringiensis* are resistant against yellow stem borer (*Scirpophaga incertulas*). *Proceedings of the National Academy of Sciences of the USA* 94, 2111–2116.

Oerke, E.C., Weber, A., Dehne, H.W. & Schönbeck, F. (1994) Conclusions and perspectives. In: *Crop Production and Crop Protection: estimated losses in major food and cash crops* (eds E.C. Oerke *et al.*), pp. 742–770. Elsevier, Amsterdam.

Ohme-Takagi, M., Taylor, C.B., Newman, T.C. & Green, P.J. (1993) The effect of sequences with high AU content on mRNA stability in tobacco. *Proceedings of the National Academy of Sciences of the USA* 90, 11811–11815.

Orr, D.B. & Landis, D.A. (1997) Oviposition of European corn borer (Lepidoptera: Pyralidae) and impact of natural enemy populations in transgenic versus isogenic corn. *Journal of Economic Entomology* 90, 905–909.

Peferoen, M., Jansens, S., Reynaerts, A. & Leemans, J. (1990) In: *Molecular and Cellular Biology of the Potato* (eds M.E. Vayda & W.C. Park), pp. 193–204. CAB International, Wallingford.

Perlak, F.J., Deaton, R.W., Armstrong, T.A. *et al.* (1990) Insect resistant cotton plants. *Bio/Technology* 8, 939–943.

Perlak, F.J., Fuchs, R.L., Dean, D.A., McPherson, S.L. & Fischhoff, D.A. (1991) Modification of the coding sequence enhances plant expression of insect control protein genes. *Proceedings of the National Academy of Sciences of the USA* 88, 3324–3328.

Perlak, F.J., Stone, T.B., Muskopf, Y.M. *et al.* (1993) Genetically improved potatoes: protection from damage by Colorado potato beetles. *Plant Molecular Biology* 22, 313–321.

Pilcher, C.D., Obrycki, J.J., Rice, M.E. & Lewis, L.C. (1997) Preimaginal development, survival and field abundance of insect predators on transgenic *Bacillus thuringiensis* corn. *Environmental Entomology* 26, 446–454.

Pimentel, D. (1991) Diversification of biological control strategies in agriculture. *Crop Protection* 10, 243–253.

Roush, R.T. (1993) Occurrence, genetics and management of insecticide resistance. *Parasitology Today* 9, 174–179.

Roush, R.T. (1997) *Bt*-transgenic crops: just another pretty insecticide or a chance for a new start in resistance management? *Pesticide Science* 51, 328–334.

Saxena, D., Flores, S. & Stotzky, G. (1999) Insecticidal toxin in root exudates from *Bt* corn. *Nature* 402, 480.

Schneph, E., Crickmore, N., Van Rie, J. *et al.* (1998) *Bacillus thuringiensis* and its pesticidal proteins. *Microbiology and Molecular Biology Reviews* 62, 775–806.

Sears, M.K., Stanley-Horn, D.E. & Mattila, H.R. (2000) http://www.cfia-acia.agr.ca/english/plaveg/pbo/btmone.shtml.

Shelton, A.M., Tang, J.D., Roush, R.T., Metz, T.D. & Earle, E.D. (2000) Field tests on managing resistance to *Bt*-engineered plants. *Nature Biotechnology* 18, 339–342.

Sims, S. & Holden, L. (1996) Insect bioassay for determining soil degradation of *Bacillus thuringiensis* subsp. *kurstaki* CryIA(b) protein in corn tissues. *Physiological and Chemical Ecology* 25, 659–664.

Sobek, E.A. & Munkvold, G.P. (1999) European corn borer (Lepidoptera: Pyralidae) larvae as vectors of *Fusarium moniliforme*, causing kernel rot and symptomless infection of maize kernels. *Journal of Economic Entomology* 92, 503–509.

Strizhov, N., Keller, M., Mathur, J. *et al.* (1996) A synthetic *cry1C* gene, encoding a *Bacillus thuringiensis* δ-endotoxin, confers *Spodoptera* resistance in alfalfa and tobacco. *Proceedings of* the *National Academy of Sciences of the USA* **93**, 15012–15017.

Tabashnik, B.E., Cushing, N.L., Finson, N. & Johnson, M.W. (1990) Field development of resistance to *Bacillus thuringiensis* in diamondback moth (Lepidoptera: Plutellidae). *Journal of Economic Entomology* **83**, 1671–1676.

Tabashnik, B.E., Malvar, T., Liu, Y.-B. *et al.* (1996) Cross-resistance of diamondback moth indicates altered interactions with domain II of *Bacillus thuringiensis* toxins. *Applied and Environmental Microbiology* **62**, 2839–2844.

Tabashnik, B.E., Liu, Y.-B., Malvar, T. *et al.* (1997) Global variation in the genetic and biochemical basis of diamondback moth resistance to *Bacillus thuringiensis*. *Proceedings of the National Academy of Sciences of the USA* **94**, 12780–12785.

Tang, J.D., Shelton, A.M., Van Rie, J. *et al.* (1996) Toxicity of *Bacillus thuringiensis* spore and crystal protein to resistant diamondback moth (*Plutella xylostella*). *Applied and Environmental Microbiology* **62**, 564–569.

Vaeck, M., Reynaerts, A., Höfte, H. *et al.* (1987) Transgenic plants protected from insect attack. *Nature* **327**, 33–37.

van Aarssen, R., Soetaert, P., Stam, M. *et al.* (1995) Cry1A(b) transcript formation in tobacco is inefficient. *Plant Molecular Biology* **28**, 513–524.

Van Hoof, A. & Green, P.J. (1997) Rare codons are not sufficient to destabilize a reporter gene transcript in tobacco. *Plant Molecular Biology* **35**, 383–387.

Van Rie, J. & Ferré, J. (2000) Insect resistance to *Bacillus thuringiensis* insecticidal crystal proteins. In: *Entomopathogenic Bacteria: from laboratory to field application* (eds J.-F. Charles *et al.*) pp. 219–236. Kluwer, Dordrecht.

Van Rie, J., McGaughey, W.H., Johnson, D.E., Barnett, B.D. & Van Mellaert, H. (1990) Mechanism of insect resistance to the microbial insecticide *Bacillus thuringiensis*. *Science* **247**, 72–74.

Warren, G.W. (1997) Vegetative insecticidal proteins: novel proteins for control of corn pests. In: *Advances in Insect Control: the role of transgenic plants* (eds N. Carozzi & M. Koziel), pp. 109–121. Taylor & Francis, London.

Warren, G.W., Carozzi, N.B., Desai, N. & Koziel, M.G. (1992) Field evaluation of transgenic tobacco containing a *Bacillus thuringiensis* insecticidal protein gene. *Journal of Economic Entomology* **5**, 1651–1659.

Wilson, F.D., Flint, H.M., Deaton, W.R. *et al.* (1992) Resistance of cotton lines containing a *Bacillus thuringiensis* toxin to pink bollworm (Lepidoptera: Gelechiidae) and other insects. *Journal of Economic Entomology* **85**, 1516–1521.

Wold, S.J., Burkness, E.C., Hutchison, W.D. & Venette, R.C. (2001) In-field monitoring of beneficial insect populations in transgenic sweet corn expressing a *Bacillus thuringiensis* toxin. *Journal of Entomological Sciences* **36**, 177–187.

Wünn, J., Klöti, A., Burkhardt, P.K. *et al.* (1996) Transgenic indica rice breeding line IR58 expressing a synthetic *cry1A(b)* gene from *Bacillus thuringiensis* provides effective insect pest control. *Bio/Technology* **14**, 171–176.

Chapter 13

Bacillus sphaericus and its Insecticidal Toxins

Fergus G. Priest

Introduction

Aerobic rod-shaped bacteria that differentiate into spherical endospores have traditionally been placed in the species *Bacillus sphaericus*. Prior to the discovery of mosquitocidal activity in some of these bacteria they had received scant attention, their major claim to fame being the presence of L-lysine or D-ornithine in the peptidoglycan of the cell wall rather than *meso*-diaminopimelic acid as is found in most other bacilli (Stackebrandt *et al.* 1987). Such chemotaxonomic markers are often indicative of phylogenetic divergence, and it was therefore reassuring that following 16S rRNA sequence analysis those organisms that produced spherical spores formed a distinctive line of descent (rRNA group 2; Ash *et al.* 1991). This phylogenetic independence is also reflected in the (for *Bacillus*) unusual physiology of these bacteria. *Bacillus sphaericus* strains lack many of the early enzymes of the Embden–Meyerhof pathway (Russell *et al.* 1989) and they do not catabolize sugars as carbon and energy sources. Instead, *B. sphaericus* and relatives use acetate, glutamate and other acids as the preferred sources of carbon, which are catabolized *via* the TCA cycle (White & Lotay 1980; Baumann *et al.* 1984; Alexander & Priest 1990). This has been used as the basis of at least two selective isolation media for *B. sphaericus* (Massie *et al.* 1985; Yousten *et al.* 1985) and it also explains why *B. sphaericus* and relatives do not grow anaerobically, not even in the presence of nitrate as electron acceptor.

The strictly aerobic lifestyle of *B. sphaericus* introduced problems for traditional taxonomists because these bacteria are negative for the various physiological tests based on carbohydrate metabolism (Gordon *et al.* 1973). As a result, all spherical spore-forming strains tended to be 'lumped' into the one species, and it was only when physiological tests appropriate to the metabolism of the bacteria were introduced in an extensive numerical study of *B. sphaericus sensu lato* that distinct taxa of spherical spore-forming bacteria could be recognized (Alexander & Priest 1990).

Insect pathogenicity in aerobic, endospore-forming bacteria

Insect pathogenicity is a fairly common feature among aerobic, endospore-forming bacteria (Priest & Dewar 2000) (table 13.1). The most widely studied

Table 13.1 Insect pathogenicity among aerobic, endospore-forming bacteria.

Bacterium	Taxonomic position	Typical insect host	Relationship with insect/disease
Bacillus thuringiensis	rRNA group 1[a]	Coleoptera, Diptera and Lepidoptera	Opportunistic pathogens, the crystal protein is the primary virulence determinant
B. sphaericus	rRNA group 2	*Anopheles* and *Culex* mosquito larvae	Opportunistic pathogen, death of host by toxicity and bacterial growth
Brevibacillus laterosporus	rRNA group 4	*Aedes, Anopheles, Culex* and *Simulium* larvae	Opportunistic pathogen, death of host by ingestion of toxin
Paenibacillus alvei	rRNA group 3	*Aedes, Anopheles, Culex* and *Simulium* larvae	Unknown
P. larvae ssp. *larvae*	rRNA group 3	Honey bee larvae	Obligate[b] pathogen, septicaemic, American foulbrood
P. lentimorbus and *P. popilliae*	rRNA group 3	Scarabaeid beetle larvae	Obligate pathogens, septicaemic, 'milky disease'

[a] rRNA groups based on the work of Ash *et al.* (1991).
[b] Obligate refers to the necessity for *in vivo* growth in the insect for completion of endospore formation.

entomopathogen is *B. thuringiensis*, characterized by synthesis of a parasporal crystal protein that is largely responsible for the toxicity to the insect (see Chapter 11). *Bacillus sphaericus, Brevibacillus laterosporus* and *Paenibacillus popilliae* strains also synthesize parasporal crystal proteins. The toxic proteins of *B. sphaericus* share no homology with those of *B. thuringiensis* (Baumann *et al.* 1991; Porter *et al.* 1993; Charles *et al.* 1996) but the crystal (Cry) protein of *P. popilliae* is closely related (40% sequence identity) to the Cry2 polypeptides of *B. thuringiensis* (Zhang *et al.* 1997) indicating that they have a common evolutionary background. Because of the separate evolutionary origins of the *B. thuringiensis* and *P. popilliae* host bacteria (figure 13.1), the *cry* gene of *P. popilliae* was presumably obtained by lateral gene transfer across the generic boundary. Nothing is known of the primary structure of the crystal proteins of *Br. laterosporus* (Favret & Yousten 1985; Orlova *et al.* 1998).

The insect pathogenic paenibacilli tend to be obligate pathogens (require growth in the larva to sporulate and thus complete the life cycle) and have fastidious nutritional requirements (Bulla *et al.* 1978). An exception is *P. alvei*, of which some strains have been reported to be pathogenic to mosquito and blackfly larvae (Singer 1996). The basis of this virulence is unknown. The aerobic endospore-forming bacteria are therefore a rich source of insect pathogens representing both toxigenic and septicaemic types of pathogenicity. Given the phylogenetic diversity of both the bacteria (figure 13.1) and their toxins, it seems likely that entomopathogenicity emerged on several occasions during the evolution of these organisms. Several of these bacteria, notably *B. thuringiensis*, *B. sphaericus* and *P. popilliae*, have been exploited for biological control purposes.

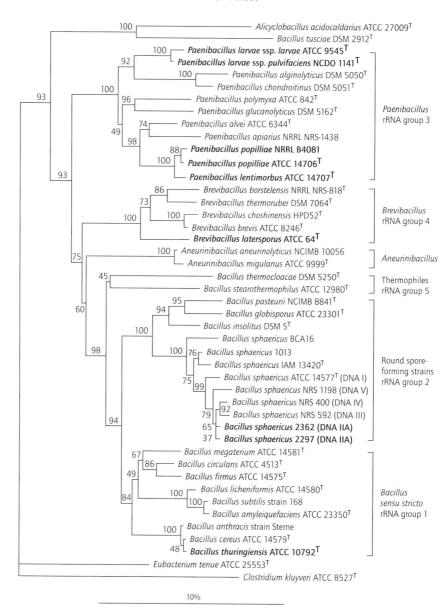

Figure 13.1 An evolutionary tree of insect pathogenic, endospore-forming bacteria derived from distance data calculated from a nonedited alignment of 16S rRNA gene sequences from insect pathogens and reference strains by using the neighbour-joining method. Insect pathogens are shown in bold (see text for detail). *Eubacterium tenue* and *Clostridium kluyveri* served as outgroups. The bootstrap values are given as the percentage times out of 500 replicates that a species or a strain to the right of the actual node occurred. The scale bar indicates nucleotide substitutions per site (adapted from Priest & Dewar 2000[T], type strain).

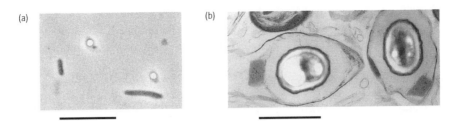

Figure 13.2 (a) Phase-contrast micrograph of *Bacillus sphaericus* strain 9002 and (b) electron micrograph of *B. sphaericus* 2362 showing insecticidal crystal proteins as dark spots associated with phase bright spores and in thin section, respectively. Bars indicate (a) 10 μm and (b) 1 μm.

Bacillus sphaericus

Mesophilic bacilli that differentiate into spherical spores are classified in *B. sphaericus*. The division of this species into six taxa according to DNA homology groups (Krych *et al.* 1980; Rippere *et al.* 1997) has now been supported by numerous taxonomic studies including 16S rRNA sequence analysis (Aquino de Muro & Priest 1993), ribotyping (Aquino de Muro *et al.* 1992; Miteva *et al.* 1998), multilocus enzyme electrophoresis (MLEE) (Zahner & Momen 1995) and pulsed field gel electrophoresis (PFGE) of chromosomal DNA restriction enzyme digests (Zahner *et al.* 1998). There is no doubt that these groups represent taxa of species rank but, despite numerous attempts, it has been impossible to find reliable diagnostic features to distinguish the molecularly defined groups. As a consequence, species status has only been afforded to the strains of DNA homology group IIB as *B. fusiformis* (Priest *et al.* 1988), and in the present chapter I shall refer to the taxa by the original DNA homology group nomenclature (Krych *et al.* 1980). Of the six DNA homology groups, mosquito pathogenicity is associated only with strains of group IIA (Rippere *et al.* 1997).

Bacillus sphaericus mosquito pathogens

The first isolation of a mosquitocidal strain of *B. sphaericus* (strain K) was achieved by Kellen *et al.* (1965) from moribund larvae of *Culiseta incidens* in California. The activity of these strains was so low that they could not be considered for vector control (Kellen *et al.* 1965) but other strains followed, and the isolation of strains 1593 in Indonesia, 2362 in Nigeria and 2297 in Sri Lanka opened up the possibilities of bacterial mosquito control because of their high toxicity (reviewed by Baumann *et al.* 1991; Charles *et al.* 1996). These highly toxic isolates all had a similar feature, the synthesis of a crystalline parasporal toxin, that was absent in the earlier, less toxic strains. The crystal is smaller than that of *B. thuringiensis*, but generally visible under a good phase-contrast microscope (figure 13.2a). Over the following 30 years numerous screening programmes resulted in the isolation of mosquitocidal *B. sphaericus* strains from dead insects, soil, mud and water from all around the globe and over 560 strains are now held by the International Entomophogenic *Bacillus* Centre at the Institut Pasteur, Paris. Few, if any, of these are more toxic than the original strains 1593, 2362 and 2297.

Bacillus sphaericus affects the larval stages of the mosquito life cycle. Larvae are filter feeders and ingest the spore–crystal complex as part of their natural diet. *Bacillus sphaericus* has no effect on adults. Mosquitoes vary in their susceptibility, and it is not possible to generalize within genera. *Culex quinquefasciatus* is highly susceptible to *B. sphaericus* (LC_{50} = *c.* 50–100 ng toxin protein/mL; Davidson 1989) which has resulted in it being used to combat transmission of filariasis and Japanese encephalitis by culicine mosquitoes (Regis *et al.* 1995; Yadav *et al.* 1997). Several anopheline species (vectors of malaria) are intermediate in their susceptibility (LC_{50} = 360–5000 ng/mL; Davidson 1989) including *Anopheles albimanus* and *A. quadrimaculatus* (Lacey & Singer 1982). Consequently, *B. sphaericus* has been considered for prevention of malaria transmission in several countries (Das & Amalraj 1997; Barbazan *et al.* 1998). *Aedes aegypti* (LC_{50} = 42 000 ng/mL; Davidson 1989) is notoriously insensitive to *B. sphaericus* but other species of *Aedes* such as *Ae. atropalpus* and *Ae. nigromaculis* are reasonably susceptible (Berry *et al.* 1993). *Bacillus sphaericus* has no toxic effect against blackflies, has little or no effect on nontarget insects, and no effect on mammals (Shadduck *et al.* 1980; Mulla *et al.* 1984; Yousten *et al.* 1995).

Upon ingestion of the spore–toxin complex by a susceptible larva, signs of intoxication appear rapidly. Ultrastructural studies show swelling of mitochondria followed by the appearance of large vacuoles in cells of the gastric caecum and posterior midgut (Davidson 1981). The midgut cells are disrupted and peristalsis ceases. The larvae are dead within 6 h (up to 48 h with low doses) and the spores germinate within the cadaver resulting in the release of some 10^5 fresh spores per larva in due course.

Virulence factors of *B. sphaericus*

The binary toxin

The major toxin of *B. sphaericus* comprises the proteins making up the parasporal crystal that accumulates within the exosporium starting around stage 3 (engulfment) of sporulation (figure 13.2b). The crystal is composed of two polypeptides with molecular weights of 42 kDa (BinA) and 51 kDa (BinB) that together form the so-called binary (Bin) toxin. All high toxicity strains contain the Bin toxin. Upon ingestion by the mosquito larva, the crystal proteins dissolve in the high pH of the insect midgut. BinA is slowly converted into a smaller protein of around 39 kDa by insect proteinases and BinB is more rapidly processed to an active form of around 43 kDa (Broadwell & Baumann 1987; Aly *et al.* 1989). Both proteins are required for optimal toxicity. Early studies were confused on this point because of the difficulties of separating the two proteins biochemically, but once individual proteins were prepared from cloned genes in *Escherichia coli*, it became apparent that BinAB comprised a true binary toxin similar in context to the diphtheria and cholera toxins (see Baumann *et al.* 1991; Priest 1992 for reviews). However, BinA alone will lyse cultured cells of *C. quinquefasciatus* (Baumann & Baumann 1991) and in large excess this protein will kill *C. quinquefasciatus* larvae (Nicolas *et al.* 1993), but the BinB protein alone, even in large excess, is not toxic.

Binary toxin complex binds to midgut cells of susceptible larvae through the BinB component allowing internalization of BinA (Davidson 1988). There is a strong and regionalized binding of the complex and BinB exhibits the same regional binding as whole binary toxin, whereas BinA shows only weak binding throughout the gut (Oei *et al.* 1992). However, in the presence of BinB, BinA becomes strongly regionalized indicating that BinB effects the specific binding. The biochemistry of the toxic action of BinA is unknown, but may involve pore formation in the midgut cells because *in vitro* assays in planar lipid bilayers and permeabilization measurements in liposomes have confirmed channel formation by BinA (C. Berry, pers. comm.). It must be emphasized that the *B. sphaericus* Bin proteins have no homology with the pore-forming toxins of *B. thuringiensis*.

BinB binds to a specific receptor in *C. quinquefasciatus* larvae (Nielsen-LeRoux & Charles 1992) and susceptible *Anopheles* larvae (Silva-Filha *et al.* 1997), but not to the resistant larvae of *Ae. aegypti*. This, and other evidence, shows that the target range is determined by the availability of receptor sites rather than by proteolytic processing in the gut. The receptor in *Culex pipiens* has now been identified as an alpha-glucosidase of around 60 kDa that is localized to the membrane of the cells by a glycosyl-phosphatidylinositol anchor (Silva-Filha *et al.* 1999). In *Anopheles gambiae*, the binding protein has not been fully identified, but the candidate site is also attached to the cell surface by a glycolipid anchor.

Variation in *bin* genes

The crystal protein genes of *B. sphaericus* are very highly conserved, especially when compared to the diversity of their *B. thuringiensis* counterparts. Toxins affecting several orders of insects and other arthropods have been found in *B. thuringiensis* strains and represent marked heterogeneity of *cry* genes. Despite intensive screening programmes in search of diversity of insect pathogenicity in *B. sphaericus*, toxicity appears to be restricted to a few mosquito genera and does not extend into the blackflies or other biting flies.

Only four variants of the *bin* genes have been detected in DNA homology group IIA strains of *B. sphaericus*, and in all cases variations in the *A* gene are matched by changes in the *B* gene. In other words, *binA1* is always associated with *binB1* and *binA2* with *binB2*, etc. We have yet to discover natural examples of mixing of the *binA* and *binB* genes (see table 13.2). The four known *bin* genes differ little (Priest *et al.* 1997; Humphreys & Berry 1998). The *binA1* gene differs from *binA2* in only two bases resulting in the glutamate residue at position 104 of BinA1 being replaced by a serine residue in BinA2. In the *binB* genes there are further differences, resulting in five amino acid changes between BinB1 and BinB2 (Humphreys & Berry 1998). Similarly, BinA3 and BinA4 are highly related, differing by only two amino acids at positions 93 and 99. These proteins both have

Table 13.2 Some representative strains of DNA homology group IIA *Bacillus sphaericus* and their characteristics.

SRP[a]	Strain	Origin	Serotype	bin[c]	mtx1	mtx2	mtx3
1	K	USA	1a	−	+	n.d.	+
1	Q	USA	1a	−	+	+	+
2	9002	India	1a	A1B1	+	n.d.	n.d.
3	SSII-1	India	2a2b	−	+	+	+
3	1883	Israel	2a2b	−	+	n.d.	n.d.
5	LP24-4	Singapore	2a2b	−	−	n.d.	n.d.
5	LP35-6	Singapore	2a2b	−	−	n.d.	n.d.
9	BDG2	France	3	−	−	n.d.	n.d.
10	IAB 881	Ghana	3	A1B1	−	n.d.	n.d.
11	LP1-G	Singapore	3	A4B4	−	n.d.	n.d.
11	LP7-A	Singapore	3	A4B4	−	n.d.	n.d.
12	1593	India	5a5b	A2B2	+	+	+
12	2362	Nigeria	5a5b	A2B2	+	+	+
12	BSE 18	Scotland	5a5b	A2B2	+	+	n.d.
13	IAB 59	Ghana	6	A1B1	+	+	+
13	IAB 481	Ghana	6	+	+	n.d.	n.d.
13	IAB 774	Ghana	6	+	+	n.d.	n.d.
15	R-1e	Brazil	6	−	−	n.d.	n.d.
15	Gt-1a	Brazil	6	−	−	n.d.	n.d.
17	2297	Sri Lanka	25	A3B3	+	+	+
17	M2-1	Malaysia	25	+	+	n.d.	n.d.
19	2377	India	26a26b	−	−	n.d.	n.d.
20	2315	Thailand	26a26b	−	−	−	n.d.
21	IMR 66.1S	Malaysia	48	−	−	n.d.	n.d.
21	Pr-1	Scotland	n.d.	A1B1	+	n.d.	n.d.
21	IAB 872	Ghana	48	A1B1	+	n.d.	n.d.

[a] SRP assignments based on PFGE of *Sma*I-digested chromosomal DNA (Zahner *et al.* 1998), see text for details.
[b] Data for toxin gene distribution taken from Thanabalu and Porter (1996), Liu *et al.* (1996) and Zahner *et al.* (1998).
[c] *bin* gene designations given where known, otherwise presence/absence indicated by + or −. n.d., not determined.

the ser-104 characteristic of BinA2. Moreover, the *binB3* and *binB4* genes differ in only four bases and the proteins in two amino acid residues. However, the greater differences between the two pairs of genes suggest two lineages, one comprising *binA1B1* and *binA2B2* and the other *binA3B3* and *binA4B4* (Zahner *et al.* 1998).

Differences in toxicity have been detected among the Bin variants. The BinA4B4 proteins from strain LP1-G confer on this strain a lower toxicity against *C. quinqefasciatus* than the other variants (Liu *et al.* 1993). The serine residue at position 93 of BinA4 is the only unique site in this toxin combination so it may be responsible for this lack of toxicity. The *bin* genes were expressed separately in *E. coli* by Berry *et al.* (1993) and examined for toxicity against three mosquito species, *C. quinqefasciatus*, *Ae. atropalpus* and *Ae. aegypti*. Bin2 toxins were the most active against *Ae. aegypti*, Bin3 toxins were weakly active and Bin1 showed no toxicity against these larvae. All three were equally effective against *C. quinquefasciatus*. Site-directed mutagenesis revealed that alterations at residues 99 and 104 of BinA were sufficient to account for these changes in toxicity (Berry *et al.* 1993). This is close to the unique ser-93 of BinA4 and together probably identifies the active site of the BinA toxin in the region 93 to 104.

Mosquitocidal (Mtx) toxins

The early mosquitocidal strains of *B. sphaericus*, such as strain SSII-1, were toxic despite the lack of a parasporal crystal. Preparation of a gene library of this strain in *E. coli* led Thanabalu *et al.* (1991) to the identification of the mosquitocidal toxin or *mtx1* gene. The Mtx1 protein of around 100 kDa accumulates in vegetative cells and, when ingested by larvae, is cleaved into two subunits of 27 and 70 kDa which resemble an ADP ribosylating toxin and a glycoprotein-binding protein, respectively (Thanabalu *et al.* 1993; Hazes & Read 1995). Mxt1 is highly toxic to mosquito larvae ($LC_{50} = 15$ ng/mL), including the usually recalcitrant *Ae. aegypti* (Thanabalu *et al.* 1992), but its contribution to the overall toxicity of the bacterium is low because of low gene expression (Ahmed *et al.* 1995) and proteolytic degradation in the vegetative cell (Wati *et al.* 1997). With enhanced gene expression in a reduced proteolytic background this could be a valuable toxin for biocontrol purposes (Thanabalu & Porter 1995).

Genes for other mosquitocidal toxins were cloned from strain SSII-1, namely *mtx2* (Liu *et al.* 1996) and *mtx3* (Thanabalu & Porter 1996). The two proteins of around 32–36 kDa are related to each other and to the epsilon toxin of *Clostridium perfringens* and the cytotoxin of *Pseudomonas aeruginosa*, both of which act by pore formation in susceptible cells. It is likely that other mosquitocidal toxins are present in *B. sphaericus* DNA homology group IIA strains because several weakly pathogenic strains, such as some strains of serotypes 2a2b, 6 and 26a26b, lack all four known toxins and yet are weakly toxic.

Strain typing and the clonal population structure of *Bacillus sphaericus* DNA homology group IIA

In any pathogenic bacterium, some means of being able to identify virulent strains is necessary for epidemiological purposes. Flagellar (H) serotyping was developed for *B. sphaericus* DNA homology group IIA prior to the allocation of nonpathogenic and pathogenic strains to different DNA homology groups. Thus, the serotyping scheme does not have consecutive numbering and the various strains of DNA group IIA are distributed in nine, nonconsecutive serotypes (table 13.2; de Barjac *et al.* 1985). The serotyping scheme was supplemented with a phage typing system by Yousten (1984) that largely corroborated the sero-groupings. Most high toxicity (Bin$^+$) strains are members of serotype 5a5b, but high toxicity strains have also been assigned to serotypes 1a, 3, 6 and 48 (table 13.2). As the number of isolates grew, it became apparent that serotype and phage type did not always concur with pathogenicity (table 13.2; for reviews see

Priest *et al.* 1994; Charles *et al.* 1996) and that some form of genomic typing may be more effective than traditional methods for intraspecific classification. Furthermore, genomic typing methods give some insight into the genetic structure of bacterial populations.

A key concept in bacterial population genetics is that of the clone. If bacteria evolve from a common ancestor in the absence of significant interstrain DNA exchange they will have highly similar genomic structures. Mutations will occur and be passed on to progeny. A mutation that allows the bacterium to exploit a particular environmental niche will allow the bacterium to outcompete other members of the species, resulting in spread of that lineage or clone. Unsuccessful clones become extinct as the successful ones become widely disseminated. This purging of diversity is known as periodic selection and results in a distinctive clonal structure to the population (Maynard Smith 1995). Clonal populations are characterized by: (i) allele linkage disequilibrium (i.e. there will be distinctive associations of alleles when analysed by a method such as MLEE); (ii) low diversity, or relatively few different genomic types within the species; and (iii) a tree-like phylogeny if the clones are classified using gene sequences. Members of clones become widely disseminated in space (particularly from endospore-forming bacteria) and over evolutionary time and can be recognized by some form of strain typing, originally MLEE (Selander *et al.* 1986) but more recently genomic typing methods such as random amplified polymorphic DNA (RAPD), PFGE of restriction enzyme-digested chromosomal DNA (Eisenstein 1990) or multilocus sequence typing (Selander *et al.* 1994; Maiden *et al.* 1998).

Bacterial species which partake in frequent DNA exchange, such as the transformable bacterium *B. subtilis*, do not evolve as clones because the genomes of clonal lines are mixed by recombination on a regular basis (Istock *et al.* 1992). The chromosomes of such species become mosaics of DNA from various sources and clonal lineages do not have the opportunity to develop in isolation. Such species are characterized by: (i) alleles in linkage equilibrium (i.e. there is a random assortment of alleles in MLEE); (ii) increased diversity as two strains are rarely identical; and (iii) a net-like phylogeny of individual strains. It is not possible to reconstruct the phylogeny of a nonclonal population using a bifurcating tree because of the extensive lateral gene transfer. Some species may have a nonclonal structure but, through the emergence of a particularly effective lineage, a clonal sub-population may emerge.

PFGE analysis of *B. sphaericus* DNA homology group IIA strains revealed a clonal structure to this taxon (Zahner *et al.* 1998) (table 13.2). The PFGE types, referred to as SRPs (*Sma*I restriction profiles), that emerged were consistent with serotypes in that on no occasion were two identical SRPs found in strains of different serotypes. However, the converse was not true, and in several instances serotypes contained strains allocated to different SRPs. Thus, the PFGE was providing a more discriminatory classification than serotyping. There is a much stronger correlation between SRP and toxicity than serotype and toxicity. For example, serotype 3 encompasses strains that are weakly toxic (no *bin* or *mtx1* genes) and those that are highly toxic (Bin⁺). These strains were allocated to different SRPs, and similar situations occurred in serotypes 1a, 2a2b and 6 (table 13.2).

Serotype 48 was inconsistent because three strains studied were assigned to the same SRP but had differences in toxicity; a Malaysian strain (IMR 66.1S) lacking *bin* and *mtx1* genes and two other strains from Ghana and Scotland with high toxicity and both genes. This suggests loss of the two toxin genes from strain IMR 66.1S or acquisition in the other lineage; whichever event occurred it must have happened recently, because there has been insufficient time for accumulation of mutations separating the genomic backgrounds and giving rise to two SRPs. Isolation and typing of more SRP 21 strains would perhaps clarify the evolution of this lineage. Just as *bin* and *mtx* genes are lost and/or acquired, so in some strains *bin* genes are duplicated. This has been observed in strains IAB 881 (SRP 10), LP1-G (SRP 11), 2297 (SRP 17) and CHE 5.1S (SRP 18). It is not clear how frequently this trait occurs, or whether it is consistent within an SRP, because bands in simple Southern blots superimpose, and recognition of the duplication requires blots of PFGE gels (Zahner *et al.* 1998).

Some comments about the evolution of entomopathogenicity in *Bacillus sphaericus*

Recognition of the clonal structure of *B. sphaericus* DNA homology group IIA provides some fascinating insight into the evolution of these bacteria. SRP 12 strains (serotype 5a5b) comprise an enormously successful lineage. Supremacy of SRP 12 is apparent from scrutiny of the holdings of the Collection of *Bacillus thuringiensis* and *Bacillus sphaericus* held at the Institut Pasteur. As the reference centre for entomopathogenic bacilli, the content of this collection reflects the frequency of isolation of mosquitocidal *B. sphaericus* strains. The 232 serotype 5a5b strains deposited from all parts of the world far outnumber the next most commonly encountered group, serotype 6, of which there are only 32 DNA homology group IIA strains, largely isolated from Ghana. Other serotypes are rarer – 30 strains of serotype 25, 29 of serotype 2a2b and fewer still of the remaining serotypes; for example only three of serotype 48 and two of serotype 9a9c. Large numbers of serotype 5a5b strains have been examined by PFGE (Zahner *et al.* 1998) and MLEE (Zahner *et al.* 1994; Zahner & Momen 1995) and all are of the high toxicity (Bin⁺) phenotype and are genomically identical (SRP 12). Where *bin* genes have been sequenced from SRP 12 strains (table 13.2), they are identical (*binA2B2*) over the complete coding sequence. Moreover, these type 2 *bin* genes are only found in this SRP (table 13.2). From wherever in the world they are isolated, it appears that SRP 12 strains are essentially identical and represent an enormously successful clonal lineage. It is likely that the emergence of SRP 12 strains results from sexual isolation giving rise to limited chromosomal DNA exchange rather than reduced contact between strains, despite the fact that these bacteria are widely disseminated in the environment and in insects. SRP 12 strains possess a highly efficient restriction and modification system based on an isoschizomer of *Hae*III while strains of most other SRPs have weak restriction systems or none at all (Zahner & Priest 1997). This will hinder gene transfer into SRP 12 strains leading to their genetic isolation. At some point, the unique combination of *binA2B2* in an SRP 12 background gave rise to an environmentally successful *B. sphaericus* strain that has become the predominant component of this taxon.

Whereas *bin2* genes are restricted to SRP 12 strains, identical *binA1B1* genes are found in various genetic backgrounds including an SRP 2 (serotype 1a) strain from India, SRP 10 (serotype 3) and SRP 13 (serotype 6) strains from Ghana and two SRP 21 (sereotype 48) strains from Ghana and Scotland. The only way that identical genes could occur in such genomically different hosts is by lateral gene transfer, an occurrence that is now recognized to be reasonably common, particularly for traits such as pathogenicity determinants (de la Cruz & Davies 2000). In many bacteria, particularly Gram-negative pathogens, genes for virulence determinants and pathogenicity are located on sections of the genome referred to as pathogenicity islands (Hacker *et al.* 1997). These large sections of the chromosome (often around 40 kb in Gram-negative bacteria but smaller in Gram-positives) often have an atypical G+C content, disclosing their foreign origin. They are, or were, weakly mobile, and in many cases resemble defective phages or the remnants of plasmids or compound transposons. The location of *bin* genes and perhaps the *mtx* genes on a pathogenicity island could explain the lateral transfer of these genes, the duplication of *bin* genes in some strains and their loss from others.

Sequence analysis of the DNA surrounding the *bin* genes of strain 2297 (GenBank accession number AJ224478) revealed an open reading frame (ORF) with significant similarity to numerous transposases downstream of the *bin* genes (C. Berry, pers. comm.). There are several frameshift mutations in the potential ORF that would render the transposase nonfunctional. In strain 1593 (GenBank accession number AJ224477) there is an insertion of 1554 bp in the putative transposase but, following the insertion, the DNA has identical sequence to strain 2297. This insertion in strain 1593 contains the *Hind*III site that results in the *binA2B2* genes, and presumably also the *binA1B1* genes of other strains, being located on a 3.5-kb *Hind*III fragment (Berry *et al.* 1989) compared to the 4.8-kb fragment of strain 2297 (*binA3B3*) and strain LP1-G (*binA4B4*) (Liu *et al.* 1993). This variation in the distribution of the insertion between the two lineages of strains supports the sequence data for the divergent evolution of the two toxin types mentioned in 'Variation in *bin* genes' above. Interestingly, *mtx2* also appears to be flanked by a disrupted transposase (C. Berry, pers. comm.). The origins of these genes and their relationships will hopefully be revealed by further chromosome sequencing and the preparation of physical genomic maps.

Ecological aspects of *Bacillus sphaericus* and the role of the crystal

All screening programmes have focused on isolating highly toxic strains of *B. sphaericus* and then characterizing them, with the result that DNA homology group IIA strains were traditionally associated with entomopathogenicity. However, when the opposite approach was used, i.e. strains of *B. sphaericus* DNA homology group IIA were isolated and then assayed for toxicity, a very different picture emerged. Of 20 isolates from tropical soils, all assigned to group

IIA by molecular typing, none were toxic to mosquito larvae and they all lacked the known toxin genes (Jahnz *et al.* 1996). This suggests that entomopathogenicity is not an essential feature of group IIA strains and that most strains of this species have disposed of the burden of crystal protein synthesis or, indeed, never gained it. And yet there must be an advantage to crystal protein production during sporulation, despite the energy expenditure, because strains of several SRPs, especially SRP 12, are invariably highly toxic. An intriguing possibility was raised by Correa and Yousten (1995) who showed that strains producing the crystal protein tend to germinate more effectively in larval guts than less toxic strains. A combination of effective larval killing and efficient germination will provide the bacterium with a selective advantage as it multiplies in the nutrient-rich cadaver before sporulation and return to the aqueous environment. In this context, it is worth emphasizing that Bin⁺ strains of *B. sphaericus* produce the crystal within the exosporium of the spore (Nicolas *et al.* 1994; figure 13.2b) and thus the larvae, when ingesting the spore, receive a deadly package of spore and toxin. The larva is killed by the Bin toxins and, as the spores germinate and the cells multiply, also by the vegetative Mtx toxin(s). As the bacteria feast on the dead larva, multiplication and dissemination of the organism is encouraged.

This recycling of the bacterium in the dead larvae is an attractive feature of the use of *B. sphaericus* for control of *Anopheles* and *Culex* mosquito populations as it tends to provide for effective killing over prolonged periods from 2 to 9 weeks depending on the nature of the breeding site (Karch *et al.* 1992; Ansari *et al.* 1995; Mulla *et al.* 1999). Treatment is not always effective (Barbazan *et al.* 1997) and new formulations that enhance persistence, increase toxicity and, perhaps, the development of genetically manipulated hosts, are needed to encourage the use of *B. sphaericus* as a mosquito control agent (Porter *et al.* 1993; Federici 1995).

A taxon without a name

Group IIA strains of *B. sphaericus* are unique among the round spore-forming bacteria in their mosquitocidal properties. In this article I have focused on the systematics of the bacteria and their mosquitocidal toxins and yet it will be apparent that I have largely avoided the question of whether these bacteria should be allocated species status. This group of bacteria is well defined in the molecular sense, but the lack of defining phenotypic properties has discouraged some of the most fervent proponents of this bacterium – including those who did much of the formative work on the biology of the organism, in particular Elizabeth Davidson, Samuel Singer and Allan Yousten – from establishing it as a new species. This has been compounded by the fact that nontoxic strains of group IIA seem to be more common in the environment than the toxic varieties. Nevertheless, *only* strains of DNA homology group IIA have the potential to be pathogenic and there are several examples of bacterial species containing both pathogenic and less, or nonpathogenic, variants, *E. coli* for example. Moreover, there are several instances of species that can only be distinguished from close relatives by genetic/molecular means and in other respects are phenotypically homogeneous; *B. mojavensis* and

B. subtilis for example (Roberts *et al.* 1994). While this may not be ideal, we should not be deterred from recognizing a distinct species that has such strong molecular identity simply because our crude phenotypic tests do not reveal distinguishing features. I therefore believe that we should follow our convictions and give these bacteria species status. Distinguishing phenotypic features may be found in due course, but if not, the popularity of molecular identification is bound to gain prominence and the lack of phenotypic diagnostics will soon become just a minor inconvenience.

Following brief discussions with E.W. Davidson and A.A. Yousten, we suggest '*Bacillus culicivorans*' for this fascinating entomopathogen with an appetite for *Culex* larvae.

Acknowledgements

I am most grateful to Colin Berry and Jean-François Charles for providing me with unpublished material and Allan Yousten for critically reading the manuscript and, with Elizabeth Davidson, arriving at the name '*Bacillus culicivorans*'.

References

Ahmed, H.K., Mitchell, W.J. & Priest, F.G. (1995) Regulation of mosquitocidal toxin synthesis in *Bacillus sphaericus*. *Applied Microbiology and Biotechnology* 43, 310–314.

Alexander, B. & Priest, F.G. (1990) Numerical classification and identification of *Bacillus sphaericus* including some strains pathogenic for mosquito larvae. *Journal of General Microbiology* 136, 367–376.

Aly, C., Mulla, M.S. & Federici, B.A. (1989) Ingestion, dissolution, and proteolysis of the *Bacillus sphaericus* toxin by mosquito larvae. *Journal of Invertebrate Pathology* 53, 12–20.

Ansari, M.A., Sharma, V.P., Mittal, P.K. & Razdan, R.K. (1995) Efficacy of two flowable formulations of *Bacillus sphaericus* against larvae of mosquitoes. *Indian Journal of Malariology* 32, 76–84.

Aquino de Muro, M. & Priest, F.G. (1993) Phylogenetic analysis of *Bacillus sphaericus* and development of an oligonucleotide probe specific for mosquito pathogenic strains. *FEMS Microbiology Letters* 112, 205–210.

Aquino de Muro, M., Mitchell, W.J. & Priest, F.G. (1992) Differentiation of mosquito pathogenic strains of *Bacillus sphaericus* from nontoxic varieties by ribosomal RNA gene restriction patterns. *Journal of General Microbiology* 138, 1159–1166.

Ash, C., Farrow, J.A., Wallbanks, S. & Collins, M.D. (1991) Phylogenetic heterogeneity of the genus

Bacillus revealed by comparative analysis of small subunit ribosomal RNA sequences. *Letters in Applied Microbiology* 13, 202–206.

Barbazan, P., Baldet, T., Darriet, F. *et al.* (1997) Control of *Culex quinquefasciatus* (Diptera: Culicidae) with *Bacillus sphaericus* in Maroua, Cameroon. *Journal of the American Mosquito Control Association* 13, 263–269.

Barbazan, P., Baldet, T., Darriet, F. *et al.* (1998) Impact of treatments with *Bacillus sphaericus* on *Anopheles* populations and the transmission of malaria in Maroua, a large city in a savannah region of Cameroon. *Journal of the American Mosquito Control Association* 14, 33–39.

Baumann, L. & Baumann, P. (1991) Effects of components of the *Bacillus sphaericus* toxin on mosquito larvae and mosquito-derived tissue culture-grown cells. *Current Microbiology* 23, 51–57.

Baumann, L., Okamoto, K., Unterman, B., Lynch, M. & Baumann, P. (1984) Phenotypic characterization of *Bacillus thuringiensis* and *Bacillus cereus*. *Journal of Invertebrate Pathology* 44, 329–341.

Baumann, P., Clark, M.A., Baumann, L. & Broadwell, A.H. (1991). *Bacillus sphaericus* as a mosquito pathogen – properties of the organism and its toxins. *Microbiological Reviews* 55, 425–436.

Berry, C., Jackson-Yap, J., Oei, C. & Hindley, J. (1989) Nucleotide sequence of two toxin genes from *Bacillus sphaericus* IAB59 – sequence comparisons between five highly toxinogenic strains. *Nucleic Acids Research* 17, 7516–7516.

Berry, C., Hindley, J., Ehrhardt, A.F., Grounds, T., Desouza, I. & Davidson, E.W. (1993) Genetic determinants of host ranges of *Bacillus sphaericus* mosquito larvicidal toxins. *Journal of Bacteriology* 175, 510–518.

Broadwell, A.H. & Baumann, P. (1987) Proteolysis in the gut of mosquito larvae results in further activation of the *Bacillus sphaericus* toxin. *Applied and Environmental Microbiology* 53, 1333–1337.

Bulla, L.A., Costilow, R.N. & Sharpe, E.S. (1978) Biology of *Bacillus popilliae*. *Advances in Applied Microbiology* 23, 1–18.

Charles, J.F., Nielsen-LeRoux, C. & Delécluse, A. (1996) *Bacillus sphaericus* toxins, molecular biology and mode of action. *Annual Review of Entomology* 41, 451–472.

Correa, M. & Yousten, A.A. (1995) *Bacillus sphaericus* spore germination and recycling in mosquito larval cadavers. *Journal of Invertebrate Pathology* 66, 76–81.

Das, P.K. & Amalraj, D.D. (1997) Biological control of malaria vectors. *Indian Journal of Medical Research* 106, 174–97.

Davidson, E.W. (1981) A review of the pathology of bacilli infecting mosquitoes, including an ultrastructural study of larvae fed *Bacillus sphaericus* 1593 spores. *Developments in Industrial Microbiology* 22, 69–81.

Davidson, E.W. (1988) Binding of the *Bacillus sphaericus* (Eubacteriales, Bacillaceae) toxin to midgut cells of mosquito (Diptera, culicidae) larvae: relationship to host range. *Journal of Medical Entomology* 25, 151–157.

Davidson, E.W. (1989) Variation in binding of *Bacillus sphaericus* toxin and wheat-germ agglutinin to larval midgut cells of six species of mosquitos. *Journal of Invertebrate Pathology* 53, 251–259.

de Barjac, H., Larget-Thiery, I., Cosmao Dumanoir, V.C. & Ripouteau, H. (1985) Serological classification of *Bacillus sphaericus* strains in relation to toxicity to mosquito larvae. *Applied Microbiology and Biotechnology* 21, 85–90.

de la Cruz, F. & Davies, J. (2000) Horizontal gene transfer and the origin of species: lessons from bacteria. *Trends in Microbiology* 8, 128–133.

Eisenstein, B.I. (1990) New techniques for microbial epidemiology and the diagnosis of infectious diseases. *Journal of Infectious Diseases* 161, 595–602.

Favret, M.E. & Yousten, A.A. (1985) Insecticidal activity of *Bacillus laterosporus*. *Journal of Invertebrate Pathology* 45, 195–203.

Federici, B.A. (1995) The future of microbial insecticides as vector control agents. *Journal of the American Mosquito Control Association* 11, 260–268.

Gordon, R.E., Haynes, W.C. & Pang, C.H.-N. (1973) *The Genus* Bacillus. Agriculture Handbook 427. US Department of Agriculture, Washington, DC.

Hacker, J., Blum-Oehler, G., Muhidorfer, I. & Tschape, H. (1997) Pathogenicity islands of virulent bacteria: structure, function and impact on microbial evolution. *Molecular Microbiology* 23, 1089–1097.

Hazes, B. & Read, R.J. (1995) A mosquitocidal toxin with a ricin-like cell-binding domain. *Structural Biology* 2, 358–359.

Humphreys, M.J. & Berry, C. (1998) Variants of the *Bacillus sphaericus* binary toxins: implications for differential toxicity of strains. *Journal of Invertebrate Pathology* 71, 184–185.

Istock, C.A., Duncan, K.E., Ferguson, N. & Zhou, X. (1992) Sexuality in a natural population of bacteria – *Bacillus subtilis* challenges the clonal paradigm. *Molecular Ecology* 1, 95–103.

Jahnz, U., Fitch, A. & Priest, F.G. (1996) Evaluation of an rRNA-targeted oligonucleotide probe for the detection of mosquitocidal strains of *Bacillus sphaericus* in soils – characterization of novel strains lacking toxin genes. *FEMS Microbiology Ecology* 20, 91–99.

Karch, S., Asidi, N., Manzambi, Z.M. & Salaun, J.J. (1992) Efficacy of *Bacillus sphaericus* against the malaria vector *Anopheles gambiae* and other mosquitos in swamps and rice fields in Zaire. *Journal of the American Mosquito Control Association* 8, 376–380.

Kellen, W., Clark, T., Lindergren, J., Ho, B., Rogoff, M. & Singer, S. (1965) *Bacillus sphaericus* Neide as a pathogen of mosquitoes. *Journal of Invertebrate Pathology* 7, 442–448.

Krych, V., Johnson, J.L. & Yousten, A.A. (1980) Deoxyribonucleic acid homologies among strains of *Bacillus sphaericus*. *International Journal of Systematic Bacteriology* 30, 476–484.

Lacey, L. & Singer, S. (1982) The larvicidal activity of new isolates of *Bacillus sphaericus* and *Bacillus thuringiensis* H14 against anopheline and culicine mosquitoes. *Mosquito News* 42, 537–545.

Liu, J.W., Hindley, J., Porter, A.G. & Priest, F.G. (1993) New high-toxicity mosquitocidal strains of *Bacillus sphaericus* lacking a 100-kilodalton-toxin gene. *Applied and Environmental Microbiology* 59, 3470–3473.

Liu, J.W., Porter, A.G., Wee, B.Y. & Thanabalu, T. (1996) New gene from nine *Bacillus sphaericus* strains encoding highly conserved 35.8-kilodalton mosquitocidal toxins. *Applied and Environmental Microbiology* 62, 2174–2176.

Maiden, M.C.J, Bygraves, J.A., Feil, E. *et al.* (1998) Multilocus sequence typing: a portable approach to the identification of clones within populations of pathogenic microorganisms. *Proceedings of the National Academy of Sciences USA* **95**, 3140–3145.

Massie, J., Roberts, G. & White, P.J. (1985) Selective isolation of *Bacillus sphaericus* from soil by use of acetate as the only major source of carbon. *Applied and Environmental Microbiology* **49**, 1478–1481.

Maynard Smith, J. (1995) Do bacteria have population genetics? In: *Population Genetics of Bacteria* (eds S. Baumberg *et al.*), pp. 1–12. Cambridge University Press, Cambridge.

Miteva, V., Gancheva, A., Mitev, V. & Ljubenov, M. (1998) Comparative genome analysis of *Bacillus sphaericus* by ribotyping, M13 hybridization, and M13 polymerase chain reaction fingerprinting. *Canadian Journal of Microbiology* **44**, 175–80.

Mulla, M.S., Darwazeh, H.A., Davidson, E.W., Dulmage, H.T. & Singer, S. (1984) Larvicidal activity and field efficacy of *Bacillus sphaericus* strains against mosquito larvae and their safety to nontarget organisms. *Mosquito News* **44**, 336–342.

Mulla, M.S., Su, T., Thavara, U., Tawatsin, A., Ngamsuk, W. & Pan-Urai, P. (1999) Efficacy of new formulations of the microbial larvicide *Bacillus sphaericus* against polluted water mosquitoes in Thailand. *Journal of Vector Ecology* **24**, 99–110.

Nicolas, L., Nielsen-LeRoux, C., Charles, J.F. & Delécluse, A. (1993) Respective roles of the 42-kDa and 51-kDa components of the *Bacillus sphaericus* toxin overexpressed in *Bacillus thuringiensis*. *FEMS Microbiology Letters* **106**, 275–280.

Nicolas, L., Regis, L.N. & Rios, E.M. (1994) Role of the exosporium in the stability of the *Bacillus sphaericus* binary toxin. *FEMS Microbiology Letters* **124**, 271–275.

Nielsen-LeRoux, C. & Charles, J.F. (1992) Binding of *Bacillus sphaericus* binary toxin to a specific receptor on midgut brush-border membranes from mosquito larvae. *European Journal of Biochemistry* **210**, 585–590.

Oei, C., Hindley, J. & Berry, C. (1992) Binding of purified *Bacillus sphaericus* binary toxin and its deletion derivatives to *Culex quinquefasciatus* gut – elucidation of functional binding domains. *Journal of General Microbiology* **138**, 1515–1526.

Orlova, M.V., Smirnova, T.A., Ganushkina, L.A., Yacubovitch, V.Y. & Azizbekyan, R.R. (1998) Insecticidal activity of *Bacillus laterosporus*. *Applied and Environmental Microbiology* **64**, 2723–2725.

Porter, A.G., Davidson, A.G. & Liu, J.W. (1993) Mosquitocidal toxins of bacilli and their effective manipulation for biological control of mosquitoes. *Microbiological Reviews* **57**, 838–861.

Priest, F.G. (1992) Biological control of mosquitoes and other biting flies by *Bacillus sphaericus* and *Bacillus thuringiensis*. *Journal of Applied Bacteriology* **72**, 357–369.

Priest, F.G. & Dewar, S.J. (2000) Bacteria and insects. In: *Applied Microbial Systematics* (eds F.G. Priest & M. Goodfellow), 165–202. Kluwer, Dordrecht.

Priest, F.G., Goodfellow, M. & Todd, C. (1988) A numerical classification of the genus *Bacillus*. *Journal of General Microbiology* **134**, 1847–1882.

Priest, F.G., Aquino de Muro, M. & Kaji, D.A. (1994) Systematics of insect pathogenic bacilli: uses in strain identification and isolation of novel strains. In: *Bacterial Diversity and Systematics* (eds F.G. Priest *et al.*), pp. 275–296. Plenum Press, New York.

Priest, F.G., Ebdrup, L., Zahner, V. & Carter, P.E. (1997) Distribution and characterization of mosquitocidal toxin genes in some strains of *Bacillus sphaericus*. *Applied and Environmental Microbiology* **63**, 1195–1198.

Regis, L., Silva-Filha, M.H., de Oliveira, C.M., Rios, E.M., da Silva, S.B. & Furtado, A.F. (1995) Integrated control measures against *Culex quinquefasciatus*, the vector of filariasis in Recife. *Memorias do Instituto Oswaldo Cruz* **90**, 115–119.

Rippere, K.E., Johnson, J.L. & Yousten, A.A. (1997) DNA similarities among mosquito-pathogenic and nonpathogenic strains of *Bacillus sphaericus*. *International Journal of Systematic Bacteriology* **47**, 214–216.

Roberts, M.S., Nakamura, L.K. & Cohan, F.M. (1994) *Bacillus mojavensis* sp. nov. distinguishable from *Bacillus subtilis* by sexual isolation, divergence in DNA sequence and differences in fatty acid composition. *International Journal of Systematic Bacteriology* **44**, 256–264.

Russell, B.L., Jelley, S.A. & Yousten, A.A. (1989) Carbohydrate metabolism In the mosquito pathogen *Bacillus sphaericus* 2362. *Applied and Environmental Microbiology* **55**, 294–297.

Selander, R.K., Cougant, D.A., Ochman, H., Musser, J.M., Gilmour, M.N. & Whittam, T.S. (1986) Methods of multilocus electrophoresis for bacterial population genetics and systematics. *Applied and Environmental Microbiology* **51**, 873–884.

Selander, R.K., Li, J., Boyd, E.F., Wang, F.-S. & Nelson, K. (1994) DNA sequence analysis of the genetic structure of populations of *Salmonella enterica* and *Escherichia coli*. In: *Bacterial Diversity and Systematics* (eds F.G. Priest *et al.*), pp. 17–49. Plenum Press, New York.

Shadduck, J.A., Singer, S. & Lause, S. (1980) Lack of mammalian pathogenicity of entomocidal isolates of *Bacillus sphaericus*. *Environmental Entomology* 9, 403–407.

Silva-Filha, M.H., Nielsen-LeRoux, C. & Charles, J.F. (1997) Binding kinetics of *Bacillus sphaericus* binary toxin to midgut brush-border membranes of *Anopheles* and *Culex* sp. mosquito larvae. *European Journal Biochemistry* 247, 754–761.

Silva-Filha, M.H., Nielsen-LeRoux, C. & Charles, J.F. (1999) Identification of the receptor for *Bacillus sphaericus* crystal toxin in the brush border membrane of the mosquito *Culex pipiens* (Diptera: Culicidae). *Insect Biochemistry and Molecular Biology* 29, 711–721.

Singer, S. (1996) The utility of strains of morphological group II *Bacillus*. *Advances in Applied Microbiology* 42, 219–261.

Stackebrandt, E., Ludwig, W., Weizenegger, M. *et al.* (1987) Comparative 16S rRNA oligonucleotide analysis and murein types of round-spore-forming bacilli and non-sporing relatives. *Journal of General Microbiology* 133, 2523–2529.

Thanabalu, T. & Porter, A.G. (1995) Efficient expression of a 100-kilodalton mosquitocidal toxin in protease-deficient recombinant *Bacillus sphaericus*. *Applied and Environmental Microbiology* 61, 4031–4036.

Thanabalu, T. & Porter, A.G. (1996) A *Bacillus sphaericus* gene encoding a novel type of mosquitocidal toxin of 31.8 kDa. *Gene* 170, 85–89.

Thanabalu, T., Hindley, J., Jackson-Yap, J. & Berry, C. (1991) Cloning, sequencing, and expression of a gene encoding a 100-kilodalton mosquitocidal toxin from *Bacillus sphaericus* SSII-1. *Journal of Bacteriology* 173, 2776–2785.

Thanabalu, T., Hindley, J. & Berry, C. (1992) Proteolytic processing of the mosquitocidal toxin from *Bacillus sphaericus* SSII-1. *Journal of Bacteriology* 174, 5051–5056.

Thanabalu, T., Berry, C. & Hindley, J. (1993) Cytotoxicity and ADP-ribosylating activity of the mosquitocidal toxin from *Bacillus sphaericus* SSII-1, possible roles of the 27-kilodalton and 70-kilodalton peptides. *Journal of Bacteriology* 175, 2314–2320.

Wati, M.R., Thanabalu, T. & Porter, A.G. (1997) Gene from tropical *Bacillus sphaericus* encoding a protease closely related to subtilisins from Antarctic bacilli. *Biochimica et Biophysica Acta* 1352, 56–62.

White, P.J. & Lotay, K.K. (1980) Minimal nutritional requirements of *Bacillus sphaericus* NCTC 9602 and 26 other strains of this species: the majority grow and sporulate with actetate as sole major source of carbon. *Journal of General Microbiology* 118, 13–19.

Yadav, R.S., Sharma, V.P. & Upadhyay, A.K. (1997) Field trial of *Bacillus sphaericus* strain B-101 (serotype H5a, 5b) against filariasis and Japanese encephalitis vectors in India. *Journal of the American Mosquito Control Association* 13, 158–163.

Yousten, A.A. (1984) Bacteriophage typing of mosquito-pathogenic strains of *Bacillus sphaericus*. *Journal of Invertebrate Pathology* 43, 124–125.

Yousten, A.A., Fretz, S.B. & Jelley, S.A. (1985) Selective medium for insect pathogenic strains of *Bacillus sphaericus*. *Applied and Environmental Microbiology* 49, 1532–1533.

Yousten, A.A., Benfield, E.F. & Genthner, F.J. (1995) *Bacillus sphaericus* mosquito pathogens in the aquatic environment. *Memorias do Instituto Oswaldo Cruz* 90, 125–129.

Zahner, V. & Momen, H. (1995) Multilocus enzyme electrophoresis study of *Bacillus sphaericus*. *Memorias do Instituto Oswaldo Cruz* 90, 65–68.

Zahner, V. & Priest, F.G. (1997) Distribution of restriction endonucleases among some entomopathogenic strains of *Bacillus sphaericus*. *Letters in Applied Microbiology* 24, 483–487.

Zahner, V., Rabinovitch, L., Cavados, C.F.G. & Momen, H. (1994) Multilocus enzyme electrophoresis on agarose gel as an aid to the identification of entomopathogenic *Bacillus sphaericus* strains. *Journal of Applied Bacteriology* 76, 327–335.

Zahner, V., Momen, H. & Priest, F.G. (1998) Serotype H5a5b is a major clone within mosquito-pathogenic strains of *Bacillus sphaericus*. *Systematic and Applied Microbiology* 21, 162–170.

Zhang, J., Hodgman, C., Krieger, L., Schnetter, W. & Schairer, H.U. (1997) Cloning and analysis of the first *cry* gene from *Bacillus popilliae*. *Journal of Bacteriology* 179, 4336–4341.

Chapter 14

The Importance of *Bacillus* Species in the Production of Industrial Enzymes

Helle Outtrup and Steen T. Jørgensen

Introduction

The aim of this chapter is to describe how valuable the different *Bacillus* species are from the viewpoint of the microbiologists working at Novo Nordisk A/S,[1] the world's largest supplier of industrial enzymes.

The market

The global market for industrial enzymes is considered to total 1.6 billion US dollars.

The market is divided as follows:

Technical enzymes   *c*. US$1 billion
Food enzymes   *c*. US$0.5 billion
Feed enzymes   *c*. US$0.1 billion.

In the first half of the 1990s, the market was characterized by oversupply and fierce competition between five major suppliers: Novo Nordisk A/S, Genencor International Inc., Gist-Brocades, Solvay and Showa Denko. During the latter half of the 1990s, the number of major suppliers was reduced through business and patent acquisitions to three: Novo Nordisk A/S, Genencor International Inc. and DSM N.V.

Genencor International Inc. is headquartered in the USA and operates predominantly in the technical and feed enzymes segments. DSM N.V. is headquartered in the Netherlands and operates in the food and feed segments.

Novo Nordisk A/S, unlike Genencor International Inc. and DSM N.V. and its other competitors, has a significant market presence in all three sectors. According to its own estimates, Novo Nordisk A/S is the largest supplier in each of the three sectors and its share of the industrial enzyme market in 1999 was between 41 and 44%. According to estimates from the same source, Genencor International Inc. and DSM N.V. had market shares of around 21% and around 8%, respectively (Anonymous 1999, 2000). The rest of the market is divided among a few smaller enzyme producers, some of which produce enzymes for their

---

[1] The industrial enzyme business of Novo Nordisk A/S is today an independent company, Novozymes A/S.

Table 14.1 The most important global industries for enzyme applications.

Category	Industry	% of global enzyme market
Technical enzymes	Detergent	37
	Starch	13
	Textile	6
Food enzymes	Baking	9
	Beverage	6
	Dairy	14
Feed enzymes	Animal feed	8
	Other	7

own use, in the USA, Canada, Europe and Japan, as well as a number of small local producers in China.

The industries

The most important industries for application of industrial enzymes today are shown in table 14.1.

Technical enzymes

The technical enzymes that accounted for 73% of Novo Nordisk A/S's net turnover in 1999 are used in the following industries: detergent, starch, fuel, alcohol, textile, leather, personal care, and pulp and paper.

The detergent industry is the largest user of industrial enzymes, accounting for more than one third of the global market, half of the sales at Novo Nordisk A/S in 1999, and around 60% of the enzyme sales from Genencor International Inc. The enzymes used in the detergent industry today are mainly proteases, which remove protein stains from the clothes, amylases for removal of starch stains, lipases for removal of lipids and cellulases, which remove particulate soiling and secure colour stability. The detergent proteases and amylases used today are all of *Bacillus* origin, whereas the lipases and cellulases are produced mostly from fungi.

Various industries have replaced processes using chemicals which have detrimental effects on the environment and on equipment, with processes using biodegradable enzymes under conditions that are less corrosive. The starch industry, for example, has replaced strong acids with enzymes in the different steps of the conversion of cereal starch into glucose, fructose or maltose.

The use of *Bacillus* α-amylases for desizing of textiles was one of the first industrial applications of enzymes. Later, other enzymes, cellulases and oxidases emerged as especially useful for the processing of denim clothes. However, although none of these are produced from *Bacillus* origin today, new textile processes are expected to emerge that will use *Bacillus* pectate lyases for biopreparation of textiles in order to make them more susceptible to dyeing and bleaching.

In the leather industry, highly alkaline *B. halodurans* proteases are used for the dehairing of hides.

Food enzymes

The food enzymes, which accounted for 22% of Novo Nordisk A/S's net turnover in 1999, are used in baking, brewing, wine, juice, alcohol, dairy, fats and oils, and other food processing industries.

For the baking industry, Novo Nordisk A/S produces a unique amylase from a thermophilic *Bacillus* species. This amylase modifies the flour starch in a way that prevents its reaction with gluten that causes hardening of the bread, thus prolonging its shelf life.

In the beverage industry, most of the enzymes used are fungal plant cell wall degrading enzymes but a few *Bacillus* enzymes are used in the brewing and potable alcohol industry. These are α-amylases for starch degradation, β-glucanases for barley β-glucan hydrolysis, proteases to release more digestible nitrogen from the cereal proteins in order to increase yeast growth, and aceto-lactate decarboxylase (ALDC) to prevent diacetyl formation during accelerated maturation of beer.

The most important enzyme used in the dairy industry is rennin, which is of either animal or fungal origin.

Feed enzymes

The enzymes used in animal feed, which accounted for 5% of Novo Nordisk A/S's net turnover in 1999, are included as ingredients in animal feed to improve feed utilization and nutrient digestion, and are almost all of fungal origin.

The Novo Nordisk A/S enzyme products of *Bacillus* origin

The *Bacillus* products sold by Novo Nordisk A/S in 1999 are listed in table 14.2. These products account for almost half of the total enzyme sales at Novo Nordisk A/S and probably account for about one third of the global market (Anonymous 1999).

Detergent proteases

The single most important enzyme application is the addition of proteases to detergents. Almost all the detergent proteases used today are of *Bacillus* origin, accounting for more than 25% of the total enzyme sales of Novo Nordisk A/S. The different ways in which people do their laundry around the world has forced enzyme manufacturers to search continuously for new and improved proteases to meet stability and performance demands in the many different detergents that are marketed today.

The preferred protease is produced by the alkaliphilic species *B. clausii* which, in the high pH powder detergents used in many countries today, has outper-

Table 14.2 *Bacillus* enzymes produced by Novo Nordisk A/S.

Enzyme	Product name	From *Bacillus* sp.	
Protease	Savinase^R	*clausii*	
Protease	Durazyme™	*clausii*	PE variant
Protease	Kannase™	*clausii*	PE variant
Protease	Everlase™	*clausii*	PE variant
Protease	Relase™	*clausii*	PE variant
Protease	Alcalase^R	*licheniformis*	
Protease	Esperase^R	*halodurans*	
Amylase	Termamyl^R	*licheniformis*	
Amylase	Duramyl^R	*licheniformis*	PE variant
Amylase	Termamyl^R LC	*licheniformis*	PE variant
Amylase	Novamy^R	New thermophile	
Amylase	Termamyl^R S	*stearothermophilus*	
Amylase	Termamyl^R SC	*stearothermophilus*	PE variant
Amylase	Aquazym^R	*amyloliquefaciens*	
Amylase	Natalase^R	*halmapalus*	PE variant
Mannanase	Mannaway^R	New alkaliphile sp.173	
Pullulanase	Promozyme^R	'acidopullulyticus'	
Pullulanase	Promozyme^{R}D	'deramificans'	
ALDC	Maturex^R	*brevis*	
Protease	Neutrase^R	*amyloliquefaciens*	
Uricase		*fastidiosus*	

formed the 'old' *B. licheniformis* enzyme and dominates the global market for detergent protease. Certain markets, however, have specific demands and up to now four different protein-engineered (PE) variants of this protease have found their way into these markets: Kannase™ gives improved washing performance at low temperature and low water hardness and is the product launched most recently; Durazyme™ and Everlase™ are suitable for bleach-containing detergents; and Relase™ gives improved washing performance in American powder detergents. More variants are expected to appear in the future.

The proteases from *B. licheniformis* and *B. halodurans* are used in certain detergent markets all over the world.

Detergent amylases

The addition of amylases to detergents is a new development. The wild type amylase Termamyl^R from *B. licheniformis* was the first on the market but this has recently been supplemented with a PE-variant, Duramyl^R. The latter meets the demand for increased stability and performance under the highly alkaline and oxidative conditions encountered with the bleach-containing detergents.

A new amylase from the alkalitolerant species *B. halmapalus*, has shown promising wash performance and Novo Nordisk A/S introduced a stabilized PE-variant, Natalase^R, in 1999.

Other detergent enzymes

The search for new cleaning enzymes has led recently to a new product, a mannanase from a new unnamed alkaliphilic *Bacillus* species. This enzyme, Mannaway[R], has a significant effect on certain stains such as chocolate ice cream that are otherwise difficult to remove.

Amylases for starch liquefaction

The first step of starch modification is liquefaction with *Bacillus* α-amylase, which takes place under steam injection at very high temperature and at a pH just below neutral. The second step, saccharification, is carried out at pH 4–5 and a temperature of about 60°C with a mixture of *Bacillus* pullulanase and fungal amyloglucosidase to produce high yields of glucose. In a third step this may be converted to high fructose syrup by glucose isomerase, which is mainly of streptomycete origin.

This process requires the use of an α-amylase active at high temperature and low pH (4.5). Furthermore, it is desirable to minimize the addition of calcium, which usually is necessary to stabilize the amylase, as this has to be removed to avoid inhibiting isomerization of glucose to fructose.

Three different *Bacillus* species, *B. amyloliquefaciens*, *B. licheniformis* and *B. stearothermophilus*, produce industrially important α-amylases. Despite high homology in their amino acid sequences, they have very different characteristics with respect to the starch hydrolysis products resulting from their action and their specific activity and stability at very high temperatures.

The α-amylase from *B. amyloliquefaciens* has high specific activity and a desirable product profile after hydrolysis but rather low thermostability. In contrast, the α-amylases from *B. licheniformis* and *B. stearothermophilus* are both sufficiently thermostable, but the *B. licheniformis* enzyme has rather low specific activity compared to the *B. stearothermophilus* enzyme. The latter, however, has a product profile that causes the production of undesired byproducts during the subsequent saccharification.

These shortcomings have led to considerable combination and engineering of these α-amylases, resulting in the introduction of several new products during the last few years. Termamyl[R] LC is a hybrid of the amylases from *B. amyloliquefaciens* and *B. licheniformis*, which is engineered to be stable at low calcium concentrations at high temperatures. Termamyl[R] SC is a variant of the amylase from *B. stearothermophilus* engineered to meet the specific demands of the fuel alcohol industry.

Pullulanases for starch debranching

In the starch saccharification process pullulanases, which cleave the α-1, 6-glucoside bonds of amylopectin, are added along with a fungal amyloglucosidase, which has optimal activity at pH 4. The *Bacillus* enzymes used today (table 14.2) have optimum activity at pH 5 but have to work at pH 4.5 as a compromise.

In this area too, there is an ongoing protein-engineering programme aimed at achieving pH and temperature optima appropriate to the conditions of the process.

The *Bacillus* species

In this chapter we regard the genus *Bacillus* as consisting of aerobic, endospore-forming, rod-shaped bacteria; this disregards new findings, which have shown that the diversity within this group is so large that it contains several different genera.

During the period of 1965–80, screening of soil samples was the method most frequently used to identify new enzyme producers. This process resulted in the isolation of a lot of useful new alkaliphilic *Bacillus* species and nine of these were recently given names that were accepted officially (Nielsen *et al.* 1995).

A phylogenetic tree based upon full 16S rRNA sequences from the ARB database (Strunk & Ludwig 1995) and from sequences determined at Novo Nordisk A/S is shown in figure 14.1. It is apparent that the species of commercial interest are distributed throughout the tree, which shows only a minor part of the *Bacillus* isolates available today. Thus, further useful products will likely emerge from other isolates not yet examined.

Many *Bacillus* enzymes that have not yet been introduced in the market place, or are only in an early phase of introduction, are identified within the alkaliphilic species. *Bacillus agaradherens*, for instance, produces a wide spectrum of different extracellular hydrolytic enzymes such as cellulase, xylanase, and protease (Outtrup *et al.* 1997), which have not yet found extensive industrial application. Similar activities are also found in other quite different *Bacillus* species.

Examples of new *Bacillus* species identified by the industrial microbiologists are the pullulanase-producing strains of '*B. acidopullulyticus*' and '*B. deramificans*', strains that are the results of intensive screening for pullulanases active at pH 4–5 and 60°C. The unique TS25 species, which was isolated in 1977 (Outtrup 1986), produces a very important industrial amylase used in the baking industry as mentioned above. This amylase, which degrades starch into α-maltose and which, unlike other maltogenic amylases, is very active also on maltotriose, has been allocated its own number in the Enzyme Commission classification of enzymes (EC 3.2.1.133) (Enzyme Nomenclature 1992).

Recombinant *Bacillus* production strains

The evolution of genetic engineering has allowed the expression of new enzyme activities from less well-known *Bacillus* species in production strains with a long history of safe industrial use. This has opened new possibilities for improvement of production strains, and it has allowed the targeted modification of enzyme proteins by protein engineering.

Figure 14.1 Phylogenetic tree based on full 16S rRNA sequences of recognized *Bacillus* species and on 16S rRNA sequences determined at Novo Nordisk A/S. The industrially important species for enzyme production are indicated in bold.

Bacillus subtilis is the preferred work-horse because of the capability of strains of this species to become competent for the uptake of plasmid as well as chromosomal DNA with high frequencies. Initially, genes encoding novel enzyme activities are cloned into competent cells of *Bacillus subtilis* on plasmids, which are maintained by the use of antibiotic resistance selection, and which frequently have a high copy number to ensure maximal production of the enzyme. The enzyme proteins are then purified and characterized.

While the use of plasmids stabilized by the addition of antibiotics is a technology well-suited to the laboratory-scale system, it is unsatisfactory for large-scale industrial production. It is frequently observed that *Bacillus* plasmids are quite unstable – with respect to segregation as well as structure – and stabilization by the addition of antibiotics to the fermentation medium is not attractive. Another solution to the stabilization of plasmids, developed in our laboratories, is the use of the *dal* gene encoding D, L-alanine racemase, which is used to convert L-alanine into D-alanine, essential for the integrity of the cell wall. As D-alanine is not present in the very rich media commonly used for industrial fermentations, plasmids containing the *dal* gene may be stabilized in a host strain lacking the chromosomal copy of this gene (Diderichsen 1986).

Chromosomal integration of the gene expressing the enzyme of interest is an attractive alternative procedure for the construction of stable, high-yielding production strains. Chromosomal integration is easily achieved directly upon transformation of competent *B. subtilis* cells, provided that the transforming DNA has sequence homology with the chromosome. Chromosomal amplification of gene copy number can occur and be selected for if the integrated gene, together with an antibiotic resistance gene, is flanked by homologous sequences. Such a strain could be constructed, for example, by the transformation of *B. subtilis* with an *Escherichia coli* replicon carrying a marker gene selectable in *B. subtilis*, the gene coding for the enzyme of interest and a DNA fragment homologous to a segment of the *B. subtilis* chromosome. Selection of strains that can grow in the presence of increasing concentrations of the relevant antibiotic, selects cells containing chromosomal gene amplifications (figure 14.2), and such cells frequently show increased enzyme productivity relative to strains containing only one copy. These chromosomal gene amplifications have proven sufficiently stable, in the absence of selective pressure, to allow the use of such strains in large-scale fermentations.

Despite the ease with which strain constructions such as those described above can be carried out in *B. subtilis*, other expression hosts like *B. licheniformis* and *B. clausii* are being used extensively in large-scale enzyme production as higher enzyme yields frequently are obtained using these host strains. They have a long history of industrial use, as enzymes like amylase and protease have been produced for decades by nonrecombinant versions of these strains. High-yielding isolates have been obtained by classical mutagenesis, and fermentation procedures have been optimized. Such strains, however, have the drawback that high-frequency competence transformation procedures are not available, which makes the introduction of DNA and manipulations of their chromosomes technically more demanding and usually results in low transformation frequencies. Procedures in use for the introduction of DNA into such strains include electroporation, transformation and regeneration of protoplasts. In addition, the use of

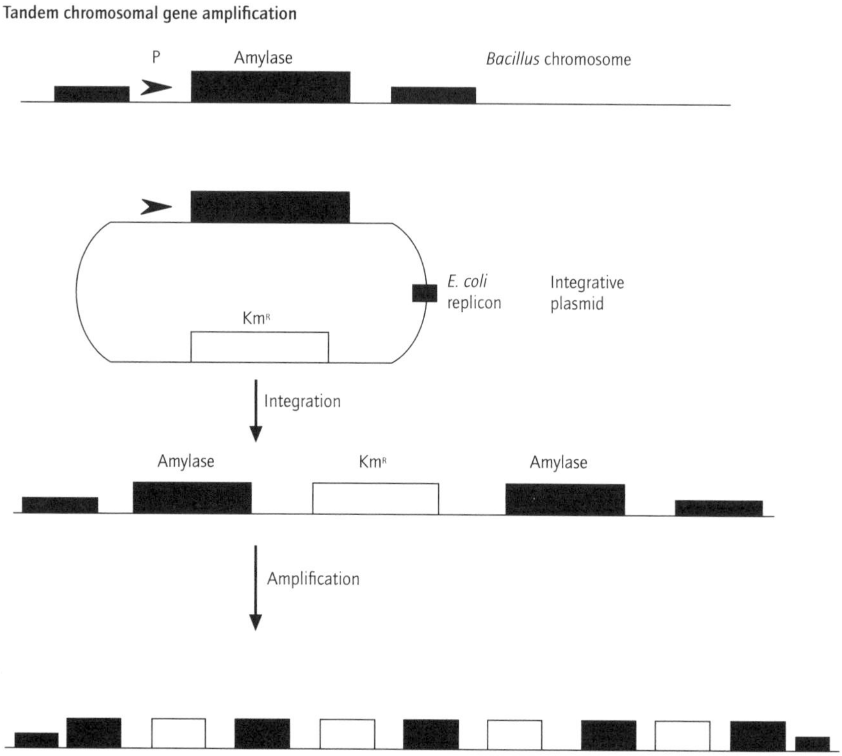

Tandem chromosomal gene amplification

Figure 14.2 Chromosomal integration in *Bacillus subtilis* of a gene expressing the enzyme of interest (e.g. amylase) by transformation with an *E. coli* replicon, followed by tandem chromosomal amplification of the gene by selection of strains growing in the presence of increasing concentrations of an antibiotic (e.g. kanamycin).

conjugation from *B. subtilis* donor strains has been developed as a means to introduce plasmids into production strains of other species (Tangney *et al.* 1998).

Subsequent to the introduction of plasmid DNA, an objective is to achieve stable chromosomal integration and amplification of the gene of interest in these non-*B. subtilis* hosts. A solution to this problem has been provided by the following method: first a plasmid molecule is constructed that will ultimately integrate into the host cell chromosome. This plasmid carries the gene of interest, an antibiotic resistance marker, a DNA segment of homology to the host chromosome, and a plasmid replication origin. Importantly, no intact gene for the replication protein – which is necessary for plasmid replication – is present on this plasmid. This first plasmid is made to replicate in a strain by the simultaneous presence of a second plasmid, which carries the intact replication protein gene, and which is selected for by the use of another antibiotic resistance marker. The replication protein is thus provided in trans. By continued propagation of such a strain, selecting only for the antibiotic resistance marker carried by the first plasmid molecule, cells can be isolated in which this first molecule carrying the gene of interest is present within the bacterial chromosome (having integrated via the segment of homology), and from which the second plasmid, not being selected for, has been lost. The integrated DNA, being devoid of active replication functions, can be maintained stably, and the copy number of the gene of interest may be increased by selection of cells able to grow in the presence of increased levels of the appropriate antibiotic, as both the gene of interest and the antibiotic resistance marker are flanked by direct repeats of the DNA segment used for chromosomal integration (Jørgensen *et al.* 1997).

It is a frequent observation that enzyme production from these non-*B. subtilis* strains is also increased if the production organism contains several copies of the enzyme-coding gene of interest, even though the well-functioning expression signals obtained from the highly expressed genes of the classical production strains are used to drive the expression in the recombinant strains.

Alternative strategies to the tandem amplification of genes by selection for increased antibiotic resistance have been devised in our laboratories. One such strategy, still resulting in tandem amplification, relies upon the use of another marker gene that can be efficiently screened for instead of the antibiotic resistance gene. Successful isolation of strains containing gene amplifications has been achieved using a gene which expresses a green fluorescent protein as a marker, and sorting of cells containing duplications, triplications, etc. can be identified by their increased content of the green fluorescent protein, using a flow cytometer with cell-sorting capability (Jørgensen & Pedersen 1999).

Another strategy produces strains in which a number of genes encoding the enzyme of interest are inserted at various locations in the chromosome of the host strain. The first step involved is the integration of the gene of interest, together with an antibiotic resistance marker gene, into the host strain chromosome by transposition from a transposon donor plasmid. In this method, recognition sites for a site-specific recombination enzyme – the resolvase from the plasmid pAMbeta1 – flank the antibiotic resistance gene and this may thus be efficiently deleted from the chromosome by the introduction of a plasmid expressing the resolvase in the second step. The use of a temperature-sensitive plasmid for this purpose ensures its easy elimination from the cell following marker deletion (figures 14.3 and 14.4). The resulting cell contains one chromosomally integrated copy of the gene of interest, but no marker gene, and may now serve as host strain for a subsequent round of gene insertion by transposition followed by marker deletion. A strain containing several inserted genes, having no retained antibiotic resistance markers, may be constructed using this iterative process (figure 14.5) (Jørgensen 1996).

Ultimately, desirable production strains would contain only one, chromosomally integrated and totally genetically stable copy of the gene of interest sufficiently well expressed to saturate the cells' capacity to produce and secrete the product, thus avoiding the presence of markers now used to ensure an increased copy number.

Current developments point towards success in this direction. Some very strong tandem promoter systems have been developed, which allow expression of enzymes in *B. subtilis* from single, chromosomally integrated gene copies at levels several-fold higher than in previously used promoter systems (Widner *et al.* 1999).

Studies of the α-amylase promoter of *B. licheniformis* have likewise resulted in the identification of improved mutant promoters. Such improved promoters,

Construction principle

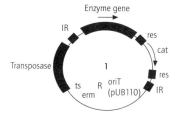

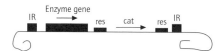

Figure 14.3 The construction of host strains containing a number of genes encoding the enzyme of interest inserted at various locations in the chromosome. The first step involves integration of the gene together with an antibiotic resistance marker (cat), flanked by resolvase recognition sites (res), by transposition from a donor plasmid containing a transposon flanked by inverted repeats (IR).

Construction principle

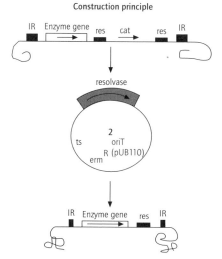

Figure 14.4 The construction of host strains containing a number of genes encoding the enzyme of interest inserted at various locations in the chromosome. The second step involves deletion of the marker gene (cat) by introduction of a temperature-sensitive plasmid expressing resolvase.

combined with a strain construction technology leading to strains which in the same locus harbour two chromosomally integrated, but oppositely oriented, copies of an expression cassette, allow higher yields from stable, marker-free strains (Jørgensen & Diderichsen 1997; Jørgensen 1999). The inverted orientation of the two neighbouring expression cassettes also prevents copy loss by homologous recombination, a possibility if orientation is parallel.

In addition to the use of gene technology for the optimization of the quality and quantity of the expression cassette encoding the desired enzyme, genes encoding undesired side activities, e.g. unwanted proteases, may be deleted from the production host strains, and the expression of 'beneficial' genes may be increased.

The range of *Bacillus* hosts used in the production of the Novo Nordisk A/S enzymes today is shown in table 14.3.

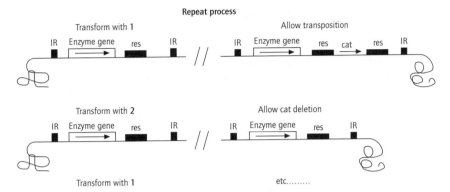

Figure 14.5 The construction of host strains containing a number of genes encoding the enzyme of interest inserted at various locations in the chromosome. After the two preceding steps (figures 14.3 and 14.4), the host strain may serve for iterative rounds of gene insertion by transposition followed by marker deletion.

Table 14.3 *Bacillus* enzymes from Novo Nordisk A/S production strains.

Enzyme	Product name	Donor strain	*Bacillus* host
Protease	Savinase[R]	*B. clausii*	*clausii*
Protease	Durazyme™	*B. clausii*	*clausii*
Protease	Kannase™	*B. clausii*	*clausii*
Protease	Everlase™	*B. clausii*	*clausii*
Protease	Relase™	*B. clausii*	*clausii*
Protease	Alcalase[R]	*B. licheniformis*	*licheniformis*
Protease	Esperase[R]	*B. halodurans*	*halodurans*
Amylase	Termamyl[R]	*B. licheniformis*	*licheniformis*
Amylase	Duramyl[R]	*B. licheniformis*	*licheniformis*
Amylase	Termamyl[R] LC	*B. licheniformis*	*licheniformis*
Amylase	Novamyl[R]	New thermophile	*subtilis*
Amylase	Termamyl[R] S	*B. stearothermophilus*	*licheniformis*
Amylase	Termamyl[R] SC	*B. stearothermophilus*	*licheniformis*
Amylase	BAN	*B. amyloliquefaciens*	*amyloliquefaciens*
Amylase	Natalase	*B. halmapalus*	*licheniformis*
Mannanase	Mannaway[R]	New alcaliphile sp.173	*licheniformis*
Pullulanase	Promozyme[R]	'*B. acidopullulyticus*'	*subtilis*
Pullulanase	Promozyme[R]D	'*B. deramificans*'	*subtilis*
ALDC	Maturex[R]	*Br. brevis*	*subtilis*
Protease	Neutrase[R]	*B. amyloliquefaciens*	*amyloliquefaciens*
Uricase		*B. fastidiosus*	*fastidiosus*
CGTase[a]	Toruzyme	*Thermoanaerobacter*	*licheniformis*

[a] Cyclomaltodextringlucanotransferase.

Conclusions

Bacillus species are the source of enzyme genes used in the production of the majority of the microbial industrial enzymes produced not only at Novo Nordisk A/S but also worldwide. The great diversity within these bacteria and the range of genetic tools available for their manipulation hold out the promise for the future of many new products arising from new or old *Bacillus* species.

References

Anonymous (1999) *Annual Report 1999.* Novo Nordisk A/S, Denmark.

Anonymous (2000) *Prospectus.* Genencor International Inc., Palo Alto, CA.

Diderichsen, B. (1986) A genetic system for stabilization of cloned genes in *Bacillus subtilis.* In: *Bacillus Molecular Genetics and Biotechnology Applications* (eds A.T. Ganesan & J.A. Hoch), pp. 35–46. Academic Press, New York.

Enzyme Nomenclature (1992) *Recommendations of the Nomenclature Committee of the International Union of Biochemistry and Molecular Biology on the Nomenclature and Classification of Enzymes.* Academic Press, New York.

Jørgensen, S.T. (1996) *DNA Integration by Transposition.* International Patent Application WO 96/23073. World Intellectual Property, Geneva.

Jørgensen, S.T. (1999) *A Prokaryotic Cell Comprising Two Copies of a Gene Transcribed in Different Directions.* International Patent Application WO 99/41358. World Intellectual Property, Geneva.

Jørgensen, S.T. & Diderichsen, B.K. (1997) Bacillus *Promoter Derived from a Variant of a* Bacillus licheniformis α-*amylase Promoter.* US Patent 5,698,415. US Patent and Trademark Office.

Jørgensen, S.T. & Pedersen, K.B. (1999) *Protein Producing Cell containing Multiple Copies of a Desired Gene and a Screenable Marker but no Selection Marker.* International Patent Application WO 99/01562. World Intellectual Property, Geneva.

Jørgensen, S.T., Jørgensen, P.L. & Diderichsen, B.K.(1997) *Stable Integrations of DNA in Bacterial Genomes.* US Patent 5,695,976. US Patent and Trademark Office.

Nielsen, P., Fritze, D. & Priest, F.G. (1995) Phenetic diversity of alkaliphilic *Bacillus* species: proposal for nine new species. *Microbiology* **141**, 1745–1761.

Outtrup, H. (1986) *Maltogenic Amylase Enzyme, Preparation and Use Thereof.* US Patent 4,604,355. US Patent and Trademark Office.

Outtrup, H., Dambmann, C., Olsen, A.A., Bisgaard-Frantzen, H. & Schülein (1997) *Alkalophilic* Bacillus *sp. AC13 and Protease, Xylanase, Cellulase Obtainable Therefrom.* US Patent 5,696,068. US Patent and Trademark Office.

Strunk, O. & Ludwig, W. (1995) *ARB: a software environment for sequence data* (retrievable from http//www.arb@mikro.biologie.tumuenchen.de). Lehrstuhl für Mikrobiologie, Technical University of Munich, Munich, Germany.

Tangney, M., Hansen, C., Pedersen, P.E., Jørgensen, P.L. & Jørgensen, S.T. (1998) *Introduction of DNA into* Bacillus *Strains by Conjugation.* US Patent 5,843,72. US Patent and Trademark Office.

Widner, W., Sloma, A. & Thomas, M.D. (1999) *Methods for Producing a Polypeptide in a* Bacillus *cell.* International Patent Application WO 99/43835. World Intellectual Property, Geneva.

Chapter 15

Plant Growth Promotion by *Bacillus* and Relatives

C.P. Chanway

Introduction

The idea that soil microorganisms can have beneficial effects on plant growth is not new. The *Rhizobium*–legume symbiosis, which results in the formation of root nodules that fix biologically significant amounts of nitrogen, was discovered over 100 years ago (Beijerinck 1888 – cited in Paul & Clark 1996). Hiltner (1904 – also cited in Paul & Clark 1996) recognized the potential importance of microbial activities associated with root systems in plant nutrition and coined the term 'rhizosphere' to describe the zone of intense microbial activity around roots of the Leguminosae. This term is now used in a more general sense to describe soil influenced physically and/or chemically by any root system.

It is not surprising that soil microorganisms may mediate plant growth when one considers their large population sizes in nature: field soil has been shown to support between 10^6 and 10^8 microorganisms per gram (dry weight) of soil (Griffin 1972), but rhizosphere soil is even more populous, with up to 3×10^9 bacterial cells per gram (dry weight) of soil (Rouatt & Katznelson 1961). It should be noted that these estimates are derived from culture-dependent, dilution plate assays. If recent estimates that only a small fraction of soil bacteria can be cultured in the laboratory are accurate (Ward *et al.* 1990; Stackebrandt 1995; Tiedje *et al.* 1999), then natural populations of these microorganisms reach truly incredible sizes. Because all soil-borne nutrients obtained by the plant root must pass through the rhizosphere, the potential for microbes to alter these compounds in a way that will affect plant growth is great.

Plant-beneficial microorganisms have been the focus of many investigations and can be further categorized into symbiotic and free-living bacteria and fungi. The distinction between symbiotic and asymbiotic microbe–plant relationships, however, is not always clear. Generally, when physical attachment of the microorganism to the plant host occurs (e.g. formation of a root nodule or infection of a root tip by a mycorrhizal fungus), the relationship is considered to be symbiotic. Bacteria which proliferate in the rhizosphere with no obvious sign of attachment to the root system or formation of special organs are considered to be associative or asymbiotic. It is difficult to classify nonpathogenic bacteria and fungi that live inside plant tissues and, in some cases, stimulate plant growth.

History of plant growth promotion by *Bacillus*

The practical significance of plant inoculation with beneficial asymbiotic rhizosphere bacteria was first demonstrated during the early and mid-twentieth century in Russia. Greenhouse and field crops were routinely inoculated with 'phosphobacterin', a preparation of '*B. megatherium* var. *phosphaticum*' and 'azotobacterin', which consisted of several species of the genus *Azotobacter* including *A. chroococcum*.

However, these Russian experiments to determine inoculation effects on crop yields were apparently not designed and analysed with the appropriate degree of statistical rigor. When such inocula were subsequently tested in carefully designed experiments and analysed with rigorous statistical tests, earlier yield claims of plant growth promotion were found to be largely unsubstantiated. The benefit in yield was observed to be only small (Mishustin & Naumova 1962) or in some cases absent (Smith *et al.* 1961; Sundara-Rao *et al.* 1963). Two independent studies assessing the overall significance of the Soviet inoculum reached similar conclusions (Cooper 1959; Mishustin 1963).

While the original claims regarding efficacy of these bacterial inoculants were seriously questioned (Brown 1974), Cakmakci *et al.* (1999) have recently demonstrated that inoculation with mixed cultures of *Paenibacillus polymyxa* and '*B. megatherium* var. *phosphaticum*', i.e. phosphobacterin, increased sugar beet root weight and barley seed yields by 19 and 25.9%, respectively, in greenhouse trials, and up to 16.5 and 18.2%, respectively, in field trials. These results suggest that the early Russian work with *Bacillus* would have been more successful if mixed cultures of *Bacillus* had been used as the inoculum. Nevertheless, the Russian work had the very positive effect of stimulating widespread interest in the use of soil microorganisms to enhance plant productivity. Following several classical studies (e.g. Rovira & Bowen 1966; Brown 1974), Kloepper and Schroth (1978) introduced the term plant growth-promoting rhizobacteria (PGPR) to describe strains of naturally occurring asymbiotic soil bacteria that have the capability to colonize plant roots and stimulate plant growth.

PGPR activity has since been reported in strains of soil bacteria belonging to several genera including *Achromobacter*, *Arthrobacter*, *Azospirillum*, *Azotobacter*, *Bacillus*, *Burkholderia*, *Clostridium*, *Hydrogenophaga*, *Microbacterium*, *Paenibacillus*, *Serratia*, *Staphylococcus* and *Streptomyces* (Rovira 1963; Kloepper *et al.* 1986, 1988, 1989; Chanway & Holl 1992; O'Neill *et al.* 1992; Kloepper 1993; Chanway *et al.* 1994; Hebbar *et al.* 1994; Bertrand *et al.* 2000). The list of plant species that have been reported to respond positively to PGPR inoculation is long, and includes cereals, oilseeds and vegetables, as well as gymnosperm and angiosperm tree seedlings (Kloepper *et al.* 1989; Zaady *et al.* 1995; Chanway 1998; Enebak *et al.* 1998).

There is little doubt that both plant and bacteria lists will grow as representatives of other genera and species are examined. Beneficial rhizobacteria may also act primarily by stimulating seedling emergence; such microorganisms are not considered to be PGPR *per se*, and have been termed emergence-promoting rhizobacteria (EPR) (Kloepper *et al.* 1986).

Ecological aspects of plant growth promotion by *Bacillus*

For some time after the Soviet work with bacterial inoculants, only 'vigorous' root colonizers were considered to be candidate PGPR strains. Spore-forming organisms such as '*B. megatherium* var. *phosphaticum*' were shown to grow less readily in the rhizosphere than other types of bacteria (Lochhead 1940), and were assumed to be ineffective candidates for PGPR. Clearly some degree of root and/or rhizosphere colonization is an obvious prerequisite for PGPR to stimulate host plant performance (Schmidt 1979). The ability to colonize roots and rhizosphere soil, however, is a strain-specific characteristic, which varies significantly within bacterial species (Kloepper *et al.* 1985). As a result, generalizations about the root-colonizing capabilities of entire genera, such as *Bacillus* and relatives, are of little value.

Furthermore, while intuitively appealing, consistent demonstration of a quantitative relationship between rhizosphere colonization by PGPR and subsequent plant growth promotion is lacking (Reddy & Rahe 1989; Chanway 1997). In one study, growth promotion of lodgepole pine (*Pinus contorta* Dougl.) by *Paenibacillus polymyxa* was inversely correlated to the size of the root-associated PGPR population (Holl & Chanway 1992).

In addition to their ability to promote plant growth, sometimes in the absence of large rhizosphere population sizes, certain *Paenibacillus* PGPR possess the intriguing ability to develop highly specific plant growth response relationships with inbred 'host' plants such as spring wheat (Rennie & Larson 1979; Chanway *et al.* 1988a) or clonal plants such as white clover and perennial ryegrass (Chanway *et al.* 1991). When clones of white clover and perennial ryegrass were inoculated in the greenhouse with *Paenibacillus* previously isolated from the rhizosphere soil of the 'parental' clover and ryegrass plants from which the clones originated, plant growth promotion was significantly greater compared to clones inoculated with *Paenibacillus* isolates they had not previously coexisted with (Chanway *et al.* 1988a,b, 1990, 1991).

Plant growth response specificity of spring wheat inoculated with *Paenibacillus* was obscured, however, when trials were conducted in small field plots instead of under controlled environmental conditions (Chanway & Nelson 1990). Indeed, the presence of only one other bacterium as a co-inoculant inhibited the PGPR strain *P. polymyxa* Pw-2 from stimulating plant growth under a controlled environment (Bent & Chanway 1998). Because plant–microbe genotype specificity is usually detected under controlled environmental conditions in the absence of other soil microorganisms, it appears that this phenomenon may be more important from an academic perspective than a practical one, as most PGPR strains do not show such close specificity in field soil (Holl *et al.* 1988; Bashan *et al.* 1989). In some cases, however, the genotypic match between plants and microbes may determine not only the magnitude but also the type of plant growth response (i.e. positive or negative) (Rovira 1963; Rennie & Larson 1979; Millet *et al.* 1985; Sumner 1990; Chanway & Holl 1992; Chanway 1995), and is interesting from ecological and short-term evolutionary perspectives.

Rennie and Larson (1979) isolated the *Paenibacillus* strain for experimentation with wheat from soil that had been used to grow wheat since 1911. The

Paenibacillus strains used by Chanway *et al.* (1988b) were isolated from a permanent pasture sown in 1939 and managed solely by adjustment of the timing and intensity of grazing. *Paenibacillus* and *Bacillus* used by Chanway *et al.* (1988a) were isolated from soil continuously cropped to spring wheat for 27 years and to the same cultivar for the 5 years immediately preceding sampling. These results indicate that plant growth response specificity involving *Bacillus* and *Paenibacillus* may develop within 5 years of coexistence with plants in the field and persist for decades.

How such plant–microbe genotype specificity develops is not known, but there are several possibilities. Rhizosphere bacteria depend on plant root exudates – which are low molecular weight, easily oxidizable compounds – for much of their nutrition (Rovira 1963). Biologically significant qualitative and quantitative differences in root exudation can occur between plants even of the same species, resulting in differential selection pressures on soil microorganisms in and around the rhizospheres of different plants (Neal *et al.* 1970, 1973). Consequently, certain *Bacillus* and *Paenibacillus* strains already present in soil may experience a competitive advantage in the rhizosphere of the 'host' plant, thereby increasing bacterial fitness so that their population(s) reach a threshold size that results in plant growth promotion.

Alternatively, it is possible that certain bacterial strains in the rhizosphere adapt to conditions near the roots of the host plant by undergoing genetic reorganization through conjugation, transduction (Vettori *et al.* 1999) or transformation. The latter process has recently been shown to be an important mechanism of horizontal gene transfer from chromosomes and plasmids in soil owing to the protective effects of sand, clay minerals and humic acids when they bind DNA (Lee & Stotzky 1999). In addition, the process of random mutation will generate several genetic variants per bacterial generation. Therefore, pre-existing genetic variability within the *Bacillus/Paenibacillus* population may not be as important as recently generated variability, and selection within the rhizosphere increases the frequency of plant growth-promoting bacterial genotypes.

Bacillus and *Paenibacillus* may also influence competitive interactions between plants of the same or different species, likely through differential effects on plant growth, as well as interactions involving other plant-associated soil microorganisms such as root nodule bacteria (Chanway *et al.* 1991) and mycorrhizal fungi (Garbaye 1994).

Mechanisms of *Bacillus* PGPR activity

Direct effects of *Bacillus* and a possible basis for plant–microbe specificity

There are several mechanisms by which PGPR, including *Bacillus* and relatives, may stimulate plant growth. These can be categorized broadly as either direct or indirect plant growth promotion mechanisms. When plant growth promotion is direct, bacteria produce a metabolite or compound that itself is stimulatory to

Table 15.1 Summary of mechanisms thought to be involved in direct plant growth promotion by soil bacteria.

Mechanism	Effect(s) on plant growth
Nitrogen fixation on or in root or shoot tissue	Increased plant biomass or nitrogen content
Production of plant growth regulators (e.g. auxin, cytokinins or giberellins)	Increased shoot or root biomass or root branching; induction of reproductive cycles
Inhibition of ethylene synthesis in inoculated plants	Increased root length
Phosphorus solubilization content	Increased plant biomass or phosphorus
Sulfur oxidation	Increased plant biomass or foliar sulfur content
Increased root permeability	Increased plant biomass and nutrient uptake
Increase nitrate reductase activity and nitrate assimilation	Increased plant biomass or nitrogen content

Table 15.2 Summary of mechanisms by which soil bacteria are thought to stimulate plant growth indirectly.

Mechanism	Effect(s) on plant growth
Increased root nodule number or size on legumes or actinorhizal plants	Increased plant biomass, nitrogen content or reproductive yield
Increased infection frequency or efficacy by endo- or ectomycorrhizal fungi	Increased plant biomass
Suppression of deleterious rhizobacteria and disease-causing microorganisms	Increased plant biomass and reduced incidence of disease symptoms and mortality
Induction of plant systemic resistance to pathogens	Increased biomass plant and reduced incidence of disease symptoms and mortality

plants, so that plant growth stimulation occurs independently of the influence of other soil microorganisms. Direct mechanisms of plant growth promotion are summarized in table 15.1. On the other hand, bacteria that stimulate plant growth indirectly do so by affecting other factors in the rhizosphere, which in turn results in enhanced plant growth (Kloepper 1993). Indirect mechanisms of plant growth promotion are summarized in table 15.2.

One of the most commonly reported direct plant growth promotion mechanism by *Bacillus* and relatives is the production of plant growth substances such as auxins, cytokinins and/or giberellins (Holl *et al.* 1988; Lebuhn *et al.* 1997). Because plant growth substances are very efficacious at extremely low concentrations, i.e. even at picomolar concentrations, bacteria do not have to produce large quantities of these compounds to affect plant growth, sometimes profoundly. This may result, in part, from the microhabitat colonized by phytohormone-producing bacteria which can be the rhizoplane (RP), i.e. the root surface, or even the root cortex (i.e. internal root tissues). Once produced, the phytohormone(s) need not diffuse great distances to affect root tissues.

This idea is supported by Mavingui *et al.* (1992), who used immunotrapping, phenotypic characterization and RFLP analysis to demonstrate the existence of genetically distinct populations of *P. polymyxa* in the rhizosphere of wheat. Genotypic diversity of *P. polymyxa* was greatest in nonrhizosphere soil (NRS) and rhizosphere soil (RS), and lowest on the RP. The major distinguishing characteristic of the RP-colonizing *P. polymyxa* population compared to the RS- and NRS-colonizing populations was its ability to metabolize sorbitol. This led the authors to hypothesize that wheat roots select a specific *Paenibacillus* population that is adapted to living on the wheat RP with the ability to grow anaerobically on highly reduced substrates such as sorbitol. These substrates would be plentiful as a consequence of the intense competition for oxygen on the RP from root respiration.

In addition, Lebuhn *et al.* (1997) demonstrated that RP-colonizing *P. polymyxa* is most likely to produce plant growth-altering substances compared to RS- and NRS-colonizing *P. polymyxa*. They also detected distinct differences in the type of plant growth-altering substances (indolic vs. phenolic compounds) *P. polymyxa* produced, depending on its specific location within the wheat rhizosphere, i.e. the NRS, RS or RP. There was a gradual decrease in the potential to produce indole-3-acetic acid (IAA) from an NRS-colonizing strain to an RP-colonizing strain, with the RS-colonizing strain able to produce an intermediate quantity of IAA. The NRS strain also possessed the greatest ability to produce more oxidized compounds such as indole-3-carboxylic acid (ICA) and benzoic acid (BA). Concomitant with the decrease in potential IAA production by strains isolated from microsites closer to the RP was an increase in the ability to produce indole-3-ethanol (TOL) and indole-3-lactic acid (ILA). The observations that RP-colonizing *P. polymyxa* ecotypes had higher dehydrogenase expression/activity compared to NRS-colonizing ecotypes, which had higher potential oxidase expression/activity, were consistent with the hypothesis proposed by Mavingui *et al.* (1992).

Paenibacillus polymyxa isolated from the RP also appear to have characteristics rendering them more likely to promote plant growth compared to NRS-colonizing isolates (Lebuhn *et al.* 1997). These include a higher ratio of potential production of TOL and ILA to IAA compared with RS and NRS isolates, as well as a higher potential for auxin catabolism to indole-3-aldehyde (Iald). Because overproduction of IAA can inhibit root growth (Patten & Glick 1996) and TOL is a plant IAA storage compound (Sandberg 1984), TOL production and auxin catabolism may be important factors in rendering RP-colonizing *Bacillus* isolates as effective plant growth promoters. In addition, *P. polymyxa* RP strains also possess higher chitinase activity and soil-borne phytopathogen antagonistic activity compared to NRS strains, which enhances their ability to stimulate plant growth in the presence of pathogens (see below).

It is also possible that colonization of plant tissues by *Paenibacillus* may stimulate the 'host' plant itself to produce plant growth substances, thereby adding to any effects caused by bacterial production or catabolism of plant growth substances. Whether the source of plant growth substances is bacterial, plant or both, production of plant growth substances in the appropriate ratios can result in enhanced shoot or root biomass, increased root branching or induction of the plant's reproductive cycle (Brown 1974).

Another novel way in which *Bacillus* and relatives might stimulate plant growth directly is by suppressing endogenous ethylene production in the 'host' plant's rhizosphere and root system. Ethylene can act as a potent phytohormone that often has negative effects on plant performance (Glick 1995). For this mechanism to work, the microorganism must possess the enzyme 1-aminocyclo-propane-1-carboxylic acid (ACC) deaminase, which hydolyses plant-produced ACC to ammonia and alpha ketoglutarate. Because ACC is the immediate precursor in ethylene synthesis (Glick *et al.* 1994), ethylene production by the plant is reduced or eliminated (at least in the vicinity of the bacteria). The end result is that the suppressive effects of ethylene are eliminated, thereby facilitating better root growth in inoculated seedlings. This mechanism has been demonstrated with *Pseudomonas* strains, but *Bacillus* has yet to be tested. Certain plant growth-promoting strains of *B. firmus*, *B. circulans* and *B. globisporus*, however, have been shown to possess ACC deaminase and can utilize ACC as their sole nitrogen source (S. Ghosh, unpubl. data), rendering it likely that certain *Bacillus* strains are also capable of promoting plant growth in this way.

Strains of many *Bacillus* species possess nitrogenase and are able to fix atmospheric nitrogen. Theoretically, such *Bacillus* strains could stimulate plant growth by colonizing plant tissues – external or internal – and providing fixed nitrogen to the host plant. In reality, nitrogen fixation by *Bacillus* has been detected in the rhizosphere of many plants, and in one case was attributed to a significant growth response in wheat (Rennie & Larson 1979). The amounts of fixed nitrogen, however, are usually too small to cause a significant increase in the host plant's biomass (Holl *et al.* 1988; Chanway & Holl 1991). Furthermore, nitrogen fixation by endophytic *Bacillus* strains remains to be demonstrated notwithstanding the isolation of strains from within plant tissues that possess nitrogenase (Chanway 1998; Shishido *et al.* 1999).

Various *Bacillus* species have also proven able to increase nutrient availability in the rhizosphere. For example, phosphorus is often abundant in soils, but in forms mostly unavailable to plants, as part of insoluble or poorly soluble inorganic or organic phosphate pools (Anderson 1976). Gerretsen (1948) argued over 50 years ago that microorganisms capable of increasing the availability of phosphorus in soil could stimulate plant growth, a point that was demonstrated to some degree by the use of 'phosphobacterin' ('*B. megatherium* var. *phosphaticum*') in Russia at about the same time. Since then, several species of *Bacillus* have been shown capable of phosphate mineralization by producing enzymes (phosphatase or phytase) which cleave phosphate groups from complex organic phosphate molecules, or by producing organic acids (e.g. 2-ketogluconic acid) (Duff *et al.* 1963) that lower the local pH and increase the solubility of inorganic phosphorus compounds (Gaur & Ostwal 1972; Kundu & Gaur 1980; Subba Rao 1982).

Other plant growth-promoting microorganisms have been shown to stimulate plant growth by enhancing sulfur availability in the rhizosphere through sulfur oxidation (Grayston & Germida 1991), increasing root permeability (Sumner 1990) or by increasing the amount of nitrate-nitrogen through the action of bacterial nitrate reductase (Kapulnik *et al.* 1985; Boddey *et al.* 1986). The role of any species of *Bacillus* or relatives in these processes has yet to be demonstrated.

Indirect effects of *Bacillus* on plant growth

There are several ways in which PGPR, including *Bacillus*, can stimulate plant growth indirectly. These include positive interactions with symbiotic bacteria and fungi. For example, *Bacillus* has been shown capable of increasing legume root nodule number and size (Chanway *et al.* 1990; Petersen *et al.* 1996), accelerating formation of endomycorrhizae (von Alten *et al.* 1993; Budi *et al.* 1999) and increasing the number of ectomycorrhizal root tips (Garbaye 1994). For enhanced root nodule formation to be of value, the host plant must be growing in a nitrogen-deficient soil and the root nodule bacteria must be effective nitrogen fixers. In the case of both endomycorrhizae and ectomycorrhizae, the fungus–root association must provide the host plant with a growth-limiting factor, such as phosphorus, or alleviate some stress such as infection by pathogens, a lack of water or an elevated heavy metal content in the soil for the mycorrhiza-stimulating bacteria to have a positive effect on plant growth. In the absence of such stresses, one could expect a mycorrhiza enhancing *Bacillus* to have a negative effect on plant growth owing to the carbon cost associated with forming the mycorrhizal root tips (Chanway & Holl 1991).

Arguably, the most important indirect mechanism of growth promotion relates to biological control of major plant pathogens, i.e. those that cause obvious symptoms of disease, and minor pathogens, which are soil microorganisms that inhibit plant growth, but do not cause symptoms of disease (Kloepper 1993). In either case, *Bacillus* can exhibit biological control activity through direct antagonism of the fungal or bacterial pathogen *in situ*. Elucidation of the precise mechanism by which *Bacillus* promotes plant growth in this case is not, however, an easy task, and can be confounded by the fact that individual strains may possess characteristics consistent with several of the aforementioned mechanisms (e.g. Holl *et al.* 1988; Kloepper 1993). Furthermore, for any given general mechanism, there are several possible specific means by which seedling growth may be promoted. Consider the case of a biological control agent known to be capable of antagonizing pathogens directly. If it is determined that a particular *Bacillus* strain stimulates plant growth by suppressing minor pathogens in the rhizosphere, then the strain may act through one or more of several possible modes of action to inhibit the activity or growth of such microorganisms in the rhizosphere (table 15.3).

These could include the production of microbial growth inhibitors as *Bacillus* and relatives are well known to produce a range of antibiotics (Katz & Demain 1977). Examples include polymyxin A, B, D and E, as well as gavaserin and saltavalin by *Paenibacillus polymyxa* (Ramachandran *et al.* 1982; Brock *et al.* 1994; Pichard & Thouvenot 1995), bacitracins A-F by *Bacillus licheniformis* (Ishihara *et al.* 1982), bacillomycin D, subtilin and iturins by *B. subtilis* as well as zwittermicin by *B. cereus* (Kurahashi *et al.* 1982; Gueldner *et al.* 1988; Krebs *et al.* 1998). In addition, several unidentified antibacterial and/or antifungal com-

Table 15.3 Summary of mechanisms by which soil bacteria may inhibit deleterious rhizosphere microorganisms and plant pathogens (adapted from Kloepper 1993).

Competition for Fe^{3+} ions through siderophore production
Competition for colonization sites and nutrients other than Fe^{3+} ions
Antibiotic production
Production of fungal cell wall lytic enzymes
Production of hydrocyanic acid
Induction of plant systemic resistance to deleterious microorganisms and pathogens

pounds are also produced by *B. cereus*, *P. polymyxa* and *B. subtilis* (Gupta & Utkhede 1987; Handelsman *et al.* 1990; Smith *et al.* 1993; Kumar 1996; Kim *et al.* 1997; Walker *et al.* 1998; Wilhelm *et al.* 1998; Mathre *et al.* 1999; Utkhede 1999; Utkhede *et al.* 1999). These compounds have been implicated in the biocontrol of fungal pathogens such as chestnut blight (*Cryphonectria parasitica*), damping off (*Alternaria brassicicola*), *Fusarium oxysporum*, *Rhizoctonia solani*, *Sclerotina sclerotiorum*, grey mould (*Botrytis cinerea*), *Pythium mamillatum*, *Py. aphanidermatum*, *Py. irregulare*, *Py. ultimum*, take-all of wheat [*Gaeumannomyces graminis* var. *tritici* (Ggt)], apple replant disease, *Phytophthora parasitica*, *Ph. cactorum*, *Ph. megasperma*, *Ph. sojae*, *Sclerotium rolfsii* and bacterial diseases such as black rot (*Xanthomonas campestris* pv. *campestris*). One *B. subtilis* strain isolated from soil in China was even shown to produce a volatile compound that inhibited the growth of Ggt (Ryder *et al.* 1999).

Antibiosis, however, is not the only mechanism by which *Bacillus* can control pathogens. As a consequence of its poor availability in oxygenated environments, the Fe^{3+} ion is often the object of intense competition between microorganisms, and the most successful scavengers of this ion are the ones able to secrete extracellular siderophores with high affinities for it. In fact, microorganisms with highly effective siderophores can inhibit the activity of, or even eliminate, competing microbes. If plant pathogens comprise part of the competing microflora, inoculation with siderophore-producing bacteria can result in biological control (Kloepper 1993). Siderophore production is usually associated with members of the genus *Pseudomonas*; however, siderophore production by *Bacillus* spp. has not been evaluated widely. In a recent study, 'moderate' siderophore production by a strain of *B. cereus* that has biocontrol capabilities against Ggt and *R. solani*, was detected (Ryder *et al.* 1999), which indicates a potential role for these compounds in biocontrol by *Bacillus*.

Other ways in which *Bacillus* may antagonize pathogens include production of extracellular chitinase, which was produced by *B. cereus* and was effective in suppressing *R. solani* on cotton (Pleban *et al.* 1997). Bacteria capable of biological control may also suppress neighbouring microorganisms by outcompeting them for nutrients other than Fe^{3+}, as well as through the production of hydrocyanic acid. While these mechanisms have been demonstrated to occur in plant growth-promoting members of other bacterial genera, some remain to be examined within the genus *Bacillus* and relatives.

Induced systemic resistance

While many *Bacillus* strains are capable of inhibiting pathogen growth or activity directly, inoculation of plants with certain strains can cause a plant response known as induced systemic resistance (ISR), in which certain plant defence mechanisms are 'turned on' so that when challenged by a disease-causing microorganism, the plant is much better able to resist infection by the pathogen.

Because bacteria capable of biocontrol are often able to suppress pathogens either directly, through antagonistic interactions with the pathogen, or indirectly, by inducing systemic resistance within the host plant, determining exactly which mechanism is at work is difficult. To unequivocally demonstrate ISR as the mechanism by which a bacterial strain controls a disease, it must be shown that no contact occurs between the inducing bacteria and the disease-causing pathogen in the plant and that no symptoms of disease develop. In addition, the possibility that biocontrol results from translocation of pathogen-inhibiting substances produced by a bacterial strain *in situ* must also be eliminated. Use of heat-killed bacteria as controls as well as purified bacterial components known to induce systemic resistance, such as lipopolysaccharides or oligopolysaccharides, have helped demonstrate ISR by endophytic bacteria. It is also reassuring if the inducing microorganism shows no sign of pathogen inhibition or antagonism *in vitro*.

The first suggestion that *Bacillus* had the ability to protect host plants through ISR was provided by Mann (1968), who drenched tobacco roots with cultures of '*B. uniflagellatus*' to induce resistance to the tobacco mosaic virus (TMV). A reduction in the number of lesions was observed and assumed to be the result of ISR. Since that initial report, there have been several demonstrations of disease control by ISR after inoculation with external or internal tissue-colonizing *Bacillus* isolates. These include: leaf spot (*Pseudomonas syringae* pv. *lachrymans*) and anthracnose (*Colletotrichum orbiculare*) on cucumber inoculated with *B. pumilis* (Wei *et al.* 1996), *Fusarium oxysporum* on tomato by *B. pumilus* (Benhamou *et al.* 1998), and *Pythium aphanidermatum* and *Phytophthora nicotianae* on tomato (*Lycopersicon esculentum* Mill.) and cucumber (*Cucumis sativus* L.) by *B. subtilis* (Grosch *et al.* 1999). More examples of *Bacillus* ISR can be found in the Proceedings of the Third (Ryder *et al.* 1994) and Fourth (Ogoshi *et al.* 1997) International Workshops on PGPR.

In addition to control of TMV, there is also evidence that ISR resulting from inoculation with endophytic bacteria affords a degree of protection from plant-parasitic nematodes (Hallmann *et al.* 1997) and cucumber beetles (Zehnder *et al.* 1997), but comparatively little work has been carried out in this area.

ISR is a particularly effective disease control mechanism for pathogens that are able to avoid triggering natural plant defence responses upon infection. Disease severity, however, can also be reduced when caused by pathogens that are able to suppress or evade plant defence responses even after they have been triggered. Bacteria that are capable of inducing systemic resistance usually do so without causing any visible symptoms of injury on the host plant, although certain anatomical changes associated with disease resistance may occur within plant tissues. These changes involve epidermal and cortical cell wall fortification through the deposition of callose and phenolic materials and may comprise an important component of the ISR response.

It is well known that the outer membrane lipopolysaccharides (LPS) of Gram-negative bacteria are effective inducers of systemic resistance in plants. Other compounds known to be important for induction of systemic resistance by rhizobacteria include siderophores, such as pyoverdin, and, in some cases, salicylic acid, although the involvement of the latter appears to vary with the specific microorganism–plant-'host' interaction. Bioactive oligosaccharides such as chitosan – a deacetylated chitin derivative that occurs in the cell wall of many fungi – have also been shown to be effective elicitors of ISR in plants. In addition, there is evidence that ethylene-dependent signalling is required at the bacterial colonization site for ISR to occur.

More recent research indicates that ISR efficacy can be significantly enhanced when plants are co-treated with endophytic bacteria and chemical elicitors of ISR (Benhamou *et al.* 1998). For example, tomato resistance against *F. oxysporum* was significantly enhanced when treated with the bacterial endophyte, *B. pumilus* SE34 in combination with chitosan. Such an approach, involving a combination of biotic and abiotic ISR elicitors, may prove to be an effective adjunct to purely biological means of pathogen control and warrants further study.

Commercial *Bacillus* inoculants

B. subtilis strain A13

Bacillus and relatives form endospores that are resistant to environmental stresses such as desiccation, heat and UV radiation (Walker *et al.* 1998). These characteristics are useful in producing commercial inocula with a long shelf-life, compared with pseudomonad-based inocula and other nonspore-forming biocontrol agents, which lose viability comparatively quickly. In view of these characteristics and the several studies that have demonstrated significant biological control or plant growth promotion by *Bacillus* and relatives, one would think that several commercial formulations of these microbes would be available for use in an agricultural context. The earliest commercial use of a *Bacillus*-based inoculant for plant growth promotion occurred in Russia when 'phosphobacterin' was applied to millions of hectares of farmland in the middle of the twentieth century. In the west, *B. subtilis* strain A13, isolated by Broadbent *et al.* (1971), was evaluated for use as a biological control agent as well as enhancing seeding vigour and yield of field-grown cereals and carrots (Merriman *et al.* 1974). However, unpredictable and large variations in seedling growth in response to inoculation with beneficial bacteria – the bane of almost all PGPR work – precluded commercial development of strain A13, notwithstanding some impressive results under controlled environments.

This setback was only temporary as Gustafson, Inc. (Plano, TX) re-evaluated strain A13 and found that biocontrol and plant growth promotion was much more consistent when it was combined with seed-treatment fungicides (Backman *et al.* 1997). Plant growth responses were large and consistent enough to warrant commercialization of strain A13 as a peanut inoculant that could promote plant growth and provide effective control of certain root pathogens such as *Rhizoctonia solani*. Strain GB03, which was obtained by host passage of strain

A13 through cotton, and a related strain GB07 were also considered to have significant commercial potential, which resulted in Gustafson, Inc., marketing these strains in 1985 under the trade names Quantum® for strain A13, Kodiak® for GB03 and Epic® for GB07. In 1996, 5 million hectares of crops including peanuts and cotton were treated with these strains (Backman *et al.* 1997) to control root diseases such as *Rhizoctonia*, *Pythium* and *Fusarium* and to stimulate root growth and plant vigour, likely resulting from the production of plant growth substances (Backman *et al.* 1994; Brannen & Backman 1994). As of today, Gustafson, Inc., appears to be marketing only Kodiak® for protection against *Fusarium*, *Rhizoctonia* and 'other' diseases of vegetables and cotton.

Yield-increasing bacteria in China

Since the early 1980s, the Chinese have been marketing yield-increasing bacteria (YIB), most of which comprise spore-forming *B. cereus*. These strains and strain mixtures are purported to colonize roots and reduce the incidence of soil-borne diseases (Chen *et al.* 1996) and aboveground plant diseases (Backman *et al.* 1997). YIB have been applied to over 20 million hectares and have resulted in average yield increases of 10% in vegetables and up to 22.5% for sweet potato (Chen *et al.* 1996). Ryder *et al.* (1999) have confirmed the plant growth-promoting effects and biocontrol activity against Ggt and *Rhizoctonia* of many YIB in glasshouse experiments with wheat. Further evaluation is required to determine whether or not YIB have a role to play in western agriculture.

Concluding remarks

In spite of the enigmatic nature of PGPR including *Bacillus* and relatives, as well as the associated complexities that result from their interaction with abiotic environmental factors such as climate, soil physical and chemical properties, and biotic factors such as relationships with plants, other soil microorganisms and soil fauna (Chanway *et al.* 1991; Fitter & Garbaye 1994), significant advances have been made in understanding how these microorganisms affect plant growth (Kloepper 1993; Ryder *et al.* 1994). Microorganisms capable of enhancing plant growth through two or more mechanisms may turn out to be the most useful from a commercial perspective. One current example is *B. cereus* strain UW85, an effective biocontrol agent (Osburn *et al.* 1995) that can also enhance root nodule number (Halverson & Handelsman 1991) and root length of soybean (*Glycine max* (L.) Merrill) (Halverson 1991). Another example is *Bacillus* sp. strain B2, which has antagonistic activity towards plant pathogens, but also stimulates endomycorrhizal fungus infection (Budi *et al.* 1999). Through continuing research, we are making important inroads to understanding the biology and ecology of *Bacillus* plant growth promoters, which may result ultimately in more commercially viable and environmentally benign biocontrol and plant growth-promoting strains or strain mixtures for use in agriculture, horticulture and forestry.

References

Anderson, G. (1976) Other organic phosphorus compounds. In: *Organic Soil Components* (ed. J.E. Geiseking), pp. 305–331. Springer, Berlin.

Backman, P.A., Wilson, M. & Murphy, J.F. (1997) Bacteria for biological control of plant diseases. In: *Environmentally Safe Approaches to Crop Disease Control* (eds N.A. Rechcigl & J.E. Rechcigl), pp. 95–109. CRC Lewis Publishers, New York.

Bashan, Y., Ream, Y., Levanony, H. & Sade, A. (1989) Nonspecific responses in plant growth, yield, and root colonization of noncereal crop plants to inoculation with *Azospirillum brasilense* Cd. *Canadian Journal of Botany* 67, 1317–1324.

Benhamou, N., Kloepper, J.W. & Tuzun, S. (1998) Induction of resistance against *Fusarium* wilt of tomato by combination of chitosan with an endophytic bacterial strain: ultrastructure and cytochemistry of the host response. *Planta* 204, 153–168.

Bent, E. & Chanway, C.P. (1998) The growth-promoting effects of a bacterial endophyte on lodgepole pine are partially inhibited by the presence of other rhizobacteria. *Canadian Journal of Microbiology* 44, 980–988.

Bertrand, H., Plassard, C., Pinochet, X., Touraine, B., Normand, P. & Cleyet-Marel, J.C. (2000) Stimulation of the ionic transport system in *Brassica napus* by a plant growth-promoting rhizobacterium (*Achromobacter* sp.). *Canadian Journal of Microbiology* 46, 229–236.

Boddey, R.M., Baldani, V.I.D., Baldani, J.I. & Dobereiner, J. (1986) Effect of inoculation of *Azospirillum* spp. on nitrogen accumulation by field grown wheat. *Plant and Soil* 95, 109–121.

Brannen, P.M. & Backman, P.A. (1994) Suppression of *Fusarium* wilt of cotton with *Bacillus subtilis* hopper box formulations. In: *Improving Plant Productivity with Rhizosphere Bacteria* (eds M.H. Ryder *et al.*), pp. 83–85. CSIRO, Adelaide.

Broadbent, P., Baker, K.F. & Waterworth, Y. (1971) Bacteria and actinomycetes antagonistic to root pathogens in Australian soils. *Australian Journal of Biological Sciences* 24, 925–944.

Brock, T.D., Madigan, M.T., Martinko, J.M. & Parker, J. (1994) *Biology of Microorganisms*, 7th edn. Prentice Hall, New Jersey.

Brown, M.E. (1974) Seed and root bacterization. *Annual Review of Phytopathology* 12, 181–197.

Budi, S.W., van Tuinen, D., Martinotti, G. & Gianinazzi, S. (1999) Isolation from the *Sorghum bicolor* mycorrhizosphere of a bacterium compatible with arbuscular mycorrhiza development and antagonistic towards soilborne fungal pathogens. *Applied and Environmental Microbiology* 65, 5148–5150.

Cakmakci, R., Kantar, F. & Algur, O.F. (1999) Sugar beet and barley yields in relation to *Bacillus polymyxa* and *Bacillus megaterium* var. *phosphaticum* inoculation. *Journal of Plant Nutrition and Soil Science* 162, 437–442.

Chanway, C.P. (1995) Differential response of western hemlock from low and high elevations to inoculation with plant growth-promoting *Bacillus polymyxa*. *Soil Biology and Biochemistry* 27, 767–775.

Chanway, C.P. (1997) Inoculation of tree roots with plant growth-promoting soil bacteria: an emerging technolgy for reforestation. *Forest Science* 43, 99–112.

Chanway, C.P. (1998) Bacterial endophytes: ecological and practical implications. *Sydowia* 50, 149–170.

Chanway, C.P. & Holl, F.B. (1991) Biomass increase and associative nitrogen fixation of mycorrhizal *Pinus contorta* Dougl. seedlings inoculated with a plant growth promoting *Bacillus* strain. *Canadian Journal of Botany* 69, 507–511.

Chanway, C.P. & Holl, F.B. (1992) Influence of soil biota on Douglas-fir (*Pseudotsuga menziesii* (Mirb.) Franco) seedling growth: the role of rhizosphere bacteria. *Canadian Journal of Botany* 70, 1025–1031.

Chanway, C.P. & Nelson, L.M. (1990) Field and laboratory studies of *Triticum aestivum* L. inoculated with growth promoting *Bacillus* strains. *Soil Biology and Biochemistry* 22, 789–795.

Chanway, C.P., Nelson, L.M. & Holl, F.B. (1988a) Cultivar specific growth promotion of spring wheat (*Triticum aestivum* L.) by co-existent *Bacillus* species. *Canadian Journal of Microbiology* 34, 925–929.

Chanway, C.P., Holl, F.B. & Turkington, R. (1988b) Genotypic coadaptation in growth promotion of forage species by *Bacillus polymyxa*. *Plant and Soil* 106, 281–284.

Chanway, C.P., Holl, F.B. & Turkington, R. (1990) Specificity of association of *Bacillus* isolates with genotypes of *Lolium perenne* L. and *Trifolium repens* L. from a grass/legume pasture. *Canadian Journal of Botany* 68, 1126–1130.

Chanway, C.P., Turkington, R. & Holl, F.B. (1991) Ecological implications of specificity between plants and rhizosphere microorganisms. *Advances in Ecological Research* 21, 121–169.

Chanway, C.P., Shishido, M. & Holl, F.B. (1994) Root-endophytic and rhizosphere plant growth promoting rhizobacteria for conifer seedlings. In: *Improving Plant Productivity with Rhizosphere Bacteria* (eds M.H. Ryder *et al.*), pp. 72–74. CSIRO, Adelaide.

Chen, Y.X., Mei, R.H., Lu, S., Liu, L. & Kloepper, J.W. (1996) The use of yield-increasing bacteria (YIB) as plant growth-promoting rhizobacteria in Chinese agriculture. In: *Management of Soil Borne Diseases* (eds R.S. Utkhede & V.K. Gupta), pp. 165–184. Kalyani, New Delhi.

Cooper, R. (1959) Bacterial fertilizers in the Soviet Union. *Soils and Fertilizers* 22, 327–333.

Duff, R.B., Webley, D.M. & Scott, R.P. (1963) Solubilization of minerals and related materials by 2-ketogluconic acid producing bacteria. *Soil Science* 95, 105–114.

Enebak, S.A., Wei, G. & Kloepper, J.W. (1998) Effects of plant growth-promoting rhizobacteria on loblolly and slash pine seedlings. *Forest Science* 44, 139–144.

Fitter, A.H. & Garbaye, J. (1994) Interactions between mycorrhizal fungi and other soil organisms. *Plant and Soil* 159, 123–132.

Garbaye, J. (1994) Helper bacteria: a new dimension to the mycorrhizal symbiosis. *New Phytologist* 128, 197–210.

Gaur, A.C. & Ostwal, K.P. (1972) Influence of phosphate dissolving bacilli on yield and phosphate uptake by wheat crop. *Indian Journal of Experimental Biology* 10, 393–394.

Gerretsen, F.C. (1948) The influence of microorganisms on the phosphate intake by the plant. *Plant and Soil* 1, 51–85.

Glick, B.R. (1995) The enhancement of plant growth by free-living bacteria. *Canadian Journal of Microbiology* 41, 109–117.

Glick, B.R., Jacobson, C.B., Schwarze, M.M.K. & Pasternak, J.J. (1994) 1-Aminocyclopropane-1-carboxylic acid deaminase mutants of the plant growth promoting rhizobacterium *Pseudomonas putida* GR12-2 do not stimulate canola root elongation. *Canadian Journal of Microbiology* 40, 911–915.

Grayston, S.J. & Germida, J.J. (1991) Sulfur-oxidizing bacteria as plant growth promoting rhizobacteria for canola. *Canadian Journal of Microbiology* 37, 521–529.

Griffin, D.M. (1972) *Ecology of Soil Fungi.* Chapman & Hall, London.

Grosch, R., Junge, H., Krebs, B. & Bochow, H. (1999) Use of *Bacillus subtilis* as a biocontrol agent. III. Influence of *Bacillus subtilis* on fungal root diseases and on yield in soilless culture. *Journal of Plant Diseases and Protection* 106, 568–580.

Gueldner, R.C., Reilly, C.C., Pusey, P.L. *et al.* (1988) Isolation and identification of iturins as antifungal peptides in biological control of peach brown rot with *Bacillus subtilis. Journal of Agricultural and Food Chemistry* 36, 366–370.

Gupta, V.K. & Utkhede, R.S. (1987) Nutritional requirements for production of antifungal substances by *Enterobacter aerogenes* and *Bacillus subtilis* antagonists of *Phytophthora cactorum. Journal of Phytopathology* 120, 143–153.

Hallmann, J., Quadt-Hallmann, A., Mahafee, W.F. & Kloepper, J.W. (1997) Bacterial endophytes in agricultural crops. *Canadian Journal of Microbiology* 43, 895–914.

Halverson, L.J. (1991) *Population biology of* Bacillus cereus *UW85: a case study for biotechnology policy.* PhD thesis, University of Wisconsin, Madison, WI.

Halverson, L.J. & Handelsman, J. (1991) Enhancement of soybean nodulation by *Bacillus cereus* UW85 in the field and in a growth chamber. *Applied and Environmental Microbiology* 57, 2767–2770.

Handelsman, J., Raffel, S., Mester, E.H., Wunderlich, L. & Grau, C.R. (1990) Biological control of damping-off of alfalfa seedlings with *Bacillus cereus* UW85. *Applied and Environmental Microbiology* 56, 713–718.

Hebbar, K.P., Martel, M.H. & Heulin, T. (1994) *Burkholderia cepacia*, a plant growth promoting rhizobacterial associate of maize. In: *Improving Plant Productivity with Rhizosphere Bacteria* (eds M.H. Ryder *et al.*), pp. 201–203. CSIRO, Adelaide.

Holl, F.B. & Chanway, C.P. (1992) Rhizosphere colonization and seedling growth promotion of lodgepole pine by *Bacillus polymyxa. Canadian Journal of Microbiology* 38, 303–308.

Holl, F.B., Chanway, C.P., Turkington, R. & Radley, R. (1988) Growth response of crested wheatgrass (*Agropyron cristatum* L.), white clover (*Trifolium repens* L.), and perennial ryegrass (*Lolium perenne* L.) to inoculation with *Bacillus polymyxa. Soil Biology and Biochemistry* 20, 19–24.

Ishihara, H., Ogawa, I. & Shimura, K. (1982) Component I protein of bacitracin synthetase: a multifunctional protein. In: *Peptide Antibiotics: Biosynthesis and Functions* (eds H. Kleinkauf & H. von Dohreren), pp. 289–296. Walter de Gruyter, Berlin.

Kapulnik, Y., Feldman, M., Okon, Y. & Henis, Y. (1985) Contribution of nitrogen fixed by *Azospirillum* to the N nutrition of spring wheat in Israel. *Soil Biology and Biochemistry* 17, 509–515.

Katz, E. & Demain, A.L. (1977) The peptide antibiotics of *Bacillus*: chemistry biogenesis and possible functions. *Bacteriological Reviews* 41, 449–474.

Kim, D.-A., Cook, J. & Weller, D.A. (1997) *Bacillus* sp. L324-92 for biological control of three root diseases of wheat grown with reduced tillage. *Biological Control* 87, 551–558.

Kloepper, J.W. (1993) Plant growth-promoting rhizobacteria as biological control agents. In: *Soil Microbial Ecology – Applications in Agricultural and Environmental Management* (ed. F.B. Metting, Jr), pp. 255–274. Marcel Dekker, New York.

Kloepper, J.W. & Schroth, M.N. (1978) Plant growth promoting rhizobacteria on radishes. In: *Proceedings of the Fourth International Conference on Plant Pathogenic Bacteria* (ed. Station de Pathologie Végétale et Phytobactériologie, INRA, Angers), Vol. 2, pp. 879–882. Gilbert Clary, Tours.

Kloepper, J.M., Scher, F.M., Laliberte, M. & Zaleska, I. (1985) Measuring the spermosphere colonizing capacity (spermosphere competence) of bacterial inoculants. *Canadian Journal of Microbiology* 31, 926–929.

Kloepper, J.W., Scher, F.M., Laliberte, M. & Tipping, B. (1986) Emergence-promoting rhizobacteria: description and implications for agriculture. In: *Iron, Siderophores, and Plant Diseases* (ed. T.R. Swinburne), pp. 155–164. Plenum Press, New York.

Kloepper, J.W., Hume, D.J., Scher, F.M. *et al.* (1988) Plant growth-promoting rhizobacteria on canola (rapeseed). *Plant Disease* 72, 42–46.

Kloepper, J.W., Lifshitz, R. & Zablotowicz, R.M. (1989) Free-living bacterial inocula for enhancing crop productivity. *Trends in Biotechnology* 7, 39–44.

Krebs, B., Hoding, B., Kubart, S. *et al.* (1998) Use of *Bacillus subtilis* as biocontrol agent. I. Activities and characterization of *Bacillus subtilis* strains. *Journal of Plant Diseases and Protection* 105, 181–197.

Kumar, B.S.D. (1996) Crop improvement and disease suppression by a *Bacillus* spp. SR2 from peanut rhizosphere. *Indian Journal of Experimental Biology* 34, 794–798.

Kundu, B.S. & Gaur, A.C. (1980) Establishment of nitrogen fixing and phosphate solubilizing bacteria in rhizosphere and their effect on yield and nutrient uptake of wheat crop. *Plant and Soil* 57, 223–230.

Kurahashi, K., Komura, S., Akashi, K. & Nishio, C. (1982) Biosynthesis of antibiotic peptides polymyxin E and gramicidin A. In: *Peptide Antibiotics: Biosynthesis and Functions* (eds H. Kleinkauf & H. von Dohreren), pp. 275–288. Walter de Gruyter, Berlin.

Lebuhn, M., Heulin, T. & Hartmann, A. (1997) Production of auxin and other indole and phenolic compounds by *Paenibacillus polymyxa* strains isolated from different proximity to plant roots. *FEMS Microbiology Ecology* 22, 325–334.

Lee, G.-H. & Stotzky, G. (1999) Transformation and survival of donor, recipient, and transformants of *Bacillus subtilis in vitro* and in soil. *Soil Biology and Biochemistry* 31, 1499–1508.

Lochhead, A.G. (1940) Qualitative studies of soil microorganisms. III. Influence of plant growth on the character of bacterial flora. *Canadian Journal of Research* 18, 42–53.

Mann, E.W. (1968) Inhibition of taobacco mosaic virus by a bacterial extract. *Phytopathology* 59, 658–662.

Mathre, D.E., Cook, R.J. & Callan, N.W. (1999) From discovery to use: traversing the world of commercializing biocontrol agents for plant disease. *Plant Disease* 83, 972–983.

Mavingui, P., Laguerre, G., Berge, O. & Heulin, T. (1992) Genotypic and phenotypic diversity of *Bacillus polymyxa* in soil and in the wheat rhizosphere. *Applied and Environmental Microbiology* 58, 1894–1903.

Merriman, P.R., Price, R.D., Kollmorgen, J.F., Piggott, T. & Ridge, E.H. (1974) Effect of seed inoculation with *Bacillus subtilis* and *Streptomyces griseus* on growth of cereals and carrots. *Australian Journal of Agricultural Research* 25, 219–226.

Millet, E., Avivi, Y. & Feldman, M. (1985) Effects of rhizospheric bacteria on wheat yield under field conditions. *Plant and Soil* 86, 347–355.

Mishustin, E.N. (1963) Bacterial fertilizers and their effectiveness. *Mikrobiologiya* 32, 911–917.

Mishustin, E.N. & Naumova, A.N. (1962) Bacterial fertilizers, their effectiveness and mode of action. *Microbiology* 31, 442–452.

Neal, J.R., Jr, Atkinson, T.G. & Larson, R.I. (1970) Changes in the rhizosphere microflora of spring wheat induced by disomic substitution of a chromosome. *Canadian Journal of Microbiology* 16, 153–158.

Neal, J.R., Jr, Larson, R.I. & Atkinson, T.G. (1973) Changes in rhizosphere populations of selected physiological groups of bacteria related to substitution of specific pairs of chromosomes in spring wheat. *Plant and Soil* 39, 209–212.

Ogoshi, A., Kobayashi, K., Homma, Y., Kodama, F., Kondo, N. & Akino, S. (eds) (1997) *Plant Growth-Promoting Rhizobacteria – Present Status and Future Prospects*. Nakanishi Printing, Sapporo.

O'Neill, G.A., Chanway, C.P., Axelrood, P.E., Radley, R.A. & Holl, F.B. (1992) Growth response specificity of spruce inoculated with coexistent rhizosphere bacteria. *Canadian Journal of Botany* 70, 2347–2353.

Osburn, R.M., Milner, J., Oplinger, E.S., Stewart Smith, R. & Handelsman, J. (1995) Effect of *Bacillus cereus* UW85 on the yield of soybean at two field sites in Wisconsin. *Plant Disease* 79, 551–556.

Patten, C.L. & Glick, B.R. (1996) Bacterial biosynthesis of indole-3-acetic acid. *Canadian Journal of Microbiology* 42, 207–220.

Paul, E.A. & Clark, F.E. (1996) *Soil Microbiology and Biochemistry*, 2nd edn. Academic Press, San Diego, CA.

Petersen, D.J., Srinivasan, M.S. & Chanway, C.P. (1996) *Bacillus polymyxa* stimulates increased *Rhizobium etli* populations and nodulation when co-resident in the rhizosphere of *Phaseolus vulgaris*. *FEMS Microbiology Letters* **142**, 271–276.

Pichard, B. & Thouvenot, D. (1995) Effect of *Bacillus polymyxa* seed treatments on control of black-rot and damping-off of cauliflower. *Seed Science and Technology* **27**, 455–465.

Pleban, S., Chernin, L. & Chet, I. (1997) Chitinolytic activity of an endophytic strain of *Bacillus cereus*. *Letters in Applied Microbiology* **25**, 284–288.

Ramachandran, L.K., Srinivasa, B.R. & Radhakrishna, G. (1982) Structural requirements for the biological activity of polymyxin B. In: *Peptide Antibiotics: Biosynthesis and Functions* (eds H. Kleinkauf & H. von Dohreren), pp. 427–443. Walter de Gruyter, Berlin.

Reddy, M.S. & Rahe, J.E. (1989) Growth effects associated with seed bacterization not correlated with populations of *Bacillus subtilis* inoculant in onion seedling rhizospheres. *Soil Biology and Biochemistry* **21**, 373–378.

Rennie, R.J. & Larson, R.I. (1979) Nitrogen fixation associated with disomic chromosome substitution lines of spring wheat. *Canadian Journal of Botany* **57**, 2771–2775.

Rouatt, J.W. & Katznelson, H. (1961) A study of bacteria on the root surface and in the rhizosphere soil of crop plants. *Journal of Applied Bacteriology* **24**, 164–171.

Rovira, A.D. (1963) Microbial inoculation of plants. I. Establishment of free-living nitrogen-fixing bacteria in the rhizosphere and their effects on maize, tomato, and wheat. *Plant and Soil* **19**, 304–314.

Rovira, A.D. & Bowen, G.D. (1966) The effects of micro-organisms on plant growth. II. Detoxification of heat-sterilized soils by fungi and bacteria. *Plant and Soil* **25**, 129–142.

Ryder, M.H., Stephens, P.M. & Bowen, G.D. (eds) (1994) *Improving Plant Productivity with Rhizosphere Bacteria*. CSIRO, Adelaide.

Ryder, M.H., Yan, Z., Terrace, T.E., Rovira, A.D., Tang, W. & Correll, R.L. (1999) Use of strains of *Bacillus* isolated in China to suppress take-all and rhizoctonia root rot, and promote seedling growth of glasshouse-grown wheat in Australian soils. *Soil Biology and Biochemistry* **31**, 19–29.

Sandberg, G. (1984) Biosynthesis and metabolism of indole-3-ethanol and indole-3-acetic acid by *Pinus sylvestris* L. needles. *Planta* **161**, 398–403.

Schmidt, E.L. (1979) Initiation of plant root–microbe interactions. *Annual Review of Microbiology* **33**, 355–376.

Shishido, M., Breuil, C. & Chanway, C.P. (1999) Endophytic colonization of spruce by plant growth-promoting rhizobacteria. *FEMS Microbiology Ecology* **29**, 191–196.

Smith, J.H., Allison, F.E. & Soulides, D.A. (1961) Evaluation of phosphobacteria as a soil inoculant. *Soil Science Society of America Proceedings* **25**, 109–111.

Smith, K.P., Havey, M.J. & Handelsman, J. (1993) Suppression of cottony leak of cucumber with *Bacillus subtilis* strain UW85. *Plant Disease* **77**, 139–142.

Stackebrandt, E. (1995) The biodiversity convention and its consequences for the inventory of prokayotes. In: *The 7th International Symposium on Microbial Ecology*, p. 1, abstract 1. Winner Graph, Sao Paulo.

Subba Rao, N.S. (1982) Phosphate solubilization by soil microorganisms. In: *Advances in Agricultural Microbiology* (ed. N.S. Subba Rao), pp. 295–303. Butterworth, London.

Sumner, M.E. (1990) Crop responses to *Azospirillum* inoculation. *Advances in Soil Science* **12**, 53–123.

Sundara-Rao, W.V.B., Bajpai, P.D., Sharma, J.P. & Subbiah, B.V. (1963) Solubilization of phosphates by phosphorus solubilizing organisms using ^{32}P as tracer and the influence of seed bacterization on the uptake by the crop. *Indian Society of Soil Science* **11**, 209–219.

Tiedje, J.M., Asuming-Brempong, S., Nusslein, K., Marsh, T. & Flynn, S.J. (1999) Opening the black box of soil microbial diversity. *Applied Soil Ecology* **13**, 109–122.

Utkhede, R.S. (1999) Biological treatments to increase apple tree growth in replant problem soil. *Allelopathy Journal* **6**, 63–68.

Utkhede, R.S., Koch, C.A. & Menzies, J.G. (1999) Rhizobacterial growth and yield promotion of cucumber plants inoculated with *Pythium aphanidermatum*. *Canadian Journal of Plant Pathology* **21**, 265–271.

Vettori, C., Stotzky, G., Yoder, M. & Gallori, E. (1999) Interaction between bacteriophage PBS1 and clay minerals and transduction of *Bacillus subtilis* by clay-phage complexes. *Environmental Microbiology* **1**, 347–355.

von Alten, H., Lindemann, A. & Schonbeck, F. (1993) Stimulation of vesicular-arbuscular mycorrhiza by fungicides or rhizosphere bacteria. *Mycorrhiza* **2**, 167–173.

Walker, R., Powell, A.A. & Seddon, B. (1998) *Bacillus* isolates from the spermosphere of peas and dwarf French beans with antifungal activity against *Botrytis cinerea* and *Pythium* species. *Journal of Applied Microbiology* **84**, 791–801.

Ward, D.M., Weller, R. & Bateson, M.M. (1990) 16S rRNA sequences reveal numerous uncultured micro-organisms in a natural community. *Nature* **345**, 63–65.

Wei, G., Kloepper, J.W. & Tuzun, S. (1996) Induced systemic resistance to cucumber diseases and increased plant growth by plant growth-promoting rhizobacteria under field conditions. *Biological Control* **86**, 221–224.

Wilhelm, E., Arthofer, W., Schafleitner, R. & Krebs, B. (1998) *Bacillus subtilis* an endophyte of chestnut (*Castanea sativa*) as antagonist against chestnut blight (*Cryphonectria parasitica*). *Plant Cell, Tissue and Organ Culture* **52**, 105–108.

Zaady, E. & Perevolotsky, A. (1995) Enhancement of growth and establishment of oak seedlings (*Quercus ithaburensis* Decaisne) by inoculation with *Azospirillum brasilense*. *Forest Ecology and Management* **72**, 81–83.

Zaady, E., Perevolotsky, A. & Okon, Y. (1993) Promotion of plant growth by inoculum with aggregated and single cell suspensions of *Azospirillum brasilense* Cd. *Soil Biology and Biochemistry* **25**, 819–823.

Zehnder, G., Kloepper, J.W., Yao, C. & Wei, G. (1997) Induction of systemic resistance in cucumber against cucumber beetles (Coleoptera: Chrysomelidae) by plant growth-promoting rhizobacteria. *Journal of Economic Entomology* **90**, 391–396.

Chapter 16

Insertion Sequence Elements and Transposons in *Bacillus*

Jacques Mahillon

Introduction

DNA molecules are constantly prone to intra- and intermolecular rearrangements, including translocation, insertion, deletion or inversion. These molecular re-assortments are essentially governed either by homologous recombination, by which two closely related DNA segments are reciprocally exchanged, or by a nonhomologous process which allows the transfer of DNA pieces between two locations sharing (site-specific), or not sharing (transpositional), some sequence similarity. Mobile elements are generally defined as those DNA entities brought about in the genome by one of these nonhomologous mechanisms (Berg & Howe 1989). Their mobility relies on the activity of one or more associated integrase/transposase enzymes whose genes are mostly included within the elements themselves. Their complexities vary enormously, from the small (less than 2.5 kb) insertion sequences (IS) encoding a single recombinase gene, to the large conjugative transposons also capable of independent intercellular transfer. In many cases, they display a modular organization, combining with their mobility additional features such as plasmid replicons, mobilization cassettes or conjugative properties (Merlin *et al*. 2000).

Transposable elements are ubiquitous and are integral parts of many bacterial genomes, where they are thought to play important roles in the genetic and genomic plasticities of their hosts. From the discovery of their first IS element (Mahillon *et al*. 1985), *Bacillus* species have since provided numerous examples of mobile elements, illustrating a wide spectrum of Gram-positive transposable elements. Moreover, judging from what is seen in other bacterial groups, discovery of a wealth of new elements is likely, in particular through the genome-sequencing projects.

These data on the genomic plasticity of *Bacillus* and relatives are expected to provide us with valuable information not only on the extent of DNA exchange among these bacteria, but also on the contribution of transposons to the *Bacillus* adaptation and evolution. In a more applied context, the recognition of several *Bacillus* species as GRAS (Generally Recognised As Safe) microorganisms is based partly on the assumption that these nonvirulent bacteria cannot easily acquire virulence genes from other opportunistic or pathogenic *Bacillus* group members. Further knowledge on the occurrence and spread of transposons among these bacteria will also bring new insights into these industrial, environmental and health issues.

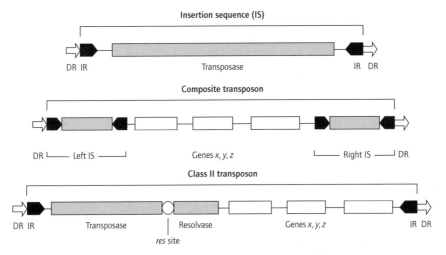

Figure 16.1 Structural features of class I and II of transposable elements. *Class I insertion sequences* (IS) contain a transposase gene (grey rectangle), encoded by one or two ORFs, and are delineated by two inverted repeat sequences (IR, black arrowheads). Upon transposition, IS generate short duplication at the target site which then flank the elements as direct repeats (DR, open arrowheads). Two IS can team up to translocate any foreign genes (*x, y, z* indicated by open rectangles). In this situation, the transposase acts on the two outermost IR and allows the entire structure – the composite transposon – to move. The new DR will then be located outside this structure. *Class II transposons* are more complex. They transpose via a two-step process. First, the transposase (large grey rectangle) acts on the IR to perform a replicative transposition. The donor and target molecules are fused, with the appearance of two copies of the transposon at their junction. In a second step, this 'cointegrate' molecule is resolved by a site-specific recombination involving a resolvase (small grey rectangle) acting on an internal resolution site (*res*, open circle). During the translocation process, DR are also generated. The main difference with class I elements is the way foreign genes are carried over. In this case, those genes are located inside the transposon. Note that the relative position of each transposition and foreign genes can vary among the various class II transposons.

This chapter will focus mainly on two groups of transposable elements which represent the essence of our current knowledge of *Bacillus* mobile elements: the IS and composite transposons, and the class II replicative (or cointegrative) transposons (figure 16.1). Recent data, however, suggest that other mobile genetic entities also exist among *Bacillus* species, such as conjugative transposons or Mobile Insertion Cassettes (MIC). A short overview of these elements will also be given. Finally, it is worth mentioning the presence of group I and II introns located into *Bacillus* transposons and phages (Goodrich-Blair *et al.* 1990; Bechhofer *et al.* 1994; Lazarevic *et al.* 1998; Huang *et al.* 1999). Although these elements are capable of intracellular translocation, they will not be considered any further here because they use RNA molecules as transfer intermediates.

Table 16.1 Transposable element occurrence among *Bacillus* species.

Bacillus species[a]	IS families (individual members)[b]	Class II elements	Conjugative Tn	Other mobile cassettes
B. anthracis	IS3 (2), IS4 (3), IS5 (1)	Tn*XO1*	–	–
B. cereus	IS4 (1), IS6 (1)	Tn*Bce1*, Tn*RC60*	–	MIC231
B. firmus	–	Tn*Bfi1*	–	–
B. halodurans	IS5 (1), IS256 (1), ISL3 (2)	–	–	–
B. megaterium	–	Tn*Bme1*, Tn*MERI1*	–	–
B. mycoides	IS6 (1)	–	–	–
B. sphaericus	–	Tn*Bsp1*	–	–
B. stearothermophilus	IS3 (1), IS4 (2), IS21 (1), IS481 (1), IS630 (1), IS982 (1), ISNCY (1)	–	–	–
B. subtilis	IS4 (1)	–	Tn916-like	–
B. thuringiensis	IS3 (2), IS4 (12), IS6 (9), IS21 (3), IS982 (2)	Tn*4430* and Tn*5401*	–	–

[a] Note that, although *B. cereus sensu stricto*, *B. thuringiensis*, *B. anthracis* and *B. mycoides* are generally recognized as forming a single species, *B. cereus sensu lato*, they are, for practical reasons, referred to here as individual entities.
[b] ISNCY (not classified yet) indicates an IS with no obvious relationship to existing families.

Bacillus and transposable elements

The first *Bacillus* transposable element to be characterized was IS*231* from the entomopathogenic bacterium *B. thuringiensis* (Mahillon *et al.* 1985). This element and its iso-forms were found to be located on large plasmids, often in close proximity to crystal toxin genes. They were also shown to be structurally associated with a cryptic transposon, Tn*4430* (Mahillon *et al.* 1987). Soon after, Tn*RC607*, another class II transposon bearing mercury resistance genes, was isolated from a natural isolate of *B. cereus* (Wang *et al.* 1989).

During the last decade, the number of mobile elements reported in *Bacillus* has reached more than 50 distinct transposable elements. Most of these are IS elements, belonging to nine families (table 16.1). They originate from ten *Bacillus* species, or subspecies. Nine class II elements are also present in six species and a putative conjugative transposon, reminiscent of Tn*916* from *Enterococcus faecalis* (Flannagan *et al.* 1994), was noted in the chromosome of strain 168 of *B. subtilis* (Kunst *et al.* 1997). A novel type of mobile DNA was also reported recently in *B. cereus*: MIC*231* (Chen *et al.* 1999), a new variation of an actual prokaryotic IS element.

The case of *B. subtilis* is particularly interesting. Determination of its chromosomal sequence revealed an apparent absence of IS elements. Yet, it was shown that this bacterium could 'sustain' the transposition activity of various transposable elements, including vectors derived from the Gram-negative IS*10* (Petit *et al.* 1990) or the Gram-positive IS*231* (Léonard *et al.* 1998). It was therefore unclear whether this lack of IS resulted from the absence of IS introduction in *B. subtilis*, its ability to 'eliminate' previous invasive elements, or was merely a peculiarity of strain 168 used for the DNA sequencing. The recent discovery of IS*4Bsu1* in *B. subtilis* (*natto*) NAF5, a starter strain for production of the fermented soybean natto (Nagai *et al.* 2000), tends to favour the last hypothesis. This 1.4-kb element belonging to the IS*4* family (table 16.2) was indeed shown to generate insertions into *comP*. This gene is part of the *comQXPA* quorum-sensing operon that controls the production of degradative enzymes, cell competence, adaptation to high cell density, and the production of capsular poly-γ-glutamate. Moreover, the IS*4Bsu1* copy number varies from 6 to 11 among *B. subtilis* (*natto*) strains (Nagai *et al.* 2000).

Insertion sequences and composite transposons

IS elements are defined as small (between 0.8 and 2.6 kb) genetic entities capable of independent translocation within and among DNA molecules. In most cases, they only code for a transposase protein, they are delineated by short (from 8 to 20 bp) inverted repeat sequences (IR) and they are flanked by direct repeats (DR) which correspond to the duplication of the target site (from 2 to 13 bp, depending on the IS). Although they do not contain 'foreign' genes inside their basic structure, the association of two IS elements can translocate any intervening DNA segments. These structures are named 'composite transposons' (figure 16.1).

Table 16.2 Insertion sequences from *Bacillus*.

Family[a]	Name	Size (bp)	DR (bp)	*Bacillus* species, serovar (sv) & strain (plasmid)	Accession no.[b]	References
IS3	IS1627L	1222	n.d.	*anthracis* Sterne (pXO1)	AF065404	Okinaka *et al.* (1999a)
	IS1627R	1282	n.d.	*anthracis* Sterne (pXO1)	AF065404	Okinaka *et al.* (1999a)
	ISBt1	999	n.d.	*thuringiensis* sv aizawai HD229	L29100	Smith *et al.* (1994)
	ISBt2	>187	n.d.	*thuringiensis* YBT-226	S68409 [P]	Hodgman *et al.* (1993)
	IST1	>1200	n.d.	*stearothermophilus* NCA1503	D10543	H. Sakoda & T. Imanaka (unpubl. data)
IS4	IS4Bsu1	1406	9	*subtilis* (natto) NAF5	AB031551	Nagai *et al.* (2000)
	IS231A	1656	10–12	*thuringiensis* sv thuringiensis berliner 1715 (p65kb)	X03397	Mahillon *et al.* (1985)
	IS231B	1643	n.d.	*thuringiensis* sv thuringiensis berliner 1715 (p65kb)	M16158	Mahillon *et al.* (1987)
	IS231C	1656	11	*thuringiensis* sv thuringiensis berliner 1715 (p65kb)	M16159	Mahillon *et al.* (1987)
	IS231D	1657	n.d.	*thuringiensis* sv finitimus T02 001	X63383	Rezsöhazy *et al.* (1992)
	IS231E	2075	n.d.	*thuringiensis* sv finitimus T02 001	X63384	Rezsöhazy *et al.* (1992)
	IS231F	1655	12	*thuringiensis* sv israelensis 4Q2-72 (p112kb)	X63385	Rezsöhazy *et al.* (1992)
	IS231G	1649	n.d.	*thuringiensis* sv darmstadiensis 73-E-10-2	M93054	Ryan *et al.* (1993)
	IS231H	>817	n.d.	*thuringiensis* sv darmstadiensis 73-E-10-2	M93054 [P]	Ryan *et al.* (1993)
	IS231M	1652	n.d.	*thuringiensis* M15	AF124259	Y.C. Jung *et al.* (unpubl. data)
	IS231N	1654	n.d.	*thuringiensis* M15	AF138876	Y.C. Jung *et al.* (unpubl. data)
	IS231R	1659	n.d.	*anthracis* Sterne (pXO1)	AF065404	Okinaka *et al.* (1999a)
	IS231S	1604	n.d.	*anthracis* Sterne (pXO1)	AF065404	Okinaka *et al.* (1999a)
	IS231T	2066	n.d.	*anthracis* Sterne (pXO1)	AF065404	Okinaka *et al.* (1999a)
	IS231V	1964	n.d.	*thuringiensis* sv israelensis 4Q2-72 (p112kb)	M86926 [V]	Rezsöhazy *et al.* (1993b)
	IS231W	1964	n.d.	*thuringiensis* sv israelensis 4Q2-72 (p112kb)	M86926	Rezsöhazy *et al.* (1993b)
	IS231Y	1645	n.d.	*cereus* CER484	n.d.	Y. Chen & J. Mahillon (unpubl. data)
	IS4712	1370	9	*stearothermophilus* PV72	AJ223150	H.C. Scholz *et al.* (unpubl. data)
	IS5377	1249	9	*stearothermophilus* CU21	X67862	Xu *et al.* (1993)
IS5	ISBan1	>1300	n.d.	*anthracis* Sterne (pXO1)	AF065404	Okinaka *et al.* (1999a)
	ISBha2	>1000	n.d.	*halodurans* C-125	AB024563	Takami *et al.* (1999b)

Family	IS	bp		Organism	Accession	Reference
IS6	IS240A	861	n.d.	*thuringiensis* sv israelensis (p112kb)	M23740	Delécluse *et al.* (1989)
	IS240B	861	n.d.	*thuringiensis* sv israelensis (p112kb)	M23741	Delécluse *et al.* (1989)
	IS240C	817	n.d.	*cereus* CER484	n.d.	Y. Chen & J. Mahillon (unpubl. data)
	IS240F	806	n.d.	*thuringiensis* sv fukuokaensis 84I113 (p197kb)	Y09946	Dunn & Ellar (1997)
	IS240J1	808	n.d.	*thuringiensis* sv jegathesan T28A001	n.d.	Delécluse *et al.* (1995), M.-L. Rosso (unpubl. data)
	IS240I2	809	n.d.	*thuringiensis* sv jegathesan T28A001	n.d.	Rosso & Delécluse (1997)
	IS240M1	861	n.d.	*thuringiensis* sv medellin T30 001	AJ251978	A. Delécluse (unpubl. data)
	IS240M2	815	n.d.	*thuringiensis* sv medellin T30 001	AJ251979	A. Delécluse (unpubl. data)
	IS240M3	862	n.d.	*thuringiensis* sv medellin T30 001	AJ251979	A. Delécluse (unpubl. data)
	IS240M4	1018	n.d.	*thuringiensis* sv medellin T30 001	AJ251979	A. Delécluse (unpubl. data)
	IS240Y	843	n.d.	*mycoides* KBS1-4	n.d.	Y. Chen & J. Mahillon (unpubl. data)
IS21	IS232A	2184	n.d.	*thuringiensis* sv thuringiensis berliner 1715 (p65kb)	M38370	Menou *et al.* (1990)
	IS232B	2200	n.d.	*thuringiensis* sv kurstaki HD73 (p75kb)	M77344 [P]	Menou *et al.* (1990)
	IS232C	2200	n.d.	*thuringiensis* sv kurstaki HD73 (p75kb)	n.d.	Menou *et al.* (1990)
	IS5376	2107	5	*stearothermophilus* CU21	X67861	Xu *et al.* (1993)
IS256	ISBha4	>800	n.d.	*halodurans* C-125	AB024564	Takami *et al.* (1999b)
IS481	ISBst1	1461	n.d.	*stearothermophilus* T-6	AF098273	Shulami *et al.* (1999)
IS630	ISBs2	>660	n.d.	*stearothermophilus* CU21	Z21626 [P]	Kiel *et al.* (1993)
IS982	IS233A	1028	n.d.	*thuringiensis* sv galleriae T05 001	n.d.	C. Léonard *et al.* (unpubl. data)
	IS233B	1026	n.d.	*thuringiensis* sv medellin T30 001	n.d.	C. Léonard *et al.* (unpubl. data)
	ISBs1	996	n.d.	*stearothermophilus* CU21	Z21626	Kiel *et al.* (1993)
ISL3	ISBha1	1384	8	*halodurans* C-125	AB011836	Takami *et al.* (1999a)
	ISBha3	>700	n.d.	*halodurans* C-125	AB024564	Takami *et al.* (1999b)
ISNCY	ISBst2	1612	8	*stearothermophilus* ATCC12980	AF162268	Egelseer *et al.* (1999)

[a] ISNCY (not classified yet) indicates an IS with no obvious relationship to existing families.
[b] [P] refers to partial DNA sequences.
n.d., not determined.

Originally discovered by Jordan *et al*. (1968) as the cause of polar mutations in the *Escherichia coli gal* operon, IS elements have now been encountered across most bacterial species, from Gram-negative bacteria to Archaea. More than 600 distinct elements have already been registered in the IS database (ISFinder at http://www-is.biotoul.fr/is.html) and a veritable explosion can be expected from the exponential number of available bacterial genome sequences (Mahillon *et al*. 1999). Attempts to classify these elements into groups sharing similar structural organization and transposition mechanism has led to the description of about 20 distinct families of prokaryotic IS elements (Mahillon & Chandler 1998).

As already mentioned, the first *Bacillus* IS elements, IS231, were isolated from the biopesticidal bacteria *B. thuringiensis* where they clustered with δ-endotoxin genes. Since then, these iso-elements belonging to the IS4 family (Rezsöhazy *et al*. 1993a) have been observed among all members of the *B. cereus sensu lato* group, including *B. anthracis* (Henderson *et al*. 1995; Okinaka *et al*. 1999a,b), *B. cereus sensu stricto* and *B. mycoides* (Léonard *et al*. 1997) (table 16.2). Preliminary PCR experiments designed to detect IS231-related elements outside the *B. cereus s.l.* group have so far been unsuccessful: no related elements could be found in 7 strains of *B. subtilis*, 12 strains of *B. pumilus*, 22 strains of *B. sphaericus*, 8 strains of *Brevibacillus laterosporus*, 4 strains of *Paenibacillus alvei*, 3 strains of *P. lentimorbus* and 12 strains of *P. popilliae* (Y. Chen & J. Mahillon, unpubl. results). This suggests that the IS231 elements have been particularly successful inside the *B. cereus* group, and/or that members of this group of bacteria are exchanging DNA material more often than with *Bacillus* outside the group, including some sharing the same soil niche. However, these data are still limited and further testing on more strains and species is required before any firm conclusions can be drawn.

Nevertheless, comparison of the various iso-IS231 elements found in *B. cereus s.l.* reinforces the idea that these bacteria share a common horizontal gene pool, as indicated by the dendrogram of their IS231 elements (figure 16.2). One striking example of this is *B. anthracis*, which contains on its pXO1 plasmid IS231R and T elements that are nearly identical to IS231E from *B. thuringiensis* serovar finitimus, and IS231S, whose closest relative, IS231Y, originates from *B. cereus*.

Two distantly related elements belonging to the same IS4 family have also been noted in *B. stearothermophilus*: IS5377 from strain CU21 (Xu *et al*. 1993) and IS4712 from strain PV72 (EMBL accession number AJ223150). The second largest IS family represented in *Bacillus* is that of IS6 (table 16.2). These small elements form a quite homogenous group of about 0.8 kb iso-IS240. Although several other members of this family were shown to duplicate 8 bp during a replicative transposition process (Mahillon & Chandler 1998), thus far, none of the *Bacillus* IS6 family members have been shown to be active (Delécluse *et al*. 1989; M.-L. Rosso & Y. Chen, unpubl. results).

Another important IS family, although less represented among *Bacillus* species, is the IS3 group. Two partial elements (ISBt1 and ISBt2) were found in *B. thuringiensis*, while two copies of another element, IS1627, were discovered flanking a 45-kb segment of the pXO1 virulence plasmid of *B. anthracis* (table 16.2). In strain NCA1503 of *B. stearothermophilus*, Sakoda and Imanaka (EMBL accession number D10543) have also recently submitted a 1.2-kb IS3-

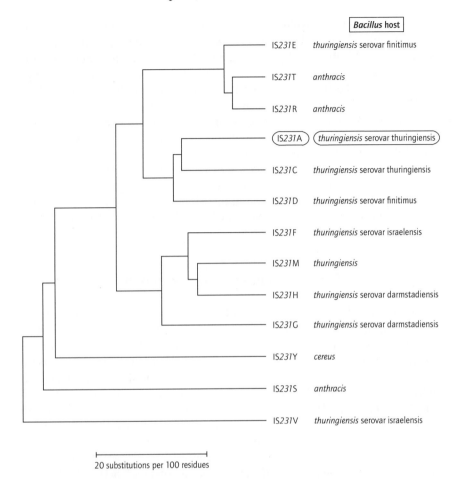

Figure 16.2 Dendrogram of IS*231* elements among *B. cereus sensu lato* group members. IS*231* DNA sequences were aligned using the PileUp program (Wisconsin Package v.10.0, Genetics Computer Group, Madison, WI) and a matrix of substitution among the different sequences calculated using the Distances program. This matrix was then used for generating a dendrogram of sequence relationship by the UPGMA (unweighted pair group method analysis) method (GrowTree). The bacterial host from which each element was isolated is indicated in the right-hand column. The original IS*231*A element from *B. thuringiensis* serovar thuringiensis is boxed.

like element they christened IST*1*. Its putative transposase is distantly related (40% identity) to that of IS*981* from *Lactococcus lactis*.

Two other IS groups have members in both *B. cereus s.l.* and *B. stearothermophilus*: the 2.2-kb IS*21*-related elements IS*232* and IS*5376* and the 1.0-kb IS*982* family members IS*233* and ISBs*1* (table 16.2). While no transposition activity could be detected for the IS*982*-like elements (Kiel *et al.* 1993; Y. Chen & J. Mahillon, unpubl. results), at least IS*232* from the first family was shown to be active in its hosts (Menou *et al.* 1990).

The remaining ISs have only been partially characterized. For instance, at the end of the glucuronic acid utilization gene cluster of *B. stearothermophilus* T-6 (Shulami *et al.* 1999), *orfC* displays striking similarities with transposases of the IS*481* family (a group recognized recently as a family on its own; Chandler & Mahillon 2002). Further analysis could identify the putative ends of this 1461-bp IS. However, the integrity of such a putative transposable element needs confirmation. A similar situation is found in *B. halodurans* where several elements were either briefly described or unnoticed in the original sequence description (Takami *et al.* 1999a,b).

Finally, preliminary examination of not-yet-annotated partial chromosomal sequences of *B. stearothermophilus* has revealed more than 25 DNA segments resembling IS elements or transposons, including putative members of the IS*3*, IS*4*, IS*21*, IS*110*, IS*605*, IS*982*, ISL*3* and Tn*916* families (M. Chandler, pers. comm.).

Class II or cointegrative transposons

Whereas class I elements are rather small, structurally simple and devoid of 'foreign' genes, class II transposons are generally large entities (from 5 to more than 25 kb), containing two transposition-related cassettes, and coding for various functions unrelated to transposition, such as antibiotic resistance or metabolic pathways (figure 16.1). They transpose in a two-step mechanism involving two independent functional cassettes. First, the transposase (TnpA) acting on the terminal repeats (IR) leads to a replicative transposition which fuses both donor and target molecules into a cointegrate molecule (hence the name 'cointegrative' transposons is sometimes used), with two copies of the transposon at their junction. This molecule is then resolved by a site-specific recombinase (TnpR or TnpI, see below) exchanging DNA strands at a resolution site, *res* (Sherratt 1989).

The archetype of these elements is the Gram-negative ampicillin resistance transposon Tn*3*, isolated from the conjugative plasmid R1 *drd*. Tn*3* harbours a 1015 amino acid TnpA and a 185 residues TnpR resolvase, is flanked by 38-bp IR and generates 5-bp duplication at the target site (Heffron *et al.* 1979). In 1986, a 4.2-kb putative class II transposon was described in *B. thuringiensis* and was shown to be active in *Escherichia coli* (Lereclus *et al.* 1986). Surprisingly, however, although this Tn*4430* element bore a TnpA-IR module resembling that of Tn*3*, its resolution cassette turned out to be completely different to that normally found in such transposons. Rather than a resolvase, Tn*4430* used a site-specific recombinase (TnpI) belonging to another family – that of the integrases – named after the enzyme involved in the integration of phage lambda DNA into the *E. coli* genome (Mahillon & Lereclus 1988). Interestingly, a similar combination was also reported for another *B. thuringiensis* class II transposon, Tn*5401* (Baum 1994, 1995; Baum *et al.* 1999). Moreover, both elements were reported as 'cryptic' as no other associated function could be found. As shown in table 16.3, this feature is rather unique among the *Bacillus* class II elements. In fact, most of them carry resistance genes to the heavy metals mercury (Hg^R) or cadmium (Cd^R). These transposons – originally reported from a soil *B. cereus* by Wang and col-

Table 16.3 Class II and conjugative transposons from *Bacillus*.

Type	Name	Size (bp)	Main feature	Species, serovar (sv) & strain (plasmid)	Accession no.[a]	References
Class II	Tn4430	4149	Cryptic	*thuringiensis* sv thuringiensis H1.1	X07651	Mahillon & Lereclus (1988)
	Tn5401	4837	Cryptic	*thuringiensis* sv morrisoni EG2158	U03554	Baum (1994)
	TnBce1	n.d.	Mercury resistance	*cereus* TA32-5 (pKLH302)	Y09204 [P]	Bogdanova et al. (1998)
	TnBfi1	>6000	Cadmium resistance	*firmus* OF4	M90749	Ivey et al. (1992)
	TnBme1	n.d.	Mercury resistance	*megaterium* MK64-1 (pKLH304)	Y18009 [P]	Bogdanova et al. (1998)
	TnBsp1	n.d.	Mercury resistance	*sphaericus* FA8-2 (pKLH301)	Y18010 [P]	Bogdanova et al. (1998)
	TnBth1	n.d.	Unknown	*thuringiensis* sv jegathesan T28A001	n.d.	M.-L. Rosso (unpubl. data)
	TnMERI1	14 500	Mercury resistance	*megaterium* MB1	AB022308 [P]	Huang et al. (1999)
	TnRC607	>10 000	Mercury resistance	*cereus* RC607	AB036431 [P]	Wang et al. (1989)
	TnXO1	8679	Spore germination response	*anthracis* Sterne (pXO1)	AF065404	Okinaka et al. (1999)
Conjugative Tn	Tn916-like	>16 000	Unknown	*subtilis* 168	Z99106	Kunst et al. (1997)

[a] [P] refers to partial DNA sequences.
n.d., not determined.

laboratories (1989) – have since been described in *B. megaterium* and *B. sphaericus* for Hg^R (Bogdanova *et al.* 1999; Huang *et al.* 1999) and in *B. firmus* for Cd^R (Ivey *et al.* 1992).

The most unusual putative class II element is probably Tn*XO1* located on the pXO1 virulence plasmid of *B. anthracis*. A detailed structural analysis of a region of this plasmid bearing a *tnpA* and a *tnpR* gene, unravelled a seemingly complete class II element (J. Mahillon, unpubl. result). This 8679-bp element is in fact delineated by two well-conserved 38-bp IR and flanked by 5-bp DR. However, the most surprising feature is the presence, inside this putative transposon (tentatively named Tn*XO1*), of three *orfs* sharing similarities with *B. subtilis* genes involved in spore germination response. Experiments are under way to demonstrate the actual integrity, transposition activity and biological functions of this element.

Transposable elements are, in essence, prone to horizontal spread. However, besides the very few examples where two nearly identical transposons are found in two distantly related bacteria, it is always difficult, and often very speculative, to interpret similarity dendrograms between transposable elements and their bacterial hosts, and to conclude on the vertical *versus* horizontal way of transmission. Notwithstanding these considerations, the data obtained with class II TnpA comparisons are particularly interesting. As shown in figure 16.3, there is a strong discrepancy between the transposase phylogram and the phylogeny of their hosts. In this simplified dendrogram focused on the *Bacillus* elements, three distinct TnpA clusters can be seen. Each of these branches bears both Grampositive and Gram-negative bacteria. In addition, the TnpA of two elements found in the same bacterial species (Tn*4430* and Tn*5401* from *B. thuringiensis*) pertain to two distinct clusters (figure 16.3). These data suggest the possible interspecies and inter-Gram transfers of class II elements (or their TnpA-IR module). The exact nature of these putative events (carrier molecule, frequency and biological relevance) still remains to be elucidated.

Other transposons and mobile elements

A third category of transposable elements encountered in several members of *Bacteria* comprises the conjugative transposons. Their archetype is the tetracycline resistant (Tet^R) Tn*916* from *Enterococcus faecalis* (Flannagan *et al.* 1994). These large mobile DNA molecules (generally >15 kb) combine the dual properties of intracellular translocation (transposition) and intercellular transfer (conjugation). Most of them carry Tet^R genes and use a site-specific recombinase to perform their strand-transfer reactions. They all depend on a carrier molecule (they do not have a replicon), but they can exist as a nonreplicative circle which is an intermediate stage in their transposition process (Scott & Churchward 1995). As indicated above, the only putative *Bacillus* conjugative element found so far is the Tn*916*-like gene cluster in the *B. subtilis* 168 genome (Kunst *et al.* 1997). In fact, a closer look at the first Tn*916*-related open reading frames (ORFs) annotated in the originally released sequence suggests that an entire transposon (with more than 15 ORFs) may be residing between coordinates 528000 and 545000 of the *B. subtilis* chromosome (J. Mahillon, unpubl. obs.). However, no Tet^R

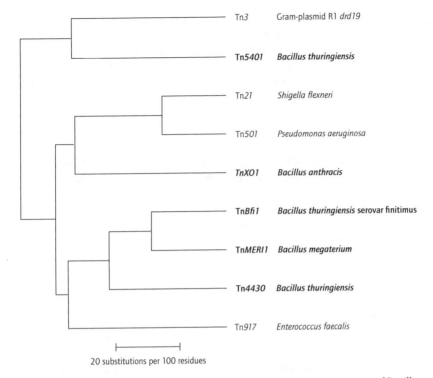

Tn3	Gram-plasmid R1 *drd19*	
Tn5401	**Bacillus thuringiensis**	
Tn21	Shigella flexneri	
Tn501	Pseudomonas aeruginosa	
TnXO1	**Bacillus anthracis**	
TnBfi1	**Bacillus thuringiensis** serovar finitimus	
TnMERI1	**Bacillus megaterium**	
Tn4430	**Bacillus thuringiensis**	
Tn917	Enterococcus faecalis	

20 substitutions per 100 residues

Figure 16.3 Phylogram of class II transposases. The transposase protein sequences of *Bacillus* class II transposons were compared to those of other representative elements from both Gram-negative and Gram-positive bacterial hosts. The phylogram was obtained using a similar approach as that described in figure 16.2, except that this transposase comparison relies on peptide sequences. The *Bacillus* transposons and their corresponding hosts are in bold characters. For Tn3, the original R1 *drd19* conjugative plasmid is widely spread among various Gram-negative bacteria.

genes could be found in the same chromosomal region (table 16.3). Confirmation of such mobile DNA inside *B. subtilis* would be of major importance in the scope of potential gene transfer.

An insertion sequence is a structural-functional module bearing a transposase and its cognate sequences, the terminal IRs. Until recently, little or no variation was observed in this simple basic structure. MIC231, a mobile insertion cassette isolated from natural isolates of *B. cereus* is an example of novel IS settings. This 1.9-kb entity consists of two 50-bp sequences, resembling the extremities of IS231, flanking a gene coding for an alkaline D-stereospecific endopeptidase (*adp*). Despite the absence of an internal transposase, this element could be translocated in the presence of a trans-acting IS231 transposase (Chen *et al.* 1999). The reason for the existence of such a functional cassette is unknown, and the exact role of the *adp* gene itself remains elusive. A recent search for other variations on the IS231 theme has suggested the presence of other MIC231 elements, either 'empty' or bearing yet other genes unrelated to transposition (C. Vermeiren & J. Mahillon, unpubl. results). It is interesting to note that in *Streptococcus pneumoniae*, a 107-bp sequence is repeated more than 100 times in its chromosome. This RUP element (Repeat Unit of *Pneumococcus*) is thought to be the remnant of an IS element (IS*630*-Spn*1*). Like the *B. cereus* MIC, RUP transposition relies on the transposase activity of its corresponding IS (Oggioni & Claverys 1999). These observations not only provide new insights into the genome potential plasticity, but also underline the necessity for in-depth structural analyses of DNA sequences.

Bacillus transposable elements: from horizontal gene transfer to bacterial adaptation and evolution

What biological roles can be attributed to the *Bacillus* transposable elements and what can be inferred from our current knowledge on such Gram-positive mobile DNA in terms of bacterial adaptation and, eventually, evolution?

Transposable elements as gene carriers

In principle, any intervening DNA fragment located between two ISs can be translocated, provided the transposase of at least one element is functional and recognizes the outermost extremities of this composite structure (figure 16.1) (Mahillon 1998). Although this type of organization has been encountered in several strains of *Bacillus*, the actual demonstration of transfer of these composite entities is still missing. This is true in the case of *B. thuringiensis* where several such arrangements have been reported in the vicinity of crystal toxin genes, without any data on their mobility (Mahillon *et al.* 1994; Rosso *et al.* 2000).

Pathogenicity Islands (PAIs) refer to chromosomal or plasmid segments containing clusters of virulence genes. They generally display G+C content different from the rest of the genome (hence the name) and therefore are thought to have a foreign origin (Hacker & Kaper 1999). In most cases, some of their structural and functional features suggest that they may originate from and/or be complemented by phages. In a few instances, IS elements were found bordering these PAIs and presumably contributed to their mobility. In the pXO1 virulence plasmid of *B. anthracis*, a 45-kb segment containing the virulence genes (and the putative spore germination response transposon Tn*XO1*; see above) is bordered by two copies of IS*1627*, an IS*3* family member (Okinaka *et al.* 1999). Moreover, this section of the pXO1 plasmid has been observed in both orientations (Thorne 1993) suggesting if not its mobility at least its local inversion. However, further data are needed in order to assess the actual integrity and mobility of this putative *B. anthracis* PAI.

Transposable elements as gene regulators

Through their insertional activity, transposable elements have the capacity to interfere with gene regulation. Besides gene inactivation, transposons can also modulate gene expression by inserting into the regulatory regions of structural genes. Examples of new promoters brought by ISs or of new hybrid promoters with a −35 box located into the transposon extremity have been reported in several bacteria (Mahillon & Chandler 1998). Although such gene expression tuning has not yet been described for *Bacillus* species, direct involvement of ISs in the appearance of spontaneous mutation of genes of interest has recently been obtained. As indicated above, insertion of the IS4*Bsu1* element into the *comP* gene of *B. subtilis* (*natto*) abolished the production of its capsular poly-γ-glutamate (Nagai *et al.* 2000). Interestingly, this transposition event was shown to be the major cause of these spontaneous γ-PGA mutants, which appear with an estimated frequency of 5×10^{-4} (event/generation). In *B. stearothermophilus*, two independent observations have also indicated that the appearance of spontaneous mutants for the production of the cell surface S-layer proteins resulted from the IS insertion in their corresponding genes, IS4712 in *sbsA* (H.C. Scholz *et al.*, unpubl. data; accession number AJ223150) and ISBst12 in *sbsC* (Egelseer *et al.* 2000) (table 16.2).

Transposable elements as genetic tools, from and for *Bacillus*

Most transposable tools were originally developed in Gram-negative bacteria. One of the major factors that has hampered the rapid development of useful genetic tools in *Bacillus* species is certainly related to the difficulties encountered in efficiently transforming plasmid-borne traits in these bacteria. The first Gram-positive transposition tools elaborated for *Bacillus* derived from the *E. faecalis* class II transposon Tn917 (Youngman *et al.* 1983; Perkins & Youngman 1986). These vectors have since been applied in various *Bacillus* species, including *B. anthracis* (Hoffmaster & Koehler 1997), *B. cereus* (Clements & Moir 1998), *B. megaterium* (Tani *et al.* 1996), *B. licheniformis* (Hertzog-Velikonja *et al.* 1994) and *Paenibacillus polymyxa* (Seldin *et al.* 1999).

In 1990, Petit *et al.* modified the well-characterized Gram-negative IS10 element from the Tet^R Tn10 transposon of *Salmonella typhimurium*. They changed its transposase promoter into a more appropriate *Bacillus*-recognized promoter and used the replication-thermosensitive plasmid pE194 as delivery vector. Later, Steinmetz and Richter (1994) improved the system by inserting, among other features, an ATS (altered target specificity) transposase variant for more random insertion.

More recently, IS231 from *B. thuringiensis* was engineered into a series of transposition vectors applicable in both Gram-negative and Gram-positive bacteria. Bearing extremely rare restriction sites, they allow rapid construction of an integrated physical and genetic map of bacterial genomes, as shown for the reference strain of *B. cereus* (Léonard *et al.* 1998).

Conjugative transposons have also been introduced in *Bacillus*. Tn916, and its close relative Tn925, both from *E. faecalis* have been used as shuttle vehicles and to transfer plasmids between *Bacillus* species, among which are *B. subtilis*, *B. pumilus* or *Paenibacillus popilliae* (Hendrick *et al.* 1991; Torres *et al.* 1991;

Dingman 1999). Similarly, Tn*1545*, a 25-kb kanamycin-, erythromycin- and tetracycline-resistant transposon from *Streptococcus pneumoniae* is self-transferable to various Gram-positive genera, including *Bacillus*, where it was shown to transpose (Courvalin & Carlier 1987).

Concluding remarks

From recent data, it is reasonable to assume that the world of transposable elements is equally well established among bacilli as it is in the other eu- and archae-bacteria. However, it remains to be seen whether the next *Bacillus* genome-sequencing projects (*B. anthracis* and *B. cereus* are already well under way) will provide a plethora of mobile elements such as found in *B. stearothermophilus* or *B. thuringiensis*, or a more modest situation as seen in *B. subtilis*.

The systematics of *Bacillus* has recently undergone important reorganization, including the recognition of new distinct groups, calling for a reviewed taxonomy. These studies are based on polyphasic approaches that combine microbiological, biochemical and molecular techniques (see Chapters 8, 9 and 10). Using the information and tools gathered from all the bacterial transposable elements, it would be particularly interesting to decipher the actual distribution and diversity of IS and transposable elements throughout the world of *Bacillus* and relatives. In particular, bacteria from exotic or unusual habitats, or those with ubiquitous niches could provide new insights into the ecology of mobile elements. This could also challenge our current view of how mobile elements have been distributed across the *Bacillus* ensemble.

The widespread commercial use of *Bacillus* strains, either as biopesticides (*B. thuringiensis* or *B. sphaericus*) or as fermentation tools, necessitate a strict control of their innocuousness. In this respect, the role of transposable and mobile elements in the possible spread of virulence or antibiotic resistance genes is of major importance. In *B. thuringiensis*, the close structural association between several mobile elements and endotoxin genes on large (conjugative) plasmids is indicative of the transposon implication in the mobility of these toxin genes, and consequently, in the virulence diversity observed among *B. thuringiensis* isolates. Similarly, in various soil species, including *B. cereus*, *B. megaterium* and *B. sphaericus*, the spread of heavy metal resistance genes has been facilitated by class II transposons. Although these examples of mobile element contribution to environmental adaptation are particularly informative, our understanding of their roles in bacterial evolution will remain elusive until extensive sets of genome sequences become available.

The world of mobile DNA is fascinating, not only in its concepts, but also in the diversity of its themes and variations. For *Bacillus*, the detailed characterization of new mobile DNA such as group I and II introns, MICs, and conjugative or integrative plasmids will certainly reveal new avenues in the understanding of their host genomics and stimulate the development of new useful genetic tools.

Acknowledgements

I would like to thank those past and present collaborators who contributed to my studies of the *B. cereus s.l.* IS and Tn elements, in particular, Pia Braathen, Yahua Chen, Kamilla Jendem, Cathy Léonard and Céline Vermeiren. Work performed in the Microbial Genetics Laboratory was supported by grants from FRIA, FNRS, Université catholique de Louvain and the EU-Biotech Concerted Action MECBAD program BIO4-CT-0099.

References

Baum, J. (1994) Tn*5401*: a new class II transposable element from *Bacillus thuringiensis*. *Journal of Bacteriology* **176**, 2835–2845.

Baum, J. (1995) TnpI recombinase: identification of sites within Tn*5401* required for TnpI binding and site-specific recombination. *Journal of Bacteriology* **177**, 4036–4042.

Baum, J., Gilmer, A.J. & Mettus, A.-M. (1999) Multiple roles for TnpI recombinase in regulation of Tn*5401* transposition in *Bacillus thuringiensis*. *Journal of Bacteriology* **181**, 6271–6277.

Bechhofer, D.H., Hue, K.K. & Shub, D.A. (1994) An intron in the thymidylate synthase gene of *Bacillus* bacteriophage β22: evidence for independent evolution of a gene, its group I intron, and the intron open reading frame. *Proceedings of the National Academy of Sciences, USA* **91**, 11669–11673.

Berg, D.E. & Howe, M.M. (eds) (1989) *Mobile DNA*. American Society for Microbiology, Washington, DC.

Bogdanova, E.S., Bass, I.A., Minakhin, L.S. *et al.* (1998) Horizontal spread of *mer* operons among Gram-positive bacteria in natural environments. *Microbiology* **144**, 609–620.

Chandler, M. & Mahillon, J. (2002) Insertion Sequences revisited. In: *Mobile DNA II* (eds N.L. Craig *et al.*), pp. 305–366. American Society for Microbiology, Washington, DC.

Chen, Y., Braathen, P., Léonard, C. & Mahillon, J. (1999) MIC*231*, a naturally occurring Mobile Insertion Cassette from *Bacillus cereus*. *Molecular Microbiology* **32**, 657–668.

Clements, M.O. & Moir, A. (1998) Role of the *gerI* operon of *Bacillus cereus* 569 in the response of spores to germinants. *Journal of Bacteriology* **180**, 6729–6735.

Courvalin, P. & Carlier, C. (1987) Tn*1545*: a conjugative shuttle transposon. *Molecular and General Genetics* **206**, 259–264.

Delécluse, A., Bourgoin, C., Klier, A. & Rapoport, G. (1989) Nucleotide sequence and characterization of a new insertion element, IS*240*, from *Bacillus thuringiensis israelensis*. *Plasmid* **21**, 71–78.

Delécluse, A., Rosso, M.-L. & Ragni, A. (1995) Cloning and expression of a novel toxin gene from *Bacillus thuringiensis* subsp. *jegathesan*, encoding a highly mosquitocidal protein. *Applied and Environmental Microbiology* **61**, 4230–4235.

Dingman, D.W. (1999) Conjugative transposition of Tn*916* and Tn*925* in *Bacillus popilliae*. *Canadian Journal of Microbiology* **45**, 530–535.

Dunn, M.G. & Ellar, D.J. (1997) Identification of two sequence elements associated with the gene encoding the 24-kDa crystalline component in *Bacillus thuringiensis* ssp. *fukuokaensis*: an example of transposable element archaeology. *Plasmid* **37**, 205–215.

Egelseer, E.M., Idris, R., Jarosch, M., Danhorn, T., Sleytr, U.B. & Sara, M. (2000) ISBst*12*, a novel type of insertion-sequence element causing loss of S-layer-gene expression in *Bacillus stearothermophilus* ATCC12980. *Microbiology* **146**, 2175–2183.

Flannagan, S.E., Zitzow, L.A., Su, Y.A. & Clewell, D.B. (1994) Nucleotide sequence of the 18 kb conjugative transposon Tn*916* from *Enterococcus faecalis*. *Plasmid* **32**, 350–354.

Goodrich-Blair, H., Scarlato, V., Gott, J.M., Xu, M.Q. & Shub, D.A. (1990) A self-splicing group I intron in the polymerase gene of *Bacillus subtilis* bacteriophage SPO1. *Cell* **63**, 417–424.

Hacker, J. & Kaper, J.B. (1999) The concept of pathogenicity islands. In: *Pathogenicity Islands and other Mobile Virulence Elements* (eds J.B. Kaper & J. Hacker), pp. 1–11. American Society for Microbiology, Washington, DC.

Heffron, F., McCarthy, B.J., Ohtsubo, H. & Ohtsubo, E. (1979) DNA sequence analysis of the transposon Tn*3*: three genes and three sites involved in transposition of Tn*3*. *Cell* **18**, 1153–1163.

Henderson, I., Dongzheng, Y. & Turnbull, P.C.B. (1995) Differentiation of *Bacillus anthracis* and other 'Bacillus cereus group' bacteria using IS231-derived sequences. *FEMS Microbiology Letters* **128**, 113–118.

Hendrick, C.A., Johnson, L.K., Tomes, N.J., Smiley, B.K. & Price, J.P. (1991) Insertion of Tn916 into *Bacillus pumilus* plasmid pMGD302 and evidence for plasmid transfer by conjugation. *Plasmid* **26**, 1–9.

Hertzog-Velikonja, B., Podlesek, Z. & Grabnar, M. (1994) Isolation and characterisation of Tn917-generated bacitracin deficient mutants of *Bacillus licheniformis*. *FEMS Microbiology Letters* **121**, 147–152.

Hodgman, T.C., Ziniu, Y., Shen, J. & Ellar, D. (1993) Identification of a cryptic gene associated with an Insertion Sequence not previously identified in *Bacillus thuringiensis*. *FEMS Microbiology Letters* **114**, 23–30.

Hoffmaster, A.R. & Koehler, T.M. (1997) The anthrax toxin activator gene *atxA* is associated with CO_2-enhanced non-toxin gene expression in *Bacillus anthracis*. *Infection and Immunity* **65**, 3091–3099.

Huang, C.-C., Narita, M., Yamagata, T., Itoh, Y. & Endo, G. (1999) Structure analysis of a class II transposon encoding the mercury resistance of the Gram-positive bacterium *Bacillus megaterium* MB1, a strain isolated from Minamata Bay, Japan. *Gene* **234**, 361–369.

Ivey, D.M., Guffanti, A.A., Shen, Z., Kudyan, N. & Krulwich, T.A. (1992) The *cadC* gene product of alkaliphilic *Bacillus firmus* OF4 partially restores Na^+ resistance to an *Escherichia coli* strain lacking an Na^+/H^+ antiporter (NhaA). *Journal of Bacteriology* **174**, 4878–4884.

Jordan, E., Saedler, H. & Starlinger, P. (1968) O° and strong polar mutations in the gal operon are insertions. *Molecular and General Genetics* **102**, 353–365.

Kiel, J.A.K.W., Boels, J.M., Ten Berge, A.M. & Venema, G. (1993) Two putative Insertion Sequences flank a truncated glycogen branching enzyme gene in the thermophile *Bacillus stearothermophilus* CU21. *Journal of DNA Sequencing and Mapping* **4**, 1–9.

Kunst, F., Ogasawara, N., Moszer, I. *et al.* (1997) The complete genome sequence of the Gram-positive bacterium *Bacillus subtilis*. *Nature* **390**, 249–256.

Lazarevic, V., Soldo, B., Düsterhöft, A., Hilbert, H., Mauël, C. & Karamata, D. (1998) Introns and intein coding sequence in the ribonucleotide reductase genes of *Bacillus subtilis* temperate bacteriophage SPβ. *Proceedings of the National Academy of Sciences, USA* **95**, 1692–1697.

Léonard, C., Chen, Y. & Mahillon, J. (1997) Diversity and differential distribution of IS231, IS232 and IS240 among *Bacillus cereus*, *B. thuringiensis*, and *B. mycoides*. *Microbiology* **143**, 2537–2547.

Léonard, C., Zekri, O. & Mahillon, J. (1998) Integrated physical and genetic mapping of *Bacillus cereus* and other Gram+ bacteria based on IS231A transposition vectors. *Infection and Immunity* **66**, 2163–2169.

Lereclus, D., Mahillon, J., Menou, G. & Lecadet, M.-M. (1986) Identification of Tn4430, a transposon of *Bacillus thuringiensis* functional in *Escherichia coli*. *Molecular and General Genetics* **204**, 52–57.

Mahillon, J. (1998) Transposons as gene haulers. *Acta Pathologica Microbiologica et Immunologica Scandinavica* **106** (Suppl. 84), 29–36.

Mahillon, J. & Chandler, M. (1998) Insertion Sequences. *Microbiology and Molecular Biology Reviews* **62**, 725–774.

Mahillon, J. & Lereclus, D. (1988) Structural and functional analysis of Tn4430: identification of an integrase-like protein involved in the co-integrase-resolution process. *EMBO Journal* **7**, 1515–1526.

Mahillon, J., Seurinck, J., Van Rompuy, L., Delcour, J. & Zabeau, M. (1985) Nucleotide sequence and structural organization of an insertion sequence element (IS231) from *Bacillus thuringiensis* strain berliner 1715. *EMBO Journal* **4**, 3895–3899.

Mahillon, J., Seurinck, J., Delcour, J. & Zabeau, M. (1987) Cloning and nucleotide sequence of different iso-IS231 elements and their structural association with the Tn4430 transposon in *Bacillus thuringiensis*. *Gene* **51**, 187–196.

Mahillon, J., Rezsöhazy, R., Hallet, B. & Delcour, J. (1994) IS231 and other *Bacillus thuringiensis* elements: a review. *Genetica* **93**, 13–26.

Mahillon, J., Léonard, C. & Chandler, M. (1999) IS elements as constituents of bacterial genomes. *Research in Microbiology* **150**, 1–13.

Menou, G., Mahillon, J., Lecadet, M.-M. & Lereclus, D. (1990) Structural and genetic organization of IS232, a new insertion sequence of *Bacillus thuringiensis*. *Journal of Bacteriology* **172**, 6689–6696.

Merlin, C., Mahillon, J., Nesvera, J. & Toussaint, A. (2000) Gene recruiters and transporters: the molecular structure of bacterial mobile elements. In: *The Horizontal Gene Pool* (ed. C. Thomas). Harwood Academic, Reading.

Nagai, T., Phan Tran, L.-S., Inatsu, Y. & Itoh, Y. (2000) A new IS4 family Insertion Sequence, IS4Bsu1, responsible for genetic instability of poly-γ-glutamic acid production in *Bacillus subtilis*. *Journal of Bacteriology* **182**, 2387–2392.

Oggioni, M.R. & Claverys, J.-P. (1999) Repeated extragenic sequences in prokaryotic genomes: a proposal for the origin and dynamics of the RUP element in *Streptococcus pneumoniae*. *Microbiology* **145**, 2647–2653.

Okinaka, R., Cloud, K., Hampton, O. *et al.* (1999a) Sequence and organization of pXO1, the large *Bacillus anthracis* plasmid harbouring the anthrax toxin genes. *Journal of Bacteriology* **181**, 6509–6515.

Okinaka, R., Cloud, K., Hampton, O. *et al.* (1999b) Sequence, assembly and analysis of pXO1 and pXO2. *Journal of Applied Microbiology* **87**, 261–262.

Perkins, J.B. & Youngman, P.J. (1986) Construction and properties of Tn*917*-*lac*, a transposon derivative that mediates transcriptional gene fusions in *Bacillus subtilis*. *Proceedings of the National Academy of Sciences, USA* **83**, 140–144.

Petit, M.A., Bruand, C., Jannière, L. & Ehrlich, S.D. (1990) Tn*10*-derived transposons active in *Bacillus subtilis*. *Journal of Bacteriology* **172**, 6736–6740.

Rezsöhazy, R., Hallet, B. & Delcour, J. (1992) IS*231*D, E and F, three new Insertion Sequences in *Bacillus thuringiensis*: extension of the IS*231* family. *Molecular Microbiology* **6**, 1959–1967.

Rezsöhazy, R., Hallet, B., Delcour, J. & Mahillon, J. (1993a) The IS4 family of insertion sequences: evidence for a conserved transposase motif. *Molecular Microbiology* **9**, 1283–1295.

Rezsöhazy, R., Hallet, B., Mahillon, J. & Delcour, J. (1993b) IS*231*V and W from *Bacillus thuringiensis* subsp. *israelensis*, two distant members of the IS*231* family of insertion sequences. *Plasmid* **30**, 141–149.

Rosso, M.-L. & Delécluse, A. (1997) Contribution of the 65-kilodalton protein encoded by the cloned gene *cry19A* to the mosquitocidal activity of *Bacillus thuringiensis* subsp. *jegathesan*. *Applied and Environmental Microbiology* **63**, 4449–4455.

Rosso, M.-L., Mahillon, J. & Delécluse, A. (2000) Genetic and genomic contexts of toxin genes. In: *Entomopathogenic Bacteria: from Laboratory to Field Application* (eds J.-F. Charles *et al.*). Kluwer, Dordrecht.

Ryan, M., Johnson, J.D. & Bulla, L.A., Jr (1993) Insertion Sequence elements in *Bacillus thuringiensis* subsp. *darmstadiensis*. *Canadian Journal of Microbiology* **39**, 649–658.

Scott, J.R. & Churchward, G.G. (1995) Conjugative transposition. *Annual Reviews in Microbiology* **49**, 367–397.

Seldin, L., de Azevedo, F.S., Alviano, D.S., Alviano, C.S. & de Freire Bastos, M.C. (1999) Inhibitory activity of *Paenibacillus polymyxa* SCE2 against human pathogenic micro-organisms. *Letters in Applied Microbiology* **28**, 423–427.

Sherratt, D. (1989) Tn*3* and related transposable elements: site-specific recombination and transposition. In: *Mobile DNA* (eds D.E. Berg & M.M. Howe), pp. 163–184. American Society for Microbiology, Washington, DC.

Shulami, S., Gat, O., Sonenshein, A.L. & Shoham, Y. (1999) The glucoronic acid utilization gene cluster from *Bacillus stearothermophilus* T-6. *Journal of Bacteriology* **181**, 3695–3704.

Smith, G.P., Ellar, D., Keeler, S.J. & Seip, C.E. (1994) Nucleotide sequence and analysis of an Insertion Sequence from *Bacillus thuringiensis* related to IS*150*. *Plasmid* **32**, 10–18.

Steinmetz, M. & Richter, R. (1994) Easy cloning of mini-Tn*10* insertions from the *Bacillus subtilis* chromosome. *Journal of Bacteriology* **176**, 1761–1763.

Takami, H., Nakasone, K., Ogasawara, N. *et al.* (1999a) Sequencing of three lambda clones from the genome of alkaliphilic *Bacillus* sp. strain C-125. *Extremophiles* **3**, 29–34.

Takami, H., Takaki, Y., Nakasone, K. *et al.* (1999b) Genetic analysis of the chromosome of alkaliphilic *Bacillus halodurans* C-125. *Extremophiles* **3**, 227–233.

Tani, K., Watanabe, T., Matsuda, H., Nasu, M. & Kondo, M. (1996) Cloning and sequencing of the spore germination gene of *Bacillus megaterium* ATCC12872: similarities to the NaH-antiporter gene of *Enterococcus hirae*. *Microbiology and Immunology* **40**, 99–105.

Thorne, C. (1993) *Bacillus anthracis*. In: Bacillus subtilis *and other Gram-positive Bacteria* (ed A.L. Sonenshein), pp. 113–124. American Society for Microbiology, Washington, DC.

Torres, O.R., Korman, R.Z., Zahler, S.A. & Dunny, G.M. (1991) The conjugative transposon Tn*925*: enhancement of conjugal transfer by tetracycline in *Enterococcus faecalis* and mobilization of chromosomal genes in *Bacillus subtilis* and *E. faecalis*. *Molecular and General Genetics* **225**, 395–400.

Wang, Y., Moore, M., Levinson, H.S., Silver, S., Walsh, C. & Mahler, I. (1989) Nucleotide sequence of a chromosomal mercury resistance determinant from a *Bacillus* sp. with broad-spectrum mercury resistance. *Journal of Bacteriology* **171**, 83–92.

Xu, K., He, Z.Q., Mao, Y.M., Sheng, R.Q. & Sheng, Z.J. (1993) On two transposable elements from *Bacillus stearothermophilus*. *Plasmid* **29**, 1–9.

Youngman, P.J., Perkins, J.B. & Losick, R. (1983) Genetic transpositional and insertional mutagenesis in *Bacillus subtilis* with *Streptococcus faecalis* transposon Tn*917*. *Proceedings of the National Academy of Sciences, USA* **80**, 2305–2309.

Chapter 17

Fingerprint Spectrometry Methods in *Bacillus* Systematics

J.T. Magee and R. Goodacre

Introduction

This chapter deals with the classification of *Bacillus* species using a variety of instrument-based physicochemical techniques for whole cell analysis (Nelson 1991; Magee 1994). Their unifying features are that: (i) sample preparation is standard, irrespective of species, rapid and simple; (ii) the analyses involve rapid, automated processing with high throughput; and (iii) they yield complex quantitative 'fingerprints' that reflect aspects of whole cell composition. These approaches derive from analytical techniques that are applied conventionally in structure determination of pure organic chemicals. Three have reached the applications stage: pyrolysis mass spectrometry (Py-MS); matrix-assisted laser desorption–ionization time of flight mass spectrometry (MALDI-TOF-MS) and Fourier transform infrared spectrometry (FT-IR). Others remain largely unexplored and so are difficult to evaluate, including: soft ionization mass spectrometry; nuclear magnetic resonance spectrometry; Raman spectrometry and UV resonance Raman spectrometry.

One aim of taxonomy is to provide a classification that maximizes the range and accuracy of the predictions that can be made from identification by *in vitro* tests. For the applied bacteriologist, the ability to deduce the pathogenic, cross-infection, spoilage or ecological properties of an isolate from species identification is a crucial feature of a classification. Many of these properties cannot be assayed in the laboratory, but their accurate prediction is often important in medicine or industry. In contrast, the academic taxonomist wishes to deduce the evolutionary lineages of the bacteria from the current group structure. The success of a single classification for both purposes depends upon the definition of homogeneous species clusters. Evolutionary bacteriologists require a realistic current group structure for their endeavours, and the reliability and range of predictions that can be made by applied bacteriologists depend on species homogeneity. There is no real conflict of interest. Controversy is inevitable, however, when one side or the other wishes to define a species that is nonhomogeneous or inadequately described (and therefore possibly not homogeneous).

The root of this controversy lies in the bad science of the quick, convenient fix. Species should be defined neither on the basis of convenient divisions based on single economically or medically important properties, nor on single currently favoured characterization techniques. By definition, homogeneous species can only be described on the basis of a thorough study of many strains in as wide a range of characterization tests as possible. This polyphasic taxonomic approach

(Colwell 1970) is designed to produce a stable classification reflecting real biological divisions, providing a firm basis for the endeavours of both sides.

Fingerprint spectrometric methods are important in this polyphasic approach. They provide data on aspects of cell composition that can otherwise be determined only by slow, costly, low-throughput conventional chemistry. Chemical composition, particularly of the polymers of the outer cell layers, is a major determinant of the interactions between bacteria and their environment, and therefore of the ecological niche(s) that a species can occupy. This niche is, essentially, what the applied bacteriologist wishes to deduce from identification. Equally, the niche moulds the genetics of occupying species by Darwinian selection, influences the potential for intra- and interspecies flow of genetic material, and so is relevant to evolutionary studies. Fingerprint spectrometric methods yield data on this important aspect of characterization and so form an essential ingredient in polyphasic studies.

Pyrolysis mass spectrometry (Py-MS)

Principle

Intact cells are heated rapidly to a fixed temperature in the range 350–1200°C in a nonoxidizing environment (vacuum or an inert gas atmosphere). Thermal intramolecular vibrations in the cell polymers exceed the strength of covalent bonds and the molecules shatter, with bonds breaking in order of their binding energies, yielding a rapidly expanding cloud of free radicals. The rapid dispersion and cooling leaves little opportunity for intermolecular recombination, and the radicals undergo internal restructuring to yield stable, moderately sized (34–500 Da), mostly volatile products.

The chemistry of pyrolysis is complex, but the quantitative and qualitative composition of the products reflects the composition of the sample and the experimental conditions. High pyrolysis temperatures produce simple compounds, carrying little information about the original sample. At lower temperatures, the yield shifts to larger products carrying more information, but condensation of poorly volatile tars in the apparatus becomes a problem. At the widely adopted compromise of 530°C, tar yield is minimal and reasonable proportions of higher molecular weight compounds are found.

The products can be separated by gas liquid chromatography (GC), and quantified with simple detector systems, e.g. flame ionization, or with fuller chemical identification by diversion of the eluted volatiles to a conventional mass spectrometer (GC-MS). However, long-term instability of the GC separation characteristics and prolonged analyses (30–60 min/specimen) have led most groups to use direct mass spectrometry. In this, the pyrolysis products flow as a molecular beam into a crossing stream of low-energy electrons. The electrons collide with the product molecules with sufficient energy to remove an electron, ionizing the molecules. Once ionized, the products are removed rapidly from the electron stream to prevent formation of multiply charged ions.

The molecular ions are then focused, separated in order of mass : charge ratio (effectively molecular weight for molecular ions with a single charge) in a mass

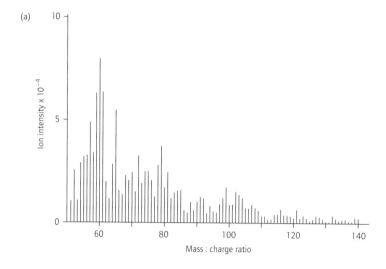

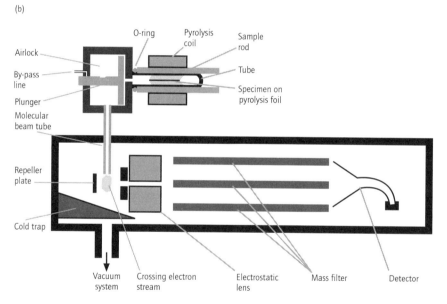

Figure 17.1 Pyrolysis mass spectrometry. (a) Pyrolysis mass spectrum of *Bacillus subtilis* grown for 24 h on nutrient agar at 30°C. The high intensities for masses 59 and 60 are typical of Gram-positive bacteria, and probably represent acetate from N-acetyl groups of peptidoglycan. The significant intensities of masses 79 and 80 are typical of endospores and may represent pyridene derived by thermal decarboxylation of dipicolinic acid. (b) Diagrammatic cross-section of a pyrolysis mass spectrometer.

spectrometer and delivered to a detector that counts the number of ions at each molecular weight. The results are acquired on a computer, and displayed as a pyrolysis mass spectrum, essentially a histogram of frequencies of ions vs. molecular weight (figure 17.1a). In the most sophisticated apparatus, specific ions can be diverted into a second mass spectrometer (pyrolysis MS-MS), to undergo high-energy ionization, producing fragmentation ions that allow fuller chemical identification.

The data so produced are complex. Quantitative reproducibility varies between products, several distinct products may arise from decomposition of a single polymer, and the same product, or distinct products with the same molecular weight, may arise from the decomposition of distinct polymers. The most successful approaches to data analysis have been multivariate statistics and neural network analysis. Both involve esoteric mathematical concepts, but, once developed, they can be packaged as fast, user-friendly 'black-box' analysis programs.

Reproducibility of the spectra hinges on biological variation between specimens, and instrument variation. Biological variation is addressed by culturing organisms under identical growth conditions. Instrument variation is complex, and difficult to disentangle from batch variation in prepared growth media. Studies have concentrated on applications where interbatch variation is irrelevant, particularly classification and typing of putative point-outbreak isolates. However, there is some promising work on corrections for interbatch and interinstrument variation (Shute *et al.* 1988; Goodacre *et al.* 1997).

Apparatus, materials and methods

Figure 17.1b shows a pyrolysis mass spectrometer. Cultures are grown concurrently on the same batch of medium, under identical conditions. A sample of growth [about 10–50 µg dry weight (dwt)] is coated onto a pyrolysis foil held in a sample tube. Samples must be fixed and dehydrated within 10 min of sampling to attain good reproducibility. Fixation is achieved by heating to 80 or 100°C for 10 min. Coated foils may be stored for up to 3 months in a vacuum desiccator, or analysed immediately. The tubes are inserted into the automated sample handling system and processed. Analysis takes about 90 s. Spectra (figure 17.1a) are acquired and stored as computer files for mathematical analysis. Replicate spectra are essential to define interspecimen and interplate reproducibility. Typically, three samples are taken from each of two replicate cultures, or four samples are taken from each culture with 10% of the isolates cultured in duplicate. The spectra are compared in computer packages and the results are expressed as a dendrogram or ordination, showing the interisolate differences in spectra (and so in whole cell composition) in diagrammatic form.

Applications

Py-MS allows rapid, inexpensive, high-throughput comparison of isolates and offers good resolution at taxonomic levels from strain, through species, to genus, but interbatch reproducibility remains a problem. It is ideal for applications where isolates must be compared rapidly (as in point-outbreak investigations), or in large numbers (as in taxonomic studies). It is one of the few methods that allow rapid, economical comparison of strains on the basis of overall chemical composition, and so it is important in polyphasic taxonomic studies. Other applications include identification, determination of metabolite levels in fermentation, antimicrobial susceptibility determination in difficult organisms, and comparison of modes of action of antimicrobials.

Much of the initial Py-MS work on the aerobic spore-forming bacilli was performed by Shute *et al.* (1984a,b, 1985, 1988). Strategies for the statistical analysis of Py-MS data were then in an early stage of development, and the published work documents the results of exploratory approaches to this problem. *Bacillus subtilis, B. pumilis, B. licheniformis* and '*B. amyloliquefaciens*' were chosen for study, as these organisms are difficult to distinguish between in conventional tests. Results for strains cultured for 16 h at 35°C, still in vegetative form, proved more discriminatory than results for extensively sporulating cultures grown for 7 days at 30°C on nutrient agar with manganese sulphate (5 mg/L). The four species showed a clear separation of species-mean spectra, but a minority of strains and spectra showed intermediate characteristics. Strains of *B. pumilis* sometimes showed poor growth on short incubation and spectra from such cultures were often atypical.

Variations in cell composition arise not only from differing genetics, but also from differences in physiology, growth rates, growth phase, growth conditions, nutrient availability and high-frequency genetic variations. Variation in several species and strains of bacteria cannot be controlled readily, causing problems in whole-cell composition approaches. Examples of rapid genetic variation are H-phase variation in salmonellae owing to mutations in sequences controlling flagellar protein synthesis, and smooth-matt colony dissociation in *Shigella sonnei* owing to plasmid loss. These variations are readily detected in Py-MS (Magee 1994). For the endospore-forming bacilli, the time of onset and extent of sporulation can vary, and has profound effects on spectra (Shute *et al.* 1988).

Within some species, the variation can be brought to a low level; for example, in *Clostridium difficile*, incubation for 48 h rather than 24 h produced a marked improvement in intrastrain reproducibility and interstrain discrimination. Variations in the intensity of masses 78 and 79 were notably reduced (J.T. Magee, unpubl. data). High intensity at these masses is highly characteristic of sporulation in *Bacillus* and *Clostridium* cultures. Mass 79 probably originates from pyrolytic decarboxylation of dipicolinic acid to form pyridine, and mass 78 may be generated from pyridine by neutral hydrogen loss. These masses are major features in Py-MS spectra of pure dipicolinic acid (Beverly *et al.* 1996), and have low intensities in nonendospore-forming species.

In taxonomic studies, variation in endospore formation may be more difficult to control across the wider range of species involved. One approach would be to select a convenient medium and incubation time, and to take the view that the extent of sporulation under these conditions is a feature of the physiology of each particular species, and so represents a valid taxonomic character. Culture on blood agar (5% horse blood in Columbia Agar base) for 24 h at 37°C or 48 h at 30°C has become a standard method in most Py-MS epidemiological studies and has been adopted successfully in several polyphasic taxonomic surveys of *Bacillus* (see below). The authors' experience is that sugar-rich media are a much greater source of problems and should be avoided. Carry-over of unmetabolized sugars in interstitial colony fluid and formation of endogenous energy storage polymers (e.g. glycogen and poly-hydroxybutyrate, which may account for a large proportion of cell mass in old cultures) yield features that can dominate spectra, obscuring discriminatory features, as was first noted in infrared spectrometry (see Magee 1994).

Later, the International Committee on Systematic Bacteriology Subcommittee on Taxonomy of the Genus *Bacillus* undertook a polyphasic taxonomic study to revise nomenclature and to establish minimal standards for the description of species, beginning with mesophilic and psychrophilic species that form swollen sporangia with ellipsoidal spores. These studies (Heyndrickx *et al.* 1995, 1996a, b) incorporated a wide variety of approaches including conventional test patterns (CTPs), whole-cell protein sodium-dodecylsulfate-polyacrylamide gel electrophoresis (SDS-PAGE), methylated fatty acid analysis (FAME), DNA base composition and hybridization, amplified rDNA restriction digest analysis (ARDRA), 16S rDNA sequencing, random amplified polymorphic DNA analysis (RAPD), amplified fragment length polymorphism (AFLP) and Py-MS.

Initial findings (Heyndrickx *et al.* 1995) for strains of *Paenibacillus gordonae* and *P. validus* showed that these two species formed a homogeneous group, and that *P. gordonae* is a later subjective synonym of *P. validus*. In Py-MS, the *P. validus* and *P. gordonae* strains formed a cluster at 89–99% similarity, with the type strains joining at 92%. This cluster was clearly distinct from strains of *P. pulvifaciens* and *P. peoriae*, which showed similarities of <60%. In CTPs with the simple matching coefficient, the two putative species clustered at >80% similarity, in FAME at a Euclidian distance of 5, in SDS-PAGE at >83%, in ARDRA at >86% and in DNA hybridization at >88%. RAPD showed two groups, each containing members of both putative species.

Similarly, a study of *P. pulvifaciens* and *P. larvae* strains resulted in a proposal that these were subspecies of a single species, *P. larvae* (Heyndrickx *et al.* 1996b). In Py-MS, strains of the subspecies formed two homogeneous subgroups with similarity >94%, with one exception, an outlying duplicate culture of ATCC9545. The subspecies clusters joined at 81% similarity to form a single species cluster. The outlying culture may represent an artefact of the clustering representation. ATCC9545 was analysed in duplicate and the duplicates were similar at 93%; however, one duplicate was placed clearly in the *P. larvae* ssp. *larvae* cluster, while one was an outlier of both subspecies groups. Outgroup strains of *P. alvei*, *P. validus*, *P. polymyxa* and *P. pabuli* showed similarities of 50% or less.

Results for other approaches divided these techniques into those that revealed the subspecies clusters, and those that did not. In CTPs the two subspecies formed homogeneous clusters at 94–95% similarity, fusing into a species cluster at 89%. In SDS-PAGE *P. larvae* ssp. *pulvifaciens* strains were characterized by a dense high molecular weight band, but were otherwise closely similar to strains of *P. larvae* ssp. *larvae* at 82%. In AFLP, the subspecies formed clusters at 88 and 95% similarity, fusing to a species cluster at 80% similarity. In RAPD, patterns were closely similar for the subspecies, but homogeneous subspecies clusters could be resolved at 80 and 90% similarity. DNA hybridization showed 99 and 95% binding within the subspecies, and 81 and 90% between the subspecies. Conversely, in ARDRA the strains of the subspecies formed a single-species cluster at 90%

similarity with no clear subspecies discrimination, reflecting similar lack of discrimination in 16S rRNA sequencing. There were problems in culturing *P. larvae* for FAME, and all strains of both subspecies were qualitatively and quantitatively similar, forming a species cluster at a Euclidean distance of 10.

A broader polyphasic study of the new genus *Paenibacillus* was undertaken with 77 strains representing ten putative species (Heyndrickx *et al.* 1996a). In Py-MS, clusters homogeneous at 95–98% were distinguished. The two subspecies of *P. larvae* were resolved, as previously. This cluster was similar to others at <60%, perhaps reflecting their distinct parasitic niche. *Paenibacillus validus* strains formed a distinct species cluster at 95% similarity in Py-MS and were resolved in a single cluster in all other methods, as were strains of *P. polymyxa*. Further species clusters of *P. macerans* (one exception), *P. amylolyticus*, *P. pabuli* (except the type strain and one other), *P. macquariensis* and *B. lautus* were evident, but showed poor interspecies separation, fusing at 90–95% similarity. *Paenibacillus alvei* strains were recovered in two distinct clusters of closely similar strain membership in Py-MS, CTPs, SDS-PAGE and RAPD with one of two primer sets, but appeared homogeneous in ARDRA, RAPD with another primer set and FAME. This pattern is reminiscent of the earlier work with *P. larvae*, suggesting the possibility of subspecies groups. Overall, the species groups of the Py-MS classification showed good concordance with those found in other approaches. *Bacillus lautus* and *B. peoria* were transferred to the genus *Paenibacillus* on the basis of this work.

This polyphasic approach was also taken up in a proposal for a new species forming highly heat-resistant endospores, *B. sporothermodurans* (Pettersson *et al.* 1996). This nonpathogenic mesophile has been isolated from UHT milk and dried milk products across Europe, and grows in milk to a maximum level of 10^5/mL without causing noticeable spoilage. PCR and sequencing of 16S rDNA indicated microheterogeneity between several gene copies, yielding multiple products differing slightly in length and sequence. These sequences showed 95–98% similarity with sequences for a group containing *B. lentus*, *B. circulans*, *B. megaterium* and similar species, and were clearly distinct from *Paenibacillus*. In Py-MS, strains of *B. sporothermodurans* were homogeneous at >90% similarity, and clearly distinct from clusters comprising strains of *B. sphaericus* serotype H5a5b (70% similarity) and *B. subtilis* (80% similarity), which showed similar intraspecies homogeneity.

These studies have wider implications outside the classification of aerobic spore-forming bacilli. Their results calibrate and define the resolution of the various approaches in taxonomic work. Those methods applied in epidemiological typing – Py-MS, SDS-PAGE and RAPD – resolved subspecies clusters, while ARDRA and FAME did not. Other Py-MS studies have shown clear separation at genus level. For example, in a polyphasic study of the Gram-positive anaerobic cocci (Murdoch *et al.* 1997), Py-MS placed strains of 'Peptostreptococcus *heliotrinreducens*' in a cluster widely separated from other species of peptostreptococci. On the basis of cell and colonial morphology, it was suggested that this species resembled members of the genus *Eubacterium*, and later studies led to a proposal for its transfer to this genus. Py-MS appears to be capable of resolving groups at levels from genus through species to subspecies, and has also proved capable of resolution at strain level in epidemiological studies.

Sisson *et al*. (1992) studied faecal and food isolates of *Bacillus* species and *Clostridium perfringens* implicated in food poisoning incidents. Py-MS clearly resolved 17 of the 18 *Bacillus* isolates into five clusters of indistinguishable isolates that coincided with their origin in five distinct food poisoning incidents. One isolate was distinct and originated from samples of lasagne, while an isolate from samples of a pizza served in the same meal and from two affected patients were indistinguishable. Conventional methods for epidemiological typing of *Bacillus* species implicated in food poisoning incidents rely on serology and are primarily targeted at flagellar antigen variation in *B. cereus*. However, serology resolves only 42 type groups, with a sizeable (10–30%) proportion of nontypeable strains, and does not resolve strains of the *B. subtilis/licheniformis* group that are also implicated in food poisoning.

Matrix-Assisted Laser Desorption–Ionization Time of Flight Mass Spectrometry (MALDI-TOF-MS)

Soft ionization methods for the analysis of biomacromolecules, such as MALDI-TOF-MS and electrospray ionization (ESI), are now important tools in protein and genomic analysis. They are particularly suited to the analysis of complex biological systems because single components can be resolved according to their mass-to-charge (m/z) values, allowing structural information to be obtained. Although Py-MS is useful for whole-cell fingerprinting, it has the disadvantage that the chemistry of pyrolysis is complex, the products are small, and so it is difficult to assign a chemical origin to particular spectrum feature. Recently, it has been demonstrated that characteristic profiles of intact microorganisms using MALDI-TOF-MS (Claydon *et al*. 1996) and ESI-MS (Goodacre *et al*. 1999) can be obtained. These approaches yield high molecular weight products that are more amenable to deduction of the chemical structure of their polymers of origin.

Principle

Cells are embedded in a matrix of a UV-absorbing chemical, by drying a cell suspension from solution, and the specimen is introduced into the mass spectrometer. Here, a small area of the sample spot is illuminated with a UV laser. The absorbed energy is converted to heat, and the matrix volatilizes, carrying with it desorbed material from the outer layers of any embedded cells in the immediate vicinity. Unlike Py-MS, the energy transferred is small, and so the desorbed material tends to comprise high molecular weight fragments. Also, volatility of the fragments is not crucial, as they are carried into the mass spectrometer with the expanding cloud of matrix material. During desorption, chemical reactions occur between the volatilized matrix and these fragments, and according to their nature, electrons are transferred, giving the fragments net positive or negative charges.

The fragment ions are extracted by repulsion down a *c*.20 Kv gradient, with a pulse of either a positive (to extract negatively charged fragments) or negative (to

(a)

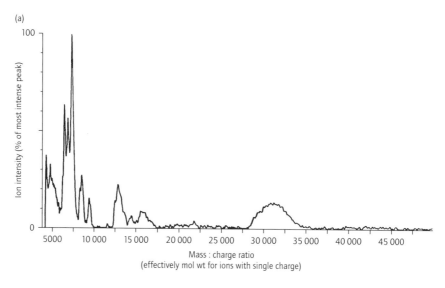

Mass : charge ratio
(effectively mol wt for ions with single charge)

(b)

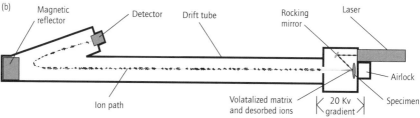

Figure 17.2 MALDI-TOF-MS. (a) MALDI-TOF spectrum of *Bacillus sphaericus* from nutrient agar in a 5-chloro-2-mercapto-benzothiazole matrix. This is scaled to show a high-mass range, but features in the lower mass range from 500 to 5000 are generally also highly discriminatory. (b) Diagrammatic cross-section of a reflectron MALDI-TOF mass spectrometer. Differential penetration of ions through the field produced by the 'magnetic mirror' or reflectron alters the transit time from specimen to detector, correcting for variations in initial ion momentum. In nonreflectron instruments, the ions travel in a straight path to a detector at the distal end of the drift tube.

extract positively charged fragments) charge. The process is designed to produce ions with a single charge, so that the mass : charge ratios of the ions are equal to their molecular weights. The ionized fragments pass into the mass spectrometer drift tube with fixed momentum (assuming that the majority of ions have a single charge), and so their velocity, and transit time down the drift tube to the detector, is a simple function of their molecular weights. Effectively, the ions are delivered sequentially to the detector in order of molecular weight. The detector records the intensity of ions, and their transit time from extraction. This process is repeated up to several hundred times, tracking the laser across the specimen, analysing the ions from each spot, and accumulating sufficient data to display the results as an averaged mass spectrum (figure 17.2a).

The data differ from those obtained in Py-MS in two respects. The region of interest is a much wider and higher range of molecular weights (600–4000 Da compared to 50–200 Da for Py-MS), and is biased strongly to reflect the composition of the outer cell layers, rather than overall cell composition. Quantitative reproducibility is poor, because a proportion of the laser pulses volatilize the matrix in an area with no embedded cells, and contribute to background rather than yielding relevant ions. The proportion of 'blank' laser pulses is strongly dependent on cell density and clumping. Accuracy of the mass scale can also be a factor in the high mass range, where an experimental error of 0.1% may shift observed mass : charge ratios over several units. Furthermore, the products are distributed over a much wider observable mass range than in Py-MS, and many more products can be observed. Most of the products above 1000 Da probably represent single chemical species, derived from a single polymer, and a good proportion of the masses show zero (or background) intensity. Overall, these differences eliminate the direct application of Py-MS data-processing packages. New data-processing techniques have had to be developed for comparison of MALDI-TOF spectra, and are now at an advanced stage.

Materials, apparatus and methods

The apparatus is illustrated in figure 17.2b. Isolates are cultured with standardized media and conditions, and a small sample of growth (*c*.10 µg dwt) or suspended cells is dried and overlaid with 1 µL of matrix solution on a clean multiwell analysis slide. Various matrices are available. For *Bacillus* spp. and other Gram-positive organisms, a matrix solution of 5-chloro-2-mercaptobenzothiazole (3 µg/L in 1 : 1 : 1 mixture of water, acetonitrile and methanol containing 0.1% formic acid) with 0.01 M 18-crown-6 ether as a molecular weight marker is suitable. The sample spots are dried in a warm air stream, and the slide is inserted into the specimen port of the mass spectrometer. Analysis is automated, and acquisition of a spectrum is accumulated from 100 laser shots taking *c*.1 min. Typically, the slide has 32 wells, and three to four replicate spots from each of two cultures are analysed for each isolate. The spectra are stored as electronic files.

Applications

Discrimination is obtained at all taxonomic levels from strain, to species, to genus (Claydon *et al.* 1996). The spectra show good long-term reproducibility, they can be interpreted visually at species level with experience, and it is possible to assign tentative chemical identities to some of the products. The known long-term reproducibility of MALDI-TOF-MS is critical to rapid identification by comparison of spectra with databases for known reference strains. Epidemiological typing applications have not been explored for *Bacillus* spp., but clear interstrain differences concordant with epidemiological information and pulse field gel electrophoresis results have been found for hospital isolates of epidemic multiresistant *Staphylococcus aureus* (Edwards-Jones *et al.* 2000).

In *Bacillus* spp., taxonomic studies are at an early stage. Detection and discrimination of pathogenic (including *B. anthracis*) and nonpathogenic bacteria has been successful (Krishnamurthy *et al.* 1996), as has identification to species level with spore suspensions (Hathout *et al.* 1999). Work on the surface layers of spore mother cells, isolated spores and purified exosporium is progressing (Prime *et al.*, pers. comm.).

Fourier transform infrared (FT-IR) spectrometry

Principle

Traditionally, sample preparation for FT-IR absorbance measurements involves grinding the dried sample to a fine powder and mixing with KBr. Alternatives suited to microbial analysis include application of samples to one of 16 ZnSe windows on a rotating disc for absorbance measurements (Naumann *et al.* 1991), or reflectance methods (Goodacre *et al.* 1996; Timmins *et al.* 1998a; Winson *et al.* 1997). Diffuse reflectance–absorbance can be achieved by applying the sample (5 µL of bacterial slurry) onto a metal plate with 100, 400 or 384 sample wells. The plate is loaded onto the motorized stage of a reflectance accessory (Glauninger *et al.* 1990), and each specimen is analysed automatically over a period of 2–4 h. Reflectance methods can be more sensitive and discriminatory than absorbance (Mitchell 1993).

FT-IR and Raman spectroscopy (dispersive, Fourier transform or UV resonance) measure the vibrations of bonds within functional chemical groups (Griffiths & de Haseth 1986; Colthup *et al.* 1990; Ferraro & Nakamoto 1994; Schrader 1995) (figure 17.3a). These functional groups absorb infrared radiation at specific wavelengths, e.g. proteins show strong amide I absorption bands at 1653/cm characteristic of stretching of C=O and C–N and bending of the N–H bond (Stuart 1996). Absorption of broadband IR radiation by biological materials in the near- (10 000–4000/cm) or mid-range (4000–600/cm) produces FT-IR spectra that reflect the proportions of functional groups present in the sample as a 'fingerprint' containing information on whole-cell composition. Raman spectroscopy differs in that it measures the exchange of energy between functional groups and monochromatic radiation (usually a near infrared diode laser emitting at 785 nm). The energy exchange effectively modulates a portion of the monochromatic light with a Raman-shifted IR spectrum. The Raman effect is

Figure 17.3 (*Opposite*) Fourier transform infrared spectrometry. (a) Spectra of *Bacillus subtilis* (faint line) and *B. cereus* (solid line). Bars indicate regions associated with specific biochemicals (FA, fatty acids; Ph, phosphates; PP, polypeptides; PS, polysaccharides). (b) FT-IR spectrometer. The beam splitter divides the IR light into two beams, one to a static reflector, and the other to a travelling reflector, then back to rejoin at the beam splitter. As the travelling mirror moves, differences in the path length from the two mirrors causes interference, ultimately yielding an inteferogram 'spectrum' of the specimen. The laser beam is also split, reflected and rejoined and interference fringes are monitored as the travelling reflector moves, giving accurate fine measurement of the path length difference. The IR beam is then directed through an absorbance cell or diffuse reflectance accessory (c) and onto a detector.

(a)

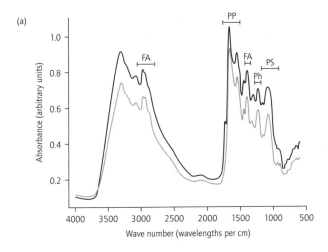

(b)

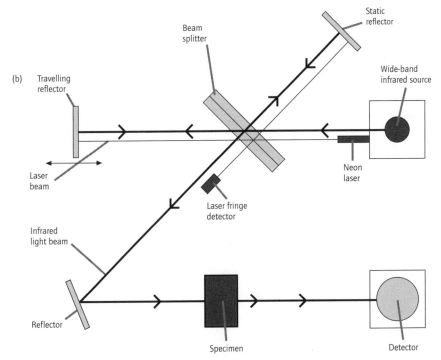

(c)

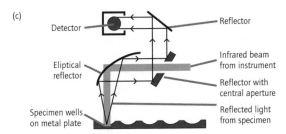

weak; only one in 10^8 photons exchange energy with a molecular bond vibration, the remainder emerge with the same frequency as the incident monochromatic light. The exchange gives two series of 'modulation' lines, called Stokes and anti-Stokes lines, resulting from subtraction or addition of energy. The Stokes shift is about ten times stronger than the anti-Stokes shift and is collected to construct a Raman 'fingerprint' of the sample. Raman data are complementary to IR absorption, but some spectral features may be more apparent in one or the other because of subtle physicochemical effects.

Materials, apparatus and methods

Figures 17.3b and 17.3c show the sample loading plate and a Bruker IFS28 FT-IR spectrometer. Cultures are grown concurrently on the same batch of medium, under identical conditions, and a sample of growth is suspended in physiological saline. The samples are spotted into the wells of an aluminium sample plate and fixed and dehydrated by heating to 50°C for 20 min within 10 min of sampling. The plate is loaded onto the motorized stage of a reflectance TLC accessory of an FT-IR instrument equipped with a mercury–cadmium–telluride detector cooled with liquid nitrogen. The instrument collects spectra from the centre of each well, over the wavenumber range 4000/cm to 600/cm at a rate of 20/s. The spectral resolution is typically 4/cm, and 256 spectra are taken and averaged, yielding a reflectance spectrum containing c.882 points. Absorbance spectra are calculated from the reflectance–absorbance data based on the Kubelka–Munk theory (Griffiths & de Haseth 1986), recorded and displayed. These can be compared in computer packages based on statistical or neural network approaches after scaling, smoothing and processing as first or second derivatives spectra using the Savitzky–Golay algorithm (Savitzky & Golay 1964).

Applications

Good discrimination is obtained at all taxonomic levels from strain to species to genus. The spectra show good long-term reproducibility, and give discrimination comparable to Py-MS and DNA analyses (Timmins *et al.* 1998a,b). For example, Naumann and co-workers (Helm *et al.* 1991; Naumann *et al.* 1991) have shown that mid-IR range FT-IR provides sufficient resolution to distinguish between strains of the same species. Analysis of spectra with neural networks allows the automatic identification of bacteria (Goodacre *et al.* 1998b), and the emergence of new chemometric techniques based on genetic algorithms (Broadhurst *et al.* 1997), genetic programmes (Gilbert *et al.* 1997; Taylor *et al.* 1998) and rule induction (Alsberg *et al.* 1998) allows the deconvolution of the FT-IR spectra in terms of biochemical structure (Goodacre *et al.* 2000; Johnson *et al.* 2000).

In *Bacillus* species, attenuated total reflectance FT-IR has been applied to real-time monitoring of germination of wild type and germination mutants of *B. subtilis,* showing compositional changes in levels of calcium dipicolinate and proteins (Cheung *et al.* 1999). Goodacre *et al.* (2000) showed that that pyridine ring vibrations at 1447–1439/cm from dipicolinate were highly characteristic spore biomarkers. Gittins *et al.* (1999) explored the potential of FT-IR in the detection of air-borne biological particles such as *Bacillus* spores and cells, which might be aerosolized as biological weapons. Most studies have concentrated on characterization of *Bacillus* spp. involved in food poisoning. A preliminary study applied FT-IR in the discrimination of *B. cereus* isolates from other *Bacillus* spp. (Lin *et al.* 1998). Beattie *et al.* (1998) discriminated successfully between *B. cereus*, *B. mycoides* and *B. thuringiensis* with FT-IR. Goodacre *et al.* (1998a) found that FT-IR-based classification of 36 strains representing *B. amyloliquefaciens*, *B. cereus*, *B. licheniformis*, *B. megaterium*, *B. subtilis* (including *B. niger* and *B. globigii*), *B. sphaericus* and *Brevi laterosporus* was congruent with the known taxonomy of the group, based on biochemical characters and phylogenetic analysis of 16S rRNA sequences.

Conclusions

These techniques offer many advantages over conventional approaches. Although instrument costs are high (£50 000–1 000 000), the simplicity of sample preparation, lack of labile reagents, speed of analysis, high throughput and low processing costs cannot be matched by any other approach that yields such comprehensively discriminatory data at genus, species and strain level.

In classification studies, these approaches can play two crucial roles.

1 They can offer confirmation that data reflecting whole-cell chemical composition is compatible with classifications based on other aspects of strain characterization. Obtaining data on chemical composition by conventional methods is so specialized, tedious, time-consuming and costly that it can be performed only on small numbers of strains. Whole-cell fingerprinting techniques eliminate these difficulties, allowing polyphasic studies that include comprehensive coverage of this important aspect of characterization.

2 The low costs, high throughput and high predictivity of consensus classification results are eminently suited to preliminary screening of strains. A collection can be rapidly sorted into presumptive clusters and wildly aberrant cultures can be identified and investigated for cryptic contaminants or incorrect inclusion. Strains that are typical of clusters can be selected for investigation in low-throughput or costly characterization methods, and imbalances in cluster membership can be corrected. Results from these methods are invaluable in the preliminary organization of a polyphasic taxonomic study, and are an essential ingredient in a comprehensive characterization process that yields stable, reliable classifications.

Convenient identification applications depend on long-term reproducibility of spectra. Although routine identification with Py-MS has been described (Wieten *et al.* 1982; Saddler *et al.* 1989), this does require the inclusion of standard reference strains with each batch, limiting the scope to readily recognized groups containing few species. Py-MS and FT-IR show sufficient instrument stability to allow long-term accumulation of extensive reference strain databases, widening the scope to routine identification across a much wider range of organisms. The development of major identification databases is inherently slow, and likely to grow from small specialist applications where speed is crucial, e.g. the emerging application of MALDI-TOF-MS in rapid differentiation and epidemiological typing of epidemic multiresistant *S. aureus*.

Epidemiological typing has been a much-favoured application for physico-chemical methods, with a preponderance of papers devoted to interstrain differentiation. These methods are ideal for investigation of putative point-outbreak situations by reason of their speed, throughput and lack of any requirement for species-specific changes in methodology. However, the medical relevance of epidemiological typing within *Bacillus* species is limited to those species involved in food poisoning, which are implicated in a fairly small proportion of outbreaks.

Fingerprint whole-cell analytical methods are tools that we can use to investigate nature. As with any other tool, they have particular properties that suit appropriate applications. In this case, these are speed, high discrimination, low specimen processing cost, automation, high throughput, broad application of a single methodology over many species, and an emerging ability to derive chemical information on cell polymers. Disadvantages are the high instrument costs and the complexity of the data. These disadvantages are more rooted in attitudes than reality. Microbiologists are accustomed to a pricing structure where costs concentrate in disposables and reagents; the one-off costs of capital items are imposing, and may not be considered in relation to the overall cost per specimen over the lifetime of the instrument. Similarly, the complexity of data analysis is the province of computers, which can now process spectrum data in fractions of a second. Designing analysis programs is undoubtedly a task for the specialist, but once this task is complete, users with little knowledge of the underlying mathematics can process results. The fingerprint whole-cell methods offer inexpensive and convenient ways of investigating and comparing cell composition and are well worth considering for studies involving classification, identification, epidemiological typing and monitoring of physiological processes.

References

Alsberg, B.K., Wade, W.G. & Goodacre, R. (1998) Chemometric analysis of diffuse reflectance-absorbance Fourier transform infrared spectra using rule induction methods: application to the classification of *Eubacterium* species. *Applied Spectroscopy* **52**, 823–832.

Beattie, S.H., Holt, C., Hirst, D. & Williams, A.G. (1998) Discrimination among *Bacillus cereus*, *B. mycoides* and *B. thuringiensis* and some other species of the genus *Bacillus* by Fourier transform infrared spectroscopy. *FEMS Microbiology Letters* **164**, 201–206.

Beverly, M.B., Basile, F., Voorhees, K.J. & Hadfield, T.L. (1996) A rapid approach for the detection of dipicolinic acid in bacterial spores using pyrolysis/mass spectrometry. *Rapid Communications in Mass Spectrometry* **10**, 455–458.

Broadhurst, D., Goodacre, R., Jones, A., Rowland, J.J. & Kell, D.B. (1997) Genetic algorithms as a method for variable selection in PLS regression, with application to pyrolysis mass spectra. *Analytica Chimica Acta* **348**, 71–86.

Cheung, H.Y., Cui, J.X. & Sun, S.Q. (1999) Real-time monitoring of *Bacillus subtilis* endospore components by attenuated total reflection Fourier-transform infrared spectroscopy during germination. *Microbiology-UK* **145**, 1043–1048.

Claydon, M.A., Davey, S.N., Edwards-Jones, V. & Gordon, D.B. (1996) The rapid identification of intact microorganisms using mass spectrometry. *Nature Biotechnology* **14**, 1584–1586.

Colthup, N.B., Daly, L.H. & Wiberly, S.E. (1990) *Introduction to Infrared and Raman Spectroscopy*. Academic Press, New York.

Colwell, R.R. (1970) Polyphasic taxonomy of bacteria. In: *Culture Collections of Microorganisms* (eds H. Iizuka & T. Hasegawa), pp. 421–436 University of Tokyo Press, Tokyo.

Edwards-Jones, V., Claydon, M.A., Evason, D.J., Walker, J., Fox, A. & Gordon D.B. (2000) Rapid discrimination between methicillin-sensitive and methicillin-resistant *Staphylococcus aureus* by intact cell mass spectrometry. *Journal of Medical Microbiology* 49, 1–6.

Ferraro, J.R. & Nakamoto, K. (1994) *Introductory Raman Spectroscopy*. Academic Press, London.

Gilbert, R.J., Goodacre, R., Woodward, A.M. & Kell, D.B. (1997) Genetic programming: a novel method for the quantitative analysis of pyrolysis mass spectral data. *Analytical Chemistry* 69, 4381–4389.

Gittins, C.M., Piper, L.G., Rawlins, W.T., Marinelli, W.J., Jensen, J.O. & Akinyemi, A.N. (1999) Passive and active standoff infrared detection of bio-aerosols. *Field Analytical Chemistry and Technology* 3, 274–282.

Glauninger, G., Kovar, K.A. & Hoffmann, V. (1990) Possibilities and limits of an online coupling of thin-layer chromatography and FTIR spectroscopy. *Fresenius Journal of Analytical Chemistry* 338, 710–716.

Goodacre, R., Timmins, É.M., Rooney, P.J., Rowland, J.J. & Kell, D.B. (1996) Rapid identification of *Streptococcus* and *Enterococcus* species using diffuse reflectance-absorbance Fourier transform infrared spectroscopy and artificial neural networks. *FEMS Microbiology Letters* 140, 233–239.

Goodacre, R., Timmins, E.M., Jones, A. *et al.* (1997) On mass spectrometer instrument standardization and interlaboratory calibration transfer using neural networks. *Analytica Chimica Acta* 348, 511–532.

Goodacre, R., Shann, B., Gilbert, R.J. *et al.* (1998a) The characterisation of *Bacillus* species from PyMS and FT IR data. In: *Proceedings of the 1997 ERDEC Scientific Conference on Chemical and Biological Defense Research*, pp. 257–265. Aberdeen Proving Ground, MD.

Goodacre, R., Timmins, É.M., Burton, R. *et al.* (1998b) Rapid identification of urinary tract infection bacteria using hyperspectral, whole organism fingerprinting and artificial neural networks. *Microbiology* 144, 1157–1170.

Goodacre, R., Heald, J.K. & Kell, D.B. (1999) Characterisation of intact microorganisms using electrospray ionization mass spectrometry. *FEMS Microbiology Letters* 176, 17–24.

Goodacre, R., Shann, B., Gilbert R.J. *et al.* (2000) The detection of the dipicolinic acid biomarker in *Bacillus* spores using Curie-point pyrolysis mass spectrometry and Fourier transform infrared spectroscopy. *Analytical Chemistry* 72, 119–127.

Griffiths, P.R. & de Haseth, J.A. (1986) *Fourier Transform Infrared Spectrometry*. Wiley, New York.

Hathout, Y., Demirev, P.A., Ho, Y.P. *et al.* (1999) Identification of *Bacillus* spores by matrix-assisted laser desorption ionization-mass spectrometry. *Applied and Environmental Microbiology* 65, 4313–4319.

Helm, D., Labischinski, H., Schallehn, G. & Naumann, D. (1991) Classification and identification of bacteria by Fourier transform infrared spectroscopy. *Journal of General Microbiology* 137, 69–79.

Heyndrickx, M., Vandemeulebroecke, K., Scheldeman, P. *et al.* (1995) *Paenibacillus* (formerly *Bacillus*) *gordonae* (Pichinoty *et al.* 1986) Ash *et al.* 1994 is a later subjective synonym of *Paenibacillus* (formerly *Bacillus*) *validus* (Nakamura 1984) Ash *et al.* 1994: emended description of *P. validus*. *International Journal of Systematic Bacteriology* 45, 661–669.

Heyndrickx, M., Vandemeulebroecke, K., Scheldeman, P. *et al.* (1996a) A polyphasic reassessment of the genus *Paenibacillus*, reclassification of *Bacillus lautus* (Nakamura 1984) as *Paenibacillus lautus* comb. nov. and of *Bacillus peoriae* (Montefusco *et al.* 1993) as *Paenibacillus peoriae* comb. nov., and emended descriptions of *P. lautus* and of *P. peoriae*. *International Journal of Systematic Bacteriology* 46, 988–1003.

Heyndrickx, M., Vandemeulebroecke, K., Hoste, B. *et al.* (1996b) Reclassification of *Paenibacillus* (formerly *Bacillus*) *pulvifaciens* (Nakamura 1984) Ash *et al.* 1994, a later subjective synonym of *Paenibacillus* (formerly *Bacillus*) *larvae* (White 1906) Ash *et al.* 1994, as a subspecies of *P. larvae*, with emended descriptions of *P. larvae* as *P. larvae* subsp. *larvae* and *P. larvae* subsp. *pulvifaciens*. *International Journal of Systematic Bacteriology* 46, 270–279.

Johnson, H.E., Gilbert, R.J., Winson, M.K. *et al.* (2000) Explanatory analysis of the metabolome using genetic programming of simple, interpretable rules. *Genetic Programming and Evolvable Machines* 1, 243–258.

Krishnamurthy, T., Ross, P.L. & Rajamani, U. (1996) Detection of pathogenic and non-pathogenic bacteria by matrix-assisted laser desorption/ionization time-of-flight mass spectrometry. *Rapid Communications in Mass Spectrometry* 10, 883–888.

Lin, S.F., Schraft, H. & Griffiths, M.W. (1998) Identification of *Bacillus cereus* by Fourier transform infrared spectroscopy (FTIR). *Journal of Food Protection* 61, 921–923.

Magee, J.T. (1994) Analytical fingerprinting methods. In: *Chemical Methods in the Classification of Prokaryotes* (eds M. Goodfellow & A.G. O'Donnell), pp. 523–553. Wiley, Chichester.

Mitchell, M.B. (1993) Fundamentals and applications of diffuse reflectance infrared Fourier transform (DRIFT) spectroscopy. *Advances in Chemical Series* **236**, 351–375.

Murdoch, D.A., Collins, M.D., Willems, A., Hardie J.M., Young K.A. & Magee, J.T. (1997) Description of three new species of the genus *Peptostreptococcus* from human clinical specimens: *Peptostreptococcus harei* sp. nov., *Peptostreptococcus ivorii* sp. nov., and *Peptostreptococcus octavius* sp. nov. *International Journal of Systematic Bacteriology* **47**, 781–787.

Naumann, D., Helm, D., Labischinski, H. & Giesbrecht, P. (1991) The characterization of microorganisms by Fourier-transform infrared spectroscopy (FT-IR). In: *Modern Techniques for Rapid Microbiological Analysis* (ed. W.H. Nelson), pp. 43–96. VCH Publishers, New York.

Nelson, W.H. (ed.) (1991) *Modern Techniques for Rapid Microbiological Analysis.* VCH Publishers, New York.

Pettersson, B., Lembke, F., Hammer, P., Stackebrandt, E. & Priest, F.G. (1996) *Bacillus sporothermodurans*, a new species producing highly heat-resistant endospores *International Journal of Systematic Bacteriology* **46**, 759–764.

Saddler, G.S., Falconer, C. & Sanglier, J.J. (1989) Preliminary experiments for the selection and identification of actinomycetes by pyrolysis mass spectrometry. *Actinomyctologia* **2**, S3–S4.

Savitzky, A. & Golay, M.J.E. (1964) Smoothing and differentiation of data by simplified least squares procedures. *Analytical Chemistry* **36**, 1627–1633.

Schrader, B. (1995) *Infrared and Raman spectroscopy: methods and applications.* Verlag Chemie, Weinheim.

Shute, L.A., Gutteridge, C.S. & Berkeley, R.C.W. (1984a) Pyrolysis mass spectrometry: a discriminatory technique with potential for rapid characterisation and identification of micro-organisms (Abstract). In: *Microbe '86: XIV International Congress of Microbiology*, p. 75.

Shute, L.A., Gutteridge, C.S., Norris, J.R. & Berkeley, R.C.W. (1984b) Curie-point pyrolysis mass spectrometry applied to characterization and identification of selected *Bacillus* species. *Journal of General Microbiology* **130**, 343–355.

Shute, L.A., Berkeley R.C.W., Norris, J.R. & Gutteridge, C.S. (1985) Pyrolysis mass spectrometry in bacterial systematics. In: *Chemical Methods in Bacterial Systematics* (eds M. Goodfellow & D.E. Minnikin), pp. 95–114. Academic Press, London.

Shute, L.A., Gutteridge, C.S., Norris, J.R. & Berkeley R.C.W. (1988) Reproducibility of pyrolysis mass spectrometry: effect of growth medium and instrument stability on the differentiation of selected *Bacillus* species. *Journal of Applied Bacteriology* **64**, 79–88.

Sisson, P.R., Kramer, J.M., Brett, M.M., Freeman, R., Gilbert, R.J. & Lightfoot, N.F. (1992) Application of pyrolysis mass spectrometry to the investigation of outbreaks of food poisoning and non-gastrointestinal infection associated with *Bacillus* species and *Clostridium perfringens*. *International Journal of Food Microbiology* **17**, 57–66.

Stuart, B. (1996) *Modern Infrared Spectroscopy.* Wiley, Chichester.

Taylor, J., Goodacre, R., Wade, W.G., Rowland, J.J. & Kell, D.B. (1998) The deconvolution of pyrolysis mass spectra using genetic programming: application to the identification of some *Eubacterium* species. *FEMS Microbiology Letters* **160**, 237–246.

Timmins, É.M., Howell, S.A., Alsberg, B.K., Noble, W.C. & Goodacre, R. (1998a) Rapid differentiation of closely related *Candida* species and strains by pyrolysis mass spectrometry and Fourier transform infrared spectroscopy. *Journal of Clinical Microbiology* **36**, 367–374.

Timmins, É.M., Quain, D.E. & Goodacre, R. (1998b) Differentiation of brewing yeast strains by pyrolysis mass spectrometry and Fourier transform infrared spectroscopy. *Yeast* **14**, 885–893.

Wieten, G., Haverkamp, J., Berwald, L.G., Groothuis, D.G. & Draper, P. (1982) Pyrolysis mass spectrometry: its application to mycobacteriology, including *Mycobacterium leprae*. *Annales de Microbiologie* **133B**, 15–27.

Winson, M.K., Goodacre, R., Woodward, A.M. *et al.* (1997) Diffuse reflectance absorbance spectroscopy taking in chemometrics (DRASTIC). A hyperspectral FT-IR-based approach to rapid screening for metabolite overproduction. *Analytica Chimica Acta* **348**, 273–282.

Chapter 18

Whole-cell Fatty Acid Analysis in the Systematics of *Bacillus* and Related Genera

Peter Kämpfer

Fatty acid analysis for bacterial identification

Very early attempts to apply cellular fatty acid analysis to bacterial identification were made in the 1950s (James & Martin 1952). In 1963, Abel *et al.* (1963) were the first to present evidence suggesting that cellular fatty acid (CFA) analysis by GLC (gas–liquid chromatography) could successfully identify bacteria. In this review, CFAs will be defined as those cellular lipid components that have carbon chain lengths of 9 to 20 atoms. This includes the majority of fatty acids located in the cell membrane as glycolipid and phospholipid. The source of fatty acids in microbial cells is lipid, primarily that of the cell membranes (e.g. phospholipid) or the lipoteichoic acid in Gram-positive bacteria.

Fatty acids can be named according to the number of carbon atoms and types of functional groups they contain, and their double bond location(s). The systematic name can be simplified by writing C followed by the number of carbon atoms to the left of a colon and the number of double bonds on the right. Different conventions are encountered for locating the double-bond positions and *cis* or *trans* isomers. ω, the lowercase letter omega of the Greek alphabet, indicates the double-bond position from the hydrocarbon end of the chain, and *c* and *t* indicate the *cis* and *trans* configurations of the hydrogen atoms. It should be mentioned that, as with other conventions, the numbering of the branched chain of cyclopropane-containing acids and of hydroxy fatty acids proceeds from the carboxyl end of the molecule. Branched fatty acids are indicated by the prefixes *iso* (*i*) or *anteiso* (*a*). These definitions permit only *iso* methyl-substituted fatty acids (an *iso* ethyl would be an *anteiso* methyl of one-higher chain number), but would allow both methyl and ethyl as *anteiso* substitutions.

The predominance of the terminally methyl-branched fatty acids – 13-methyltetradecanoic (*i*15:0), 12-methyltetradecanoic (*a*15:0) and 15-methylhexadecanoic (*i*17:0) acid – in *Bacillus subtilis* was first reported by Saito (1960). Extensive studies by Kaneda (1977, and references therein) led to the discovery of three additional branched-chain fatty acids; 12-methyltridecanoic (*i*14:0), 14-methylpentadecanoic (*i*16:0) and 14-methylhexadecanoic (*a*17:0) acid in the lipids of *B. subtilis*.

The biosynthesis of *iso* and *anteiso* fatty acids is linked closely with the biosynthesis of branched-chain amino acids, and it was suggested that these fatty acids may have played an important role at an early stage of the evolution of life (Kaneda 1977).

Analysis of cellular fatty acids (CFAs)

The study of branched-chain fatty acids depends on careful analytical work to separate the various branched-chain compounds from other closely related fatty acids and to obtain unambiguous identification and accurate quantitative measurements on the separated fractions (Kaneda 1977). The main requirement of CFA analysis is suitable instrumentation. Technological development by industry in the production of chromatographs and capillary columns has made it possible for microbiologists to use gas–liquid chromatography (GLC) more routinely for a variety of purposes. Computer-aided interpretation of the data obtained has also facilitated numerous applications of GLC in microbiology.

In the early days of GLC for fatty acid analysis three techniques were often used in the sequence: thin-layer chromatography for class separation; gas–liquid chromatography for individual separation and tentative identification; and mass spectrometry for structural identification (Kaneda 1977). The advantages and disadvantages of these techniques have been reviewed excellently by Kaneda (1977). However, as already pointed out by Welch (1991), the understanding of the molecular basis and development of the analytic instrumentation is central to an appreciation of their potential uses and limitations. The CFA composition of many microorganisms has been studied, but a fair degree of caution must be exercised in examining the earlier literature and comparing the results of different studies. In addition to changes in technique with respect to chromatographic equipment, the CFA compositions of cells vary quantitatively, according to culture conditions and various other influencing factors (Welch 1991). It was shown, for example, that the addition of the amino acid isoleucine increased the proportion of *anteiso* acids, and that valine and leucine enhanced the production of *iso* fatty acids in *B. subtilis* (Kaneda 1966) and also in a thermophilic species of *Bacillus* (Daron 1973). Furthermore, it was found that if unnatural ethyl and dimethyl branched precursors were added to growing cultures of *B. subtilis*, corresponding branched fatty acids were produced in substantial proportions (Kaneda 1977).

Willecke and Pardee (1971) studied the effect of substrate on fatty acid composition using a mutant of *B. subtilis* which could not produce branched-chain α-keto acid dehydrogenase and therefore required short-branched fatty acids for growth. The wild type strain produced *a*15:0 in high amounts and in addition substantial amounts of *a*17:0, but smaller amounts of 14:0 to 17:0 *iso* and straight-chain fatty acids. A supplement of 2-methylbutyrate (precursor of odd-numbered *anteiso* acids) to the mutant resulted in a fatty acid pattern consisting of comparable amounts of *a*15:0 and 17:0 acids, and minor proportions (<10%) of straight-chain 14:0 and 16:0 acids. Growth of the mutant on the *iso* acid precursors isobutyrate and isovalerate gave simple patterns consisting mainly of *i*14:0 and *i*16:0 acids, respectively (Goodfellow & Minnikin 1981). Weerkamp and Heinen (1972) studied the effect of temperature on the fatty acid composition of the thermophiles '*B. caldolyticus*' and '*B. caldotenax*' (now grouped into rRNA group 5 of Ash *et al.* 1991). With increasing temperature, they found a decrease of *i*15:0 and an increase of *i*17:0 fatty acids, as well as a decrease of *i*16:0 and an increase of 16:0. The effect of a moderate halophile medium on fatty acid composition was studied for strains of the genera *Gracilibacillus* and *Salibacillus* (Waino *et al.* 1999) (for details see table 18.10).

In general, the culture media and growth conditions should be exactly identical for the purpose of comparing fatty acid patterns from different strains. But, as already pointed out by Welch (1991), choice of medium, length and atmosphere of incubation, and temperature generally have little impact on the qualitative detection of CFAs.

At present, the only commercially available GLC system developed for the identification of bacteria and yeasts by CFA analysis is the Microbial Identification System (MIDI, Newark, DE). The system combines a gas chromatograph equipped with a flame ionization detector, a 5% methylphenyl silicone fused-silica capillary column (25 m by 0.2 mm), an automatic sampler, integrator and computer, and a printer. Software libraries for the identification of a large number of aerobic and anaerobic bacteria, mycobacteria and yeasts have been developed and updated in conjunction with this equipment. Upon injection of a sample into the column containing a specified flow of hydrogen (carrier) gas, the fatty acids are separated because of different retention times under conditions of increasing temperature. A computer-controlled temperature program begins at 170°C and is gradually increased to 270°C at a rate of 5°C/min. When the methyl ester derivatives reach the end of the column, a signal from the flame ionization detector is recorded as a peak by the integrator. The areas under the peaks reflect the relative amounts of individual fatty acids. The retention time of a mixture of known fatty acids is used by the computer – or by the individual if performed manually – to calculate an equivalent chain length for the molecule. The equivalent chain length is equal to the number of carbon atoms of a straight-chain saturated fatty acid or to a number that can be calculated by interpolation with a mathematical formula for other fatty acids. The accuracy of naming fatty acid peaks by comparing retention times with those of a known mixture is high when the computer is used, but definitive identification can be made only by mass spectrometry.

For most applications, however, mass spectrometry is unnecessary. The amounts of CFAs detected are calculated as a percentage of the total amount, and a summary can be printed at the end of each run to show the names and amounts of the CFAs and, optionally, the most likely identification according to similarity to entries in the database. The Microbial Identification System is based on a strict guideline for highly standardized cultivation and extraction procedures and provides a species-specific fatty acid database. It must be stressed, however, that individual species identification within the whole group of *Bacillus sensu lato* is often impossible. Additional phenotypic and/or genotypic methods are often indispensable for a correct species allocation.

Early investigations of fatty acid composition in
Bacillus sensu lato

The study of CFA patterns of strains belonging to the genus *Bacillus* has a long tradition. The extensive work of Kaneda (1977), in particular, led to the discovery of the major occurrence of the terminally methyl-branched fatty acids and their biosynthetic pathways. On the basis of the fatty acid composition of 19 *Bacillus* species studied by Kaneda (1977), six groups were found (Kaneda groups A–F), all of which, except group D, contained major amounts of branched chain acids.

Within groups A–C, only insignificant amounts (<3%) of unsaturated fatty acids are found. These three groups could be separated on the basis of their predominant fatty acids. Group A ('*B. alvei*', '*B. brevis*', *B. circulans*, *B. licheniformis*, '*B. macerans*', *B. megaterium*, *B. pumilus* and *B. subtilis*) contains *a*14:0 (26–60%) and *i*15:0 (13–30%) acids. The chain lengths of the acids ranged between 14 and 17.

In group B ('*B. polymyxa*', '*B. larvae*', '*B. lentimorbus*', '*B. popilliae*') *a*15:0 acid predominated (39–62%). Again, fatty acids of chain lengths between 14 and 17 were detected.

Kaneda group C (*B. stearothermophilus*, '*B. caldolyticus*', '*B. caldotenax*') were found to be different from A and B in having *i*15:0 acid as the predominant fatty acid. In group D ('*B. acidocaldarius*'), Kaneda (1977) found a unique fatty acid pattern. Here, the main compounds (up to 70%) were cyclohexane fatty acids with chain lengths of 17–19 (De Rosa *et al.* 1971).

Group E contained *B. anthracis*, *B. cereus* and *B. thuringiensis*. In contrast to the other groups, small proportions (7–12%) of unsaturated fatty acids were found always to be present. Also in this group, the predominant fatty acids (19–21%) were of the *i*15:0 type.

Only in the psychrophilic organisms *B. globisporus* and *B. insolitus* (group F) were large proportions of unsaturated fatty acids (17–28%) found. The predominant branched-chain fatty acids in these two species were *a*15:0 acids.

Changes in systematics of *Bacillus* and related genera

Within the last 10 years, the taxonomy of the genus *Bacillus* has undergone several substantial changes. As defined in *Bergey's Manual of Systematic Bacteriology* (Claus & Berkeley 1986), the genus *Bacillus* s. l. encompasses a diverse assemblage of Gram-positive, aerobic or facultatively anaerobic, spore-forming rod-shaped, phenotypically heterogeneous organisms that exhibit an extremely wide range of nutritional and growth requirements, and metabolic diversity. The wide range of DNA base compositions of *Bacillus* strains (32–69 mol% G+C) (Norris *et al.* 1981; Claus & Berkeley 1986) supports the subdivision of the genus *Bacillus* into several, more homogeneous, genera. At present, the more detailed results of analyses of genomic data provide a rational basis for subdivision of the genus *Bacillus* s. l. Comparisons of rRNA sequences of the type strains of many *Bacillus* species show clear phylogenetic relationships within the genus (Ash *et al.* 1991; Rössler *et al.* 1991). In the study of Ash *et al.* (1991), the small-subunit rRNA sequences of type strains of 51 species were determined and comparative analysis of the sequence data revealed five phylogenetically distinct clusters. Group 1 (*Bacillus sensu stricto*) includes *B. subtilis* (the type species of the genus) and 27 other species. Group 2 consists of *B. sphaericus*, *B. fusiformis*, *B. insolitus*, *B. pasteurii*, *B. psychrophilus* and *Sporosarcina ureae*. The species *B. globisporus*, *B. pasteurii* and *B. psychrophilus* were reclassified into the genus *Sporosarcina*, together with *Sporosarcina ureae* and a new species, *S. aquimarina* (Yoon *et al.* 2001a).

Group 3 consists of a phylogenetically coherent group of ten species: *B. alvei*, *B. amylolyticus*, *B. azotofixans*, *B. gordonae*, *B. larvae*, *B. macerans*, *B. macquariensis*, *B. pabuli*, *B. polymyxa* and *B. pulvifaciens*. Group 4 comprises *B. brevis* and *B. laterosporus*, and group 5 contains *B. kaustophilus*, *B. stearothermophilus* and *B. thermoglucosidasius*. The strains representing *B. alcalophilus*, *B. aneurinilyticus*, *B. cycloheptanicus* and *Sporolactobacillus inulinus* form distinct lines of descent. These results concurred essentially with those obtained by Rössler *et al.* (1991). In subsequent studies it has been found that strains formerly labelled as *B. circulans* but comprising three species: *B. amylolyticus*, *B. pabuli* and *B. validus* (Nakamura 1984), also belong to group 3 of Ash *et al.* (1991). This group of organisms was reclassified into the new genus *Paenibacillus* (Ash *et al.* 1993). Petterson *et al.* (1999) found that *B. popilliae* and *B. lentimorbus*, initially grouped into rRNA group 1 of Ash *et al.* (1991), were more closely related to species of rRNA group 3 and, hence, belong to the genus *Paenibacillus*. Several additional species of the genus *Paenibacillus* were proposed subsequently.

The name *Brevibacillus* was proposed by Shida *et al.* (1996) for rRNA group 4 of Ash *et al.* (1991). This genus contains the mesophilic species *Br. brevis*, *Br. agri*, *Br. centrosporus*, *Br. choshinensis*, *Br. parabrevis*, *Br. reuszeri*, *Br. formosus*, *Br. borstelensis*, *Br. laterosporus* and the thermophilic species *Br. thermoruber*. These changes in the taxonomy of the *Br. brevis* group resulted also in the proposal of the genus *Aneurinibacillus*, containing the species 'Br. aneurinilyticus', 'Br. migulanus' (Shida *et al.* 1996) and 'Br. thermoaerophilus' (Heyndrickx *et al.* 1997).

The species *Br. pantothenticus*, originally placed in rRNA group 1 (Ash *et al.* 1991), was subsequently placed in a higher taxon and, as a consequence, the genus *Virgibacillus* was created to accommodate *V. pantothenticus* (Heyndrickx *et al.* 1998); an additional new species, *V. proomii*, was described later (Heyndrickx *et al.* 1999). The species *Bacillus salexigens* – described by Garabito *et al.* (1997) – was also found to constitute a new genus, for which the name *Salibacillus* was proposed (Wainö *et al.* 1999). Wainö *et al.* (1999) also proposed that the halotolerant species *G. halotolerans* and *G. dipsosauri* be accomodated in the genus *Gracilibacillus*. Furthermore, in the phylogenetic study of alkaliphilic bacilli by Nielsen *et al.* (1994) a sixth group was defined on the basis of 16S rRNA sequence data.

For species grouped into rRNA group 5 of Ash *et al.* (1991) the genus name *Geobacillus* was proposed (Nazina *et al.* 2001). This genus contains the species *G. stearothermophilus*, *G. thermooleovorans*, *G. thermocatenulatus*, *G. kaustophilus*, *G. thermoglucosidasius*, *G. thermodenitrificans*, *G. subterraneus* and *G. uzenensis*. For rRNA group 6 (Nielsen *et al.* 1994), no formal nomenclatural changes were proposed; however, on the basis of 16S rRNA sequences, it seems evident that this group merits generic status.

The additional genera *Jeotgalibacillus* and *Marinibacillus* were described for halophilic spore-forming bacteria (Yoon *et al.* 2001b).

Because of the substantial changes outlined above, the role of fatty acid patterns for differentiation needs to be re-evaluated at both the genus and species levels.

Bacillus sensu stricto: rRNA group 1

Ash *et al.*'s (1991) group 1 (*Bacillus s. s.*) includes *B. subtilis*, the type species of the genus, and several other species, including *B. amyloliquefaciens*, *B. anthracis*, *B. atrophaeus*, *B. azotoformans*, *B. badius*, *B. benzoevorans*, *B. cereus*, *B. circulans*, *B. coagulans*, *B. fastidiosus*, *B. firmus*, *B. lentus*, *B. licheniformis*, '*B. maroccanus*', *B. medusa* (a subjective synonym of *B. cereus*), *B. megaterium*, *B. mycoides*, *B. pantothenticus* (now reclassified as *V. pantothenticus*), *B. psychrosaccharolyticus*, *B. simplex*, *B. simplex*, *B. sonorensis*, *B. smithii* and *B. thuringiensis* and some others.

Several of these species were studied in the early fatty acid analyses of Kaneda (1977). Species in Kaneda group A (*B. circulans*, *B. licheniformis*, *B. megaterium*, *B. pumilus* and *B. subtilis*) produced major amounts of *a*14:0 (26–60%) and *i*15:0 (13–30%) and only small amounts (<3%) of unsaturated fatty acids. Further species were allocated to Kaneda group E (*B. anthracis*, *B. cereus* and *B. thuringiensis*) revealing also *i*15:0 as the predominant fatty acid and also small proportions (7–12%) of unsaturated fatty acids. These results were confirmed for several other species belonging to *Bacillus s. s.* in the comprehensive study of Kämpfer (1994); however, with respect to the levels of taxonomic resolution, the fatty acid profiles were found to vary largely among the species. Details of the fatty acid compositions of species of the genus *Bacillus* can be found in table 18.1. The major CFA components of most species of the genus *Bacillus* belonging to rRNA group 1 of Ash *et al.* (1991) [with the exception of *B. badius*, the *B. cereus* group (explained below), *B. circulans*, *B. coagulans*, *B. simplex* and *B. smithii*] (with ranges as percentage of total given in parentheses) are *a*15:0 (25–66%), *i*15:0 (22–47%) and *a*17:0 (2–12%). For species of the *B. cereus* group (*B. anthracis*, *B. cereus*, *B. mycoides*, *B. pseudomycoides* and *B. thuringiensis*), the amounts of *a*15:0 were lower (3–7%) (tables 18.1 and 18.2) and the amounts for unsaturated fatty acids were generally higher (>10%, see table 18.1). On the basis of a reliable database containing information on genomically homogeneous strains, fatty acid patterns can be used for species identification (Nakamura & Jackson 1995; see table 18.2). Additional species were described recently: *B. sonorensis*, which is most closely related to *B. licheniformis* (Palmisano *et al.* 2001), and *B. jeotgali* (Yoon *et al.* 2001c).

Bacillus smithii and *B. coagulans* showed higher amounts of *a*17:0 (17–42% and lower amounts of *a*15:0 and *i*15:0; table 18.1). They form a separate lineage within *Bacillus* rRNA group 1 (Ash *et al.* 1991). Also *B. badius* revealed low amounts of *a*15:0 (<10%) and relatively high amounts of unsaturated acids (table 18.1). Very low amounts of *i*15:0 were detected in *B. circulans*. These results indicate that rRNA group 1 of Ash *et al.* (1991) is still heterogeneous, and it can be expected that further taxonomic rearrangements will lead to a clear definition of *Bacillus s. s.*

Bacillus rRNA group 2

The species *B. sphaericus*, *B. fusiformis*, *B. insolitus*, *B. pasteurii*, *B. psychrophilus* and *Sporosarcina ureae* were grouped into rRNA group 2 (Ash *et al.* 1991). The species *B. globisporus*, *B. pasteurii* and *B. psychrophilus* were reclassified into the genus *Sporosarcina*, together with *Sporosarcina ureae* and *S. aquimarina* (Yoon *et al.* 2001a). It is obvious from 16S rRNA sequence data that this group is phylogenetically different from *Bacillus* species of rRNA group 1. On the basis of the study of Kaneda (1977) the psychrophilic organisms *B. globisporus*, *B. psychrophilus* and *B. insolitus* were allocated to fatty acid group F. The predominant branched chain fatty acids in these species were *a*15:0 acids. In addition, a large proportion of unsaturated fatty acids (17–28%) was detected. These findings were essentially confirmed by Kämpfer (1994), who found that *B. sphaericus* produced large amounts of unsaturated fatty acids. Similar to the *Bacillus* species of rRNA group 1, the fatty acid profiles were found to vary largely among the species. Details of the fatty acid compositions of species of rRNA group 2 are summarized in table 18.3. With respect to fatty acid composition, two groups can be detected in rRNA group 2. The first is composed of *S. globisporus* and *S. psychrophila* (table 18.3). Both species produce low amounts of *i*15:0 (3–8%) and high amounts of *a*15:0 (47–65%). Species of this group have been reclassified into the genus *Sporosarcina* (Yoon *et al.* 2001a). In contrast, *B. insolitus*, *B. sphaericus* and *B. silvestris* reveal larger amounts of *i*15:0 (>20%) and smaller amounts of *a*15:0 (6–23%). *Bacillus sphaericus* is heterogeneous. DNA–DNA similarity studies have resulted in the recognition of at least five groups within this species (Rippere *et al.* 1997). As with *Bacillus* rRNA group 1, group 2 is also heterogeneous and further taxonomic changes can be expected.

Bacillus rRNA group 3 – *Paenibacillus*

The genus *Paenibacillus* was proposed for species grouped into *Bacillus* rRNA group 3 of Ash *et al.* (1991, 1993). Comparisons of rRNA sequences of the type strains of many strains belonging to species of the genus *Bacillus s. l.* led to the conclusion that *B. alvei*, *B. amylolyticus*, *B. azotofixans*, *B. gordonae*, *B. larvae*, *B. macerans*, *B. macquariensis*, *B. pabuli*, *B. polymyxa* and *B. pulvifaciens* fall into this phylogenetically distinct cluster. In a subsequent comprehensive polyphasic study (Heyndrickx *et al.* 1996a), these results were largely confirmed and it was shown that *Paenibacillus alvei*, *P. amylolyticus*, *P. azotofixans*, *P. gordonae* (which was shown to be a later subjective synonym of *P. validus*) (Heyndrickx *et al.* 1995), *P. larvae* (reclassified as *P. larvae* ssp. *larvae*)

Table 18.1 Fatty acid composition of several species of the genus *Bacillus* rRNA group 1 of Ash *et al.* (1991).[a]

Species	Saturated acids								Saturated iso-branched acids							Saturated anteiso-branched acids					Unsaturated acids					Other acids	Reference
	10:0	12:0	13:0	14:0	15:0	16:0	17:0	18:0	i12:0	i13:0	i14:0	i15:0	i16:0	i17:0	i18:0	a13:0	a14:0	a15:0	a16:0	a17:0	i16:1	a16:1	i17:1	a17:1	i18:1		
Group 1[b]																											
B. cereus				3	Tr	3				10	2	55	3					3		3	7	9		Tr			Kämpfer (1994)
B. cereus		1		3	5	2				8	2	49	3	6				4		1	4	3		3		8	Shida et al. (1997a)
B. thuringiensis				3		4				8	4	32	5	1				5		1	14	10		3		9	Kämpfer (1994)
Group 2																											
B. amyloliquefaciens				Tr	Tr	3				Tr	1	34	3	10				36		7	1			2		3	Kämpfer (1994)
B. amyloliquefaciens (mean of 3 strains)						3					1	25	4	14				38		12	1			2			Roberts et al. (1996)
B. atrophaeus (mean of 5 strains)						5					1	31	4	9				36		7	3	2					Roberts et al. (1996)
B. firmus				1	1	1					3	32	1	2				45		2	5	1		3		3	Kämpfer (1994)
B. flexus				2		3					4	30	Tr	2				47		4	2	2		Tr		4	Kämpfer (1994)
B. fumarioli (mean of 20 strains)												51	6	15				6	4	14		2					Logan et al. (2000)
B. jeotgali				1		3					2	49	2	4				9		4	5	7				11	Yoon et al. (2001c)
B. lentus				2		4					7	30	3					44		5	5						Kämpfer (1994)
B. licheniformis				Tr		2					1	34	3	11				27		9	1			3		9	Kämpfer (1994)
B. licheniformis (mean of 5 strains)						4					1	29	4	7				38		11	5	1					Roberts et al. (1996)
B. megaterium				2		2				Tr	6	28	Tr	1				49		3	4			1		3	Kämpfer (1994)
B. megaterium				2	2	2					8	33	1	Tr				45		1						6	Shida et al. (1997a)
B. mojavensis (mean of 22 strains)				1		2					1	22	3	9				43		13	3			3			Roberts et al. (1996)
B. oleronius				1		2						47	3	5				25		16							Kunigk et al. (1995)

B. pumilus	1		1		Tr	1	43			37	5	2			3	7	Kämpfer (1994)
B. sonorensis (mean of 8 strains)			5			Tr	30	3	9	37	12	1		Tr			Palmisano *et al.* (2001)
B. subtilis	Tr	Tr	1	Tr		1	27	1	8	39	10	1		3	1	8	Kämpfer (1994)
B. subtilis	Tr	2	1				28	2	6	45	7			Tr		9	Shida *et al.* (1997a)
B. subtilis (mean of 5 strains)			3			1	29	2	10	40	9	2		2			Roberts *et al.* (1996)
B. vallismortis (mean of 5 strains)			3			1	25	4	14	37	12	1		2			Roberts *et al.* (1996)
Other species																	
B. badius	2	3	3			4	41	1		10	4	5	16	7	4		Kämpfer (1994)
B. badius	2	10	2			2	60	3	2	8	2					9	Shida *et al.* (1997a)
B. badius	2	2	3			2	56	5	3	6	2	4		6	2	7	Heyndrickx *et al.* (1997)
B. circulans	3	1	3			4	14			47	7	6		1		14	Kämpfer (1994)
B. circulans	4	2	4			4	10	3		57	3					13	Shida *et al.* (1997a)
B. coagulans	2		2			1	10	4	Tr	63	17	3					Kämpfer (1994)
B. coagulans	1	5	1				2	Tr		66	23					2	Shida *et al.* (1997a)
B. simplex	1	1	4			3	10		1	69	3	6		1			Kämpfer (1994)
B. smithii			8				19	6	13	12	42					2	Andersson *et al.* (1995)

[a] Data are given for the type strains unless stated otherwise. The data for some unsaturated and branched fatty acids were combined in this table. Data are given as percentage of total cellular fatty acids. Tr, traces (<1%). Because some of the values were summed off, the total sum of fatty acids is not in all cases 100%. More details about the heterogeneity of fatty acids within some species can be found in the study of Kämpfer (1994). *Bacillus cohnii*, grouped into rRNA group 1 (Nielsen *et al.* 1994), showed 46–54% *i*15:0 and *a*15:0, and a proportion of unsaturated fatty acids from 22 to 27% (Spanka & Fritze 1993).

[b] For *B. anthracis*, fatty acid profiles were given by Lawrence *et al.* (1991). When grown on RCM-medium (a complex medium), *B. anthracis* produced (similar to *B. cereus*) high amounts of 16:0 (>50%) and only 7% of *i*15:0. A cultivation on a synthetic medium (RM-medium) increased the amount of *i*15:0 to 16%. Details are given by Lawrence *et al.* (1991).

Table 18.2 Fatty acid profiles of *Bacillus cereus*, *B. mycoides* and *B. pseudomycoides* as defined by Nakamura and Jackson (1995) and Nakamura (1998).

Species	No. of strains	Saturated acids			Saturated *iso*-branched acids						Saturated *anteiso*-branched acids		
		12:0	14:0	16:0	i12:0	i13:0	i14:0	i15:0	i16:0	i17:0	a13:0	a15:0	a17:0
B. cereus	8	0.2±0.2[a]	3.1±0.7	3.9±1.6	0.9±0.3	11.5±3.7	5.2±2.0	30.7±2.8	5.3±1.5	7.2±1.7	1.6±0.4	5.5±1.6	1.1±0.3
B. mycoides	9	1.7±0.7	5.3±2.3	8.3±2.1	2.9±1.0	23.1±6.8	5.3±1.7	22.9±2.9	4.7±1.8	7.1±3.0	2.3±0.5	3.1±0.8	0.9±0.4
B. pseudomycoides	7	1.3±0.4	4.4±0.8	7.9±1.7	7.2±1.8	17.0±4.4	3.5±1.8	19.3±3.2	4.6±1.2	8.8±3.0	6.1±2.1	3.4±0.8	1.6±0.5

[a] Mean values ± SD. **Bold** entries indicate differences between species.

Table 18.3 Fatty acid composition of the genus *Sporosarcina* and further species of the *Bacillus* rRNA group 2 of Ash *et al.* (1991).[a]

Species	Saturated acids								Saturated iso-branched acids							Saturated anteiso-branched acids					Unsaturated acids					Other acids	Reference
	10:0	12:0	13:0	14:0	15:0	16:0	17:0	18:0	i12:0	i13:0	i14:0	i15:0	i16:0	i17:0	i18:0	a13:0	a14:0	a15:0	a16:0	a17:0	i16:1	a16:1	i17:1	a17:1	i18:1		
Sporosarcina																											
S. aquimarina				2		4		Tr			4	5	2	2				77		4	Tr				Tr		Yoon *et al.* (2001a)
S. globisporus											5	7	4	1		1		56		9	2			1		14	Kämpfer (1994)
S. globisporus				Tr	Tr	Tr					3	4	1	Tr				62		7	8					15	Yoon *et al.* (2001a)
'B.' globisporus W25				Tr	1	3					1	3	2					47		15	9	4		15			Kaneda *et al.* (1983)
S. pasteurii				Tr	3	5					15	7	7					49		4	6				Tr	3	Yoon *et al.* (2001a)
S. psychrophila				1		1					4	7	2	6				65			4					9	Kämpfer (1994)
S. psychrophila				Tr	Tr			Tr		Tr	4	6	2	Tr				68		7	4		Tr		Tr	8	Yoon *et al.* (2001a)
S. psychrophila				1		2					2	5	2	Tr				59		12	4			10		14	Kaneda *et al.* (1983)
S. ureae		Tr		Tr	Tr	2		Tr		Tr	2	7	1	Tr		Tr		69		9	3				Tr	5	Yoon *et al.* (2001a)
'Bacillus' group 2																											
'B.' insolitus											13	40	5	4		1		23		2	1			1		10	Kämpfer (1994)
'B.' insolitus W16b			Tr	8		8		Tr		Tr	5	20	3	Tr	Tr		Tr	22		Tr	3			1		21	Kaneda *et al.* (1983)
'B.' silvestris				3		6		1			3	44	6	5	Tr			6		3	19	3	8	3			Rheims *et al.* (1999)
Other species																											
'B.' sphaericus				1	7	1					5	65	2	2				15		1	5					4	Andersson *et al.* (1995)
'B.' sphaericus					7	1					6	53	7	2				9		1						13	Shida *et al.* (1997a)
'B.' sphaericus DSM 463					10	10					3	10	33	6				3		5	13					8	Andersson *et al.* (1995)
'B.' sphaericus DSM 461					2	3					9	52	23	5				1			2					4	Andersson *et al.* (1995)

[a] Data are given for the type strains unless stated otherwise. The data for some unsaturated and branched fatty acids were combined in this table. Data are given as percentage of total cellular fatty acids. Tr, traces (<1%). Because some of the values were summed off, the total sum of fatty acids is not in all cases 100%. The fatty acid profile of the type strain of *'B.' sphaericus* indicates a grouping within group 2.

(Heyndrickx *et al.* 1996a), *P. macerans*, *P. macquariensis*, *P. pabuli*, *P. polymyxa*, *P. pulvifaciens* (reclassified as a subspecies of *P. larvae*) (Heyndrickx *et al.* 1996b), *P. validus*, *B. peoriae*, *B. lautus*, *B. longisporus* and *P. durum* (formerly *Clostridium durum*) clearly belonged to this new genus. In addition, Nakamura (1996) proposed the new species *P. apiarius*. Shida *et al.* (1997a) determined the taxonomic status of a further six '*Bacillus*' species (*B. alginolyticus*, *B. chondroitinus*, *B. curdlanolyticus*, *B. glucanolyticus*, *B. kobensis* and *B. thiaminolyticus*) by 16S rRNA gene sequence comparison. The use of a *Paenibacillus*-specific PCR primer designed for differentiating organisms of the genus *Paenibacillus* from other members of the family Bacillaceae, revealed that the amplified 16S rRNA gene fragment of the six '*Bacillus*' species was identical to those of the other members of the genus *Paenibacillus*. A further three species – *P. amylolyticus*, *P. illinoisensis* and *P. chibensis* – were described by Shida *et al.* (1997b) and Petterson *et al.* (1999) proposed the transfer of *B. lentimorbus* and *B. popilliae* into the genus *Paenibacillus* on the basis of 16S rRNA sequencing studies. Additional species were described recently: *P. borealis* (Elo *et al.* 2001), *P. jamilae*, for which *a*15:0 was the predominant fatty acid (Aguilera *et al.* 2001), and *P. koreensis* (Chung *et al.* 2000).

In the comprehensive studies of Heyndrickx *et al.* (1995, 1996a,b) and Shida *et al.* (1997a,b), fatty acid profiles were determined to find distinguishing characters between the species. Heyndrickx *et al.* (1996a) came to the conclusion that analysis of CFA could be very useful for obtaining descriptive information and can be used to cluster large sets of strains. In connection with the use of an appropriate database, FAME analysis data can result in identification of new isolates to at least the genus level for many taxa. However, in agreement with the study of Kämpfer (1994), it was concluded that the levels of taxonomic resolution can vary greatly between taxa. Consistent with the results of Kämpfer (1994), who studied more than 300 strains, the results of the numerical analysis of the FAME data assigned the strains mainly to species groups. Details of the fatty acid composition of species of the genus *Paenibacillus* can be found in table 18.4. The major CFA components of the genus *Paenibacillus* (with ranges as percentages of total given in parentheses) are *a*15:0 (36–80%), *i*16:0 (0.5–6.6%), *i*15:0 (1–12%) and *a*17:0 (2–21%) (Heyndrickx *et al.* 1995, 1996a,b; Shida *et al.* 1997a,b).

Bacillus rRNA group 4 – *Brevibacillus*

The genus *Brevibacillus* was proposed (Shida *et al.* 1996) to accommodate species previously assigned to the *B. brevis* group (rRNA group 4 of Ash *et al.* 1991) which was divided into eight species, namely: *B. brevis*, *B. agri*, *B. centrosporus*, *B. choshinensis*, *B. parabrevis*, *B. reuszeri*, *B. formosus* and *B. borstelensis* (Nakamura 1991, 1993; Takagi *et al.* 1993; Shida *et al.* 1994a,b, 1995). On the basis of nucleotide sequence comparisons of the 16S rRNA genes of the type strains, clearly these species belonged together with *B. laterosporus* and the thermophilic species *B. thermoruber*, which was described by Manachini *et al.* (1985) as a member of rRNA group 4 of Ash *et al.* (1991). *Bacillus galactophilus* has

Table 18.4 Fatty acid composition of species of the genus *Paenibacillus*.[a]

Species	Saturated acids								Saturated iso-branched acids							Saturated anteiso-branched acids					Unsaturated acids					Reference
	10:0	12:0	13:0	14:0	15:0	16:0	17:0	18:0	i12:0	i13:0	i14:0	i15:0	i16:0	i17:0	i18:0	a13:0	a14:0	a15:0	a16:0	a17:0	i16:1	a16:1	i17:1	a17:1	i18:1	
P. alginolyticus				Tr	1	3					1	5	10	2				70		6	4					Shida et al. (1997a)
P. alvei				2	1	9					1	11	6	5				58		8	4					Shida et al. (1997a)
P. alvei (mean of 6 strains)				2	1	8					Tr	11	5	4				57		7	3					Nakamura (1996)
P. amylolyticus				3	Tr	13					2	2	9	3				46		2	Tr		Tr			Shida et al. (1997b)
P. apiarius				1	2	5					Tr	8	4	5				60		16	1		Tr			Shida et al. (1997a)
P. apiarius (mean of 6 strains)				1	1	4					Tr	14	5	6				52		10	3		2			Nakamura (1996)
P. azoreducens				3	Tr	22					Tr	6	9	6				34		20	Tr					Meehan et al. (2001)
P. azotofixans				2	Tr	18					1	2	7	1		Tr		62		5	1		Tr			Shida et al. (1997a)
P. borealis				18	Tr	10					5	11	10	5				35		2	1					Elo et al. (2001)
P. chondroitinus				1	2	6					2	2	10	1		Tr		70		3						Shida et al. (1997a)
P. chibensis				Tr	1	5					1	4	12	3				58		14						Shida et al. (1997b)
P. curdlanolyticus				1	1	7					2	2	24	1				56		3						Shida et al. (1997a)
P. dendritiformis				1	3	6						6	5	8				43		21	3		1			Tcherpakov et al. (1999)
P. glucanolyticus				2	1	11					1	3	14	2				56		8	Tr		Tr			Shida et al. (1997a)
P. illinoisensis				1	1	24					2	1	6	1				57		5						Shida et al. (1997b)
P. kobensis					3	12						2	9	Tr				65		3						Shida et al. (1997a)
P. koreensis				3	2	28							21					51								Chung et al. (2000)
P. larvae (mean of 6 strains)						28					1	11	3	7				28		9						Nakamura (1996)
P. larvae ssp. *pulvifaciens*				1	1	6						10	5	5				49		21	1					Shida et al. (1997a)
P. lautus				1	Tr	20					Tr	3	4	6				37		11	Tr		1			Shida et al. (1997b)
P. lautus				2		7					2	6	9					58		9	4					Kämpfer (1994)
P. lautus				Tr	Tr	16					1	1	7	1				57		10	2		Tr			Shida et al. (1997a)
P. macerans				4	Tr	18					8	3	16	1				36		12	Tr		Tr			Shida et al. (1997a)
P. macquariensis				1	1	3					1	5	3	Tr				81		1	1		Tr			Shida et al. (1997a)
B. pabuli				2	Tr	5					5	5	9					60		3	2					Kämpfer (1994)
P. pabuli				1	Tr	10						2	5	1				74		4	Tr					Shida et al. (1997a)
P. peoriae				1	Tr	11					1	8	7	5				55		10	Tr		Tr			Shida et al. (1997a)
P. polymyxa				Tr	Tr	9					Tr	1	6	2				63		17	5					Shida et al. (1997a)
P. thiaminolyticus (mean of 6 strains)				2	Tr	12					Tr	6	3	4				47		16	5		1			Nakamura (1996)
P. thiaminolyticus				1	Tr	11						11	6	6				45		16	5		2			Shida et al. (1997a)
P. validus				1	1	11					1	4	12	3				57		7	Tr		Tr			Shida et al. (1997a)

[a] Data are given for the type strains unless stated otherwise. The data for some unsaturated and branched fatty acids were combined in this table. Data are given as percentage of total cellular fatty acids. Tr, traces (<1%). Because some of the values were summed off, the total sum of fatty acids is not in all cases 100%. For *P. campinasensis* (Yoon et al. 1998), 53% a15:0 were reported as the predominant fatty acid.

been recognized to be a synonym of *B. agri* (Shida *et al.* 1994). The major fatty acids of the genus *Brevibacillus* are *i*15:0 and *a*15:0, and the differences between the species are summarized in table 18.5. With the exception of *B. reuszeri*, all *Brevibacillus* species produced *i*15:0 in amounts ranging from 18 to 42%; fatty acid *a*15:0 was produced in amounts ranging from 32 to 72%. Because only few strains of each species were studied for fatty acid composition (Kämpfer 1994; Shida *et al.* 1995), it is difficult to assess the differentiation potential of fatty acid analyses at the species level.

Bacillus rRNA group 5 – *Geobacillus*

This group is a phylogenetically coherent group of thermophilic bacilli (*G. stearothermophilus*, *G. thermocatenulatus*, *G. thermoleovorans*, *G. kaustophilus*, *G. thermodenitrificans*, *G. subterraneus* and *G. uzenensis* showing a high similarity of 16S rRNA sequences (ranging from 98.5 to 99.2%) (Ash *et al.* 1991; Rainey *et al.* 1994; Nazina *et al.* 2001). The taxonomic position of the closely related species '*B.*' *caldolyticus*, '*B.*' *caldotenax*, '*B.*' *caldovelox*, '*B.*' *caldoxylolyticus* and '*B.*' *thermoantarcticus* and the thermophilic asporogenous species *Saccharococcus thermophilus*, should be investigated further (Rainey *et al.* 1994; Nazina *et al.* 2001). The fatty acid patterns of some of these species have been published by Kämpfer (1994), Andersson *et al.* (1995) and Nazina *et al.* (2001). Details on the fatty acid composition of species of the '*Bacillus*' rRNA group 5 (*Geobacillus*) are given in table 18.6. The major CFA components are *i*15:0, *i*16:0 and *i*17:0, which account for 60–80%. As minor components, *a*15:0 and *a*17:0 are detected.

Bacillus rRNA group 6

A comprehensive phylogenetic study of alkaliphilic endospore-forming bacteria revealed a sixth rRNA group within the genus *Bacillus* (Nielsen *et al.* 1994). On the basis of rRNA sequence data and phenotypic data, nine new species were described: *B. agaradhaerens*, *B. clarkii*, *B. clausii*, *B. gibsonii*, *B. halmapalus*, *B. halodurans*, *B. horikhoshii*, *B. pseudalcalophilus* and *B. pseudofirmus* (Nielsen *et al.* 1994, 1995). These species, together with *B. alkalophilus* and the recently descibed species *B. horti* (Yumoto *et al.* 1999), form a distinct phylogenetic group. Data on fatty acid compositions have been published only for *B. alkalophilus* and *B. hortii* (Clejan *et al.* 1986; Yumoto *et al.* 1999). Those studies show that the major CFA components are *a*15:0 and *i*15:0 (table 18.7).

Genus *Alicyclobacillus*

Organisms belonging to the strictly aerobic, slightly thermophilic or thermophilic bacteria of the genus *Alicyclobacillus* were initially placed in the genus *Bacillus* because of their ability to form endospores. 16S rRNA gene sequencing

Table 18.5 Fatty acid composition of species of the genus *Brevibacillus*.[a]

Species	Saturated acids								Saturated *iso*-branched acids							Saturated *anteiso*-branched acids					Unsaturated acids	Reference
	10:0	12:0	13:0	14:0	15:0	16:0	17:0	18:0	i12:0	i13:0	i14:0	i15:0	i16:0	i17:0	i18:0	a13:0	a14:0	a15:0	a16:0	a17:0		
Br. agri				1	Tr	3					1	41	3	4				40		4	2	Shida *et al.* (1995)
Br. borstelensis				Tr	Tr	1					1	42	2	4				32		3	10	Shida *et al.* (1995)
Br. brevis				Tr	1	3					1	18	3	5				54		10	4	Shida *et al.* (1995)
Br. centrosporus				1	Tr	2						23	3	2				62		4	1	Shida *et al.* (1995)
Br. choshinesis				1	1	3					1	12	2	1				72		4	1	Shida *et al.* (1995)
Br. formusus					Tr	2					Tr	25	2	9				44		4	14	Shida *et al.* (1995)
Br. laterosporus (mean of 6 strains)				1							7	34		1				48		1	2	Nakamura (1996)
Br. laterosporus				2		Tr					3	31	1	1		Tr		58		1	1	Kämpfer (1994)
Br. reuszeri				1		3					1	9	3	1				64		5	2	Shida *et al.* (1995)
Br. parabrevis				1	1	2					1	32	2	3				53		4	1	Shida *et al.* (1995)

[a] Data are given for the type strains unless stated otherwise. The data for some unsaturated and branched fatty acids were combined in this table. Data are given as percentage of total cellular fatty acids. Tr, traces (<1%). Because some of the values were summed off, the total sum of fatty acids is not in all cases 100%. For *Br. thermoruber*, Manachini *et al.* (1985) reported 80–90% i15:0, a15:0, a17:0 and a17:0 with a proportion of 44.6% i15:0.

Table 18.6 Fatty acid composition of species of the genera *Ureibacillus*, *Geobacillus* and further species of the *Bacillus* rRNA group 5 of Ash *et al.* (1991).[a]

Species	Saturated acids								Saturated *iso*-branched acids							Saturated *anteiso*-branched acids					Other acids	Reference
	10:0	12:0	13:0	14:0	15:0	16:0	17:0	18:0	i12:0	i13:0	i14:0	i15:0	i16:0	i17:0	i18:0	a13:0	a14:0	a15:0	a16:0	a17:0		
Ureibacillus																						
U. thermosphaericus DSM 10633					3	6						14	59	13							1	Fortina *et al.* (2001)
U. terrenus DSM 12654						12						12	19	43						13		Fortina *et al.* (2001)
'Bacillus'																						
'B.' caldolyticus					1	5	1					22	37	22	2			1		8		Andersson *et al.* (1995)
'B.' caldotenax					3	3	2					29	31	21	1			2		7	1	Andersson *et al.* (1995)
'B.' caldovelox					2	3	1					27	26	27				1		11	1	Andersson *et al.* (1995)
Geobacillus																						
G. stearothermophilus						3						28	6	17				8		39		Shida *et al.* (1997a)
G. stearothermophilus				1	1	9					Tr	40	21	17		5		6		13	1	Kämpfer (1994)
G. thermodenitrificans DSM 465					2	3		2				33	14	34				2		10		Andersson *et al.* (1995)
G. thermodenitrificans DSM 466					2	5		3				35	10	35				1		6		Andersson *et al.* (1995)
G. thermocatenulatus						6						44	9	31				2		8		Andersson *et al.* (1995)
G. uzenensis U1					Tr	4	Tr					21	17	37				2		19		Nazina *et al.* (2001)
G. subterraneus 34					1	2	Tr				3	38	29	18	Tr			2		6		Nazina *et al.* (2001)
G. thermoglucosidasius				Tr		11	1	Tr				22	10	30				2		16	5	Kämpfer (1994)

[a] Data are given for the type strains unless stated otherwise. The data for some unsaturated and branched fatty acids were combined in this table. Data are given as percentage of total cellular fatty acids. Because some of the values were summed off, the total sum of fatty acids is not in all cases 100%.
Tr, traces (<1%).

Table 18.7 Fatty acid composition of species of the genera Amphibacillus, Jeotgalibacillus, Marinibacillus, Virgibacillus, Thermobacillus, Sporolactobacillus, some species of rRNA group 6 of Nielsen et al. (1994) and several other Bacillus species.[a]

Species	Saturated acids								Saturated iso-branched acids							Saturated anteiso-branched acids					Unsaturated acids					Other acids	Reference
	10:0	12:0	13:0	14:0	15:0	16:0	17:0	18:0	i12:0	i13:0	i14:0	i15:0	i16:0	i17:0	i18:0	a13:0	a14:0	a15:0	a16:0	a17:0	i16:1	a16:1	i17:1	a17:1	i18:1		
Amphibacillus																											
A. xylanus				11		20					11	10	19	1				21		5							Niimura et al. (1990)
Bacillus species of rRNA group 6[b]																											
B. horti	1						2					38	5	1	2			30	4		5					8	Yumoto et al. (1999)
B. alcalophilus				1		7		1	1			21	5	12			2	29		1				2	2	13	Clejan et al. (1986)
B. haloalkaliphilus[b] (mean of 10 strains)												48		6				11		11				2		19	Fritze (1996)
Virgibacillus																											
V. pantothenticus				Tr		Tr		2			Tr	11	5	8				34		31							Wainø et al. (1999)
V. pantothenticus (group 1, 10 strains)				1							4	16	8	3		Tr		50		14							Heyndrickx et al. (1998)
V. pantothenticus (group 2, 2 strains)				2						1	4	35	4	5		1		34		6							Heyndrickx et al. (1998)
V. proomii (4 strains)						8				2	6	33	5	4		1		33		6							Heyndrickx et al. (1999)
Sporolactobacillus																											
S. inulinus					5	1						2	1					53		35							Shida et al. (1997a)
Jeotgalibacillus																											
J. alimentarius				Tr		4		3			2	48	2	6				15		5	5			4		3	Yoon et al. (2001b)
Marinibacillus																											
M. marinus						1					5	22	5	3				47		11	3					2	Yoon et al. (2001b)
Thermobacillus																											
T. xylanilyticus				1	2	21		1			1	1	48	2	1			9		1							Touzel et al. (2000)
Other 'Bacillus' species																											
B. thermoleovorans						7		3				34	8	29						12						8	Duffner et al. (1997)
B. ehimensis				1		16					1	6	6	8				50		11						2	Kuroshima et al. (1996)
B. chitinolyticus				2		21					2	4	4	3				58		8						1	Kuroshima et al. (1996)

[a] Data are given for the type strains unless stated otherwise. The data for some unsaturated and branched fatty acids were combined in this table. Data are given as percentage of total cellular fatty acids. Tr, traces (<1%). Because some of the values were summed off, the total sum of fatty acids is not in all cases 100%.

[b] Only few fatty acid data on 'Bacillus' species belonging to rRNA group 6 are published. For B. cohnii, Spanka and Fritze (1993) reported 46–54% iso-C15:0 and anteiso-C15:0 and a proportion of unsaturated fatty acids ranging from 22 to 27%. The phylogenetic position of B. haloalkaliphilus (Fritze 1996) has not been investigated.

and phylogenetic analyses showed that the species of the genus *Alicyclobacillus* belonged to a distinct line of descent within the low G+C Gram-positive lineage of the bacteria that also included the closely related, facultatively autotrophic species of the genus *Alicyclobacillus* (Wisotzkey *et al.* 1992; Durand 1996). Four validly named species were initially described: *A. acidocaldarius* (Darland & Brock 1971), *A. acidoterrestris* (Deinhard *et al.* 1987a), *A. cycloheptanicus* (Deinhard *et al.* 1987b) and *A. hesperidum* (Albuquerque *et al.* 2000). The most distinctive characteristic of these organisms is the presence of ω-cyclohexyl or ω-cycloheptyl fatty acids; the species *A. acidoterrestris* and *A. acidocaldarius* have relatively large proportions of ω-cyclohexyl fatty acids (De Rosa *et al.* 1971; Oshima & Ariga 1975), whereas ω-cycloheptyl fatty acids are the predominant fatty acids of *A. cycloheptanicus* (Allgaier *et al.* 1985). Only a few other bacteria examined have ω-cyclic fatty acids, among them *Sulfobacillus* spp. (Golovacheva & Karavaiko 1978), which also possess ω-cyclohexyl fatty acids (Dufresne *et al.* 1996). For this reason, these fatty acids are excellent markers for these acid-ophilic spore-forming bacteria (table 18.8). The main differences in fatty acid compositions between the four species are shown in table 18.8; the quantitative differences shown can be used, in combination with other markers, for species identification.

Genus *Amphibacillus*

The genus *Amphibacillus* was described by Niimura *et al.* (1990) for facultatively anaerobic, xylan-digesting organisms. 5S rRNA gene sequencing and phyloge-netic analyses showed that these organisms are different from representatives of the previously described genera *Bacillus*, *Clostridium* and *Sporolactobacillus*. On the basis of 16S rRNA sequence comparison, *Amphibacillus* seems to be most closely related to representatives of the genus *Halobacillus* (P. Kämpfer, unpubl. results). The major fatty acids of *Amphibacillus* were *a*15:0 (15–28%) and *i*16:0 (7–22%). High amounts of 16:0 (14–36%) were also reported (Niimura *et al.* 1990). Details are given in table 18.7.

Genus *Halobacillus*

The genus *Halobacillus* was described by Spring *et al.* (1996) for moderately halophilic organisms. 16S rRNA gene sequencing and phylogenetic analyses showed that these organisms are different from representatives of the previously described genera *Bacillus*. The major fatty acids of *Halobacillus* species were *a*15:0 (> 45%) and *i*17:0 (7–25%). High amounts of *i*15:0 (5–23%) are also found in some strains (P. Kämpfer, unpubl. results)

Genus *Aneurinibacillus*

Along with modifications of the taxonomy of the *B. brevis* group, which resulted in the proposal of the genus *Brevibacillus*, it became obvious that on the basis of a

Table 18.8 Fatty acid compositions of *Alicyclobacillus*[a] spp. according to Albuquerque *et al.* (2000).

Species	a14:0	i15:0	a15:0	i16:0	16:0	i17:0	a17:0	ω-Cyclohexyl 17:0	ω-Cyclohexyl 19:0
A. acidocaldarius		2.2		1.1	0.7	10.8	3.0	48.9	33.3
Genomic species 1		1.7	0.8	1.4	0.9	8.3	8.4	51.8	26.7
A. aciditerrestris	1.0				1.1		0.6	71.6	25.8
A. hesperidium		5.4	6.6	0.9	2.1	4.9	10.3	56.8	13.3

[a] For *A. cycloheptanicus* >90% ω-cyclohexyl-fatty acids are reported (Wisotzkey *et al.* 1992).

16S rDNA gene sequence analysis (of the type strains only), closely related species such as *B. aneurinilyticus* and *B. migulanus* should be transferred to a new genus, for which the name *Aneurinibacillus* was proposed (Shida *et al.* 1996). In a further study, a new thermophilic species *B. thermoaerophilus* was described (Meier-Stauffer *et al.* 1996) that was closely related to species of the genus *Aneurinibacillus*. The analyses of the fatty acids of the type strains revealed some interesting results. It could be shown that *B. badius* (belonging to rRNA group 1 of Ash *et al.* 1991) showed a quite similar fatty acid profile to *A. aneurinilyticus* and *A. migulanus*, whereas that of *B. thermoaerophilus* was different (Meier-Stauffer *et al.* 1996). This result can be at least partly explained by the difference in cultivation temperature; *B. thermoaerophilus* strains could not be grown at the normal mesophilic temperature of 28°C and therefore was cultivated at 55°C. A shift in fatty acid composition toward longer-chain fatty acids and lower ratios of unsaturated fatty acids to saturated fatty acids for the thermophilic strains compared to the mesophilic strains has been reported as being linked with higher growth temperatures (Suzuki *et al.* 1993). The major cellular fatty acid components of the genus *Aneurinibacillus* [ranges (percentage of total) are given in parentheses] are *i*15:0 (41.9–66.8%), *i*16:0 (0.5–6.6%), 16:0 (1.8–8.5%) and *i*17:0 (1.0–23.8%) (Shida *et al.* 1996; Heyndrickx *et al.* 1997). All species can be differentiated on the basis of qualitative and quantitative differences in fatty acid composition (table 18.9).

Genus *Thermobacillus*

Recently, the genus *Thermobacillus*, with the species *T. xylanilyticus*, has been described (Touzel *et al.* 2000). It comprises aerobic, thermophilic, xylanolytic spore-forming bacteria related to *B. viscosus*, *Paenibacillus curdlanolyticus* and *P. popillae*, with 16S rRNA sequence similarities of 91.2, 90.1 and 91%, respectively. Its major cellular fatty acids are *i*16:0 (48%), 16:0 (21.4%) and *a*17:0 (table 18.7). These features allowed a clear differentiation from species of the genus *Paenibacillus*, for which *a*15:0 was found to be predominant.

Genus *Ureibacillus*

The genus *Ureibacillus*, with the species *U. thermosphaericus* and *U. terrenus*, has been described recently (Fortina *et al.* 2001). It comprises aerobic, thermophilic, spore-forming bacteria moderately related to the genus *Geobacillus* (Andersson *et al.* 1995; Fortina *et al.* 2001). Its major cellular fatty acids are *i*16:0 (58–61%), when grown on TSA. The fatty acids *i*15:0, *i*17:0 and 16:0 occurred in smaller amounts (table 18.6).

Table 18.9 Fatty acid compositions of species of the genus *Aneurinibacillus* according to Heyndrickx *et al.* (1997a).

	Saturated acids			Saturated iso-branched acids				Saturated anteiso-branched acids		Unsaturated acids								Summed features[a]		
	14:0	15:0	16:0	i14:0	i15:0	i16:0	i17:0	a15:0	a17:0	15:1 ω6c	16:1 ω7C alcohol	i16:1 (H)	16:1 ω11c	16:1 ω5c	i17:1 ω10c	17:1 ω6c	2[b]	4[c]	5[d]
A. aneurinilyticus	2.6	0.8	5.0	3.7	57.2	2.3	2.0	1.3					10.7	0.8	2.6		1.4	6.0	2.1
A. migulanus	1.9	2.8	3.5	9.0	48.6	6.5	2.1	1.3		0.9	1.4	1.5	6.5	1.1	1.9	1.6	1.4	4.9	2.4
A. thermoaerophilus		2.2	2.2		58.5	3.8	23.6	5.1	4.2										

[a] Fatty acids were detected by using the MIDI System.
[b] Summed feature 2: 15:1 *iso* H, 15:1 *iso* I and/or 13:0 3-OH.
[c] Summed feature 4: 16:1 ω7c and/or 15:0 *iso* 2-OH.
[d] Summed feature 5: 17:1 *iso* I and/or 17:1 *anteiso* B.

Genera *Virgibacillus, Gracilibacillus, Salibacillus, Jeotgalibacillus* and *Marinibacillus*

The genus name *Virgibacillus* was proposed by Heyndrickx *et al.* (1998) for organisms formerly classified as '*Bacillus*' *pantothenticus*. On the basis of 16S rRNA sequence data, this species originally was allocated to rRNA group 1 of Ash *et al.* (1991); however, on the basis of a polyphasic study, Heyndrickx *et al.* (1998) provided clear evidence that these organisms should be transferred to a new genus. A second species, *V. proomii* was described recently by Heyndrickx *et al.* (1999). The major cellular fatty acids of *Virgibacillus* are *a*15:0 (24–52%), *i*15:0 (12–40%), 16:0 (3–11%) and *a*17:0 (1–19%) (table 18.10). On the basis of the *i*15:0/*a*15:0 ratios, a differentiation between the two species is possible: in *V. pantothenticus* the ratio is about 1 : 3, but in *V. proomii* it is 1 : 1 (Heyndrickx *et al.* 1999). (Details can be found in tables 18.7 and 18.10.) The genus *Gracilibacillus* was described by Wainö *et al.* (1999) for an extremely halotolerant bacterium isolated from the Great Salt Lake, UT. On the basis of morphological, physiological and 16S rDNA sequence data, this organism was associated with *Bacillus* rRNA group 1, but showed the greatest degree of sequence similarity to strains belonging to *Halobacillus* and the species *Marinococcus albus, V. pantothenticus*, '*B.*' *salexigens* and '*B.*' *dipsosauri*. In this polyphasic study it was also proposed that '*B.*' *dipsosauri* should be transferred to this genus as *G. dipsosauri* and that '*B.*' *salexigens* be transferred to the genus *Salibacillus* as *S. salexigens*. Again, *a*15:0 was the predominant fatty acid in *Gracilibacillus* and *Salibacillus* (30–56%) and *i*15:0 was found in amounts ranging from 4 to 28% depending on the cultivation conditions. Relatively high amounts of *a*17:0 could also be detected (table 18.10). Fatty acid compositions for the genera *Jeotgalibacillus* and *Marinibacillus* are also shown in table 18.7.

Conclusions

At present, the taxomomy of the genus *Bacillus s. l.* is in a state of transition. Sequence analyses of rRNA have provided (and still provide) a firm basis for the division of *Bacillus* into several phylogenetically distinct genera. This procedure is still under way. However, fatty acid data provide a new basis with which to re-evaluate our differentiation of taxa.

In the past, fatty acid data have often been used for classification purposes (see Kaneda 1977). Recently published numerical analyses of fatty acid data, however, concluded that with respect to the levels of taxonomic resolution, fatty acid profiles vary largely within the species (Kämpfer 1994; Heyndrickx *et al.* 1996a). Within a polyphasic taxonomy, CFA analysis is at present often used as a rapid and fairly inexpensive screening and identification method (Vandamme *et al.* 1996). It is, however, important to determine the resolution level of this technique for every taxon under investigation. With respect to *Bacillus s. l.* and the related genera covered here, individual species within the genera often cannot be differentiated, especially in cases when large numbers of strains are examined. Even a clear differentiation of genera (table 18.11) is not always possible, with a

Table 18.10 Fatty acid compositions of *Gracilibacillus*, *Salibacillus* and *Virgibacillus* spp. according to Wainö *et al.* (1999) under different cultivation conditions.[a]

	Culture medium	Saturated acids				Saturated *iso*-branched acids				Saturated *anteiso*-branched acids		Unsaturated acids	
		14:0	15:0	16:0	18:0	i14:0	i15:0	i16:0	i17:0	a15:0	a17:0	16:1	Isomers of 18:1
Gracilibacillus halotolerans	MB	Tr	6.61	14.9	Tr	4.1	8.2	7.1		41.0	9.9	3.0	5.2
	10% MH	Tr	Tr	9.0	Tr	9.0	4.2	7.0		56.1	14.6	Tr	Tr
G. disposauri	MB	Tr	2.8	16.3	Tr	Tr	27.8	3.3	6.7	29.7	13.3	Tr	
	10% MH	Tr	Tr	7.2	Tr	Tr	17.6	4.7	7.2	39.1	24.2		
Salibacillus salexigens	MB	Tr	Tr	19.4	Tr	3.0	22.2	8.1	5.4	30.1	11.6		
	10% MH	Tr	Tr	1.7	Tr	3.9	26.0	11.2	6.1	33.8	17.3		
Virgibacillus pantothenticus	MB	Tr	Tr	9.7	1.5	Tr	11.2	4.8	7.6	34.3	30.8		
	10% MH					1.7	5.1	9.4	2.0	49.9	31.9		

[a] Data are given for the type strains unless stated otherwise. Data are given as percentage of total cellular fatty acids. Tr, traces (<1%). Because some of the values were summed off, the total sum of fatty acids is not in all cases 100%.
MB, Bacto Marine Broth; 10% MH, 10% moderate halophile medium (Garabito *et al.* 1997).

Table 18.11 Important fatty acids for differentiation of *Bacillus* and related genera (for details see tables 18.1 to 18.10).[a]

Genus (group)	Fatty acids important for differentiation at the group or genus level						
	a15:0	i15:0	i16:0	16:0	a17:0	Other acids	Remarks
Alicyclobacillus	0–2	0–5	0–2	Tr–2	1–10	**ω-alicyclic acids (>60%)**	Only the genus *Sulfobacillus* also produces ω-alicyclic acids
Amphibacillus	3–6	**30–55**	3–6	3–9	1–3		Results based on few strains
Aneurinibacillus	2–6	**48–59**	2–7	2–6	0–4		
Gracilibacillus	**29–58**	4–28	3–7	7–16	**9–24**		
Salibacillus	**30–33**	**22–26**	8–11	2–19	**11–18**		Results based on few strains
Sporolactobacillus	**53**	2	1	1	**35**		Results based on few strains
Ureibacillus	0–2	**12–41**	5–61	5–23	1–13		Results based on few strains
Virgibacillus	**33–50**	**11–36**	4–8	5–10	6–31		
Thermobacillus	9	1	**48**	21	1		Results based on few strains
Bacillus (rRNA group 1)							
Group 1 (*B. cereus* group)	3–6	**30–55**	3–6	3–9	1–3		
Group 2	**25–59**	**25–47**	Tr–4	1–5	2–16		
B. badius	6–10	**40–60**	1–5	2–3	2–4		
B. circulans	**47–57**	10–14	Tr–3	3–4	3–7		
B. smithii	12	19	6	8	**42**		
B. coagulans	**63–66**	2–10	Tr–4	1–2	**17–23**		
'Bacillus' (rRNA group 2)							
Group 1 (*Sporosarcina*)	**47–77**	3–8	2–7	Tr–5	9–15		
Group 2	6–31	**20–44**	3–6	Tr–8	Tr–4		
Paenibacillus (rRNA group 3)	**36–70**	2–12	3–24	3–24	2–21		
Brevibacillus (rRNA group 4)	**32–72**	9–42	1–3	Tr–3	1–10		
Geobacillus (rRNA group 5)	1–8	**22–44**	**6–42**	3–11	5–39	i17:0 (7–35%)	
Bacillus (rRNA group 6)	**29–30**	**21–38**	Tr–5	1–7	Tr–1		Results based on few strains

[a] Data are summarized from tables 18.1 to 18.10. Data are given as percentage of total cellular fatty acids. Tr, traces (<1%). Important fatty acids are marked in bold. Because only the important fatty acids are shown, the total sum of fatty acids is not in all cases 100%;

few exceptions (e.g. the separation of *Alicyclobacillus* and *Sulfobacillus* from all other genera). In combination with other methods, however, the relative amounts of important fatty acids (table 18.11) allow a presumptive identification at the genus level or group level within a genus group.

On the other hand, a differentiation of individual species, or even subspecies in certain cases, can be possible if small numbers of strains are studied (because possible intraspecific variation is not detected and hence cannot influence the interpretation of the results), or in cases of studying genomically well-characterized groups of strains (for example, see table 18.2). It must be stressed that it is very important that strains are grown under highly standardized conditions in order to obtain reproducible results. In the future, a further subdivision of the genus *Bacillus s. l.* can be expected. This will offer a new basis for re-assessing the value of CFA analyses at different taxonomic ranks.

References

Abel, K., deSchmertzing, H. & Peterson, J.I. (1963) Classification of microorganisms by analysis of chemical composition. Feasibility of utilizing gas chromatography. *Journal of Bacteriology* 85, 1039–1044.

Aguilera, M., Monteoliva-Sánchez, M., Suárez, A. et al. (2001) *Paenibacillus jamilae* sp. nov., an exopolysaccharide-producing bacterium able to grow in olive-mill waste water. *International Journal of Systematic and Evolutionary Microbiology* 51, 1687–1692.

Albuquerque, L., Rainey, F.A., Chung, A.P. et al. (2000) *Alicyclobacillus hesperidium* sp. nov. and a related genomic species from solfataric soils of Sao Miguel in the Azores. *International Journal of Systematic and Evolutionary Microbiology* 50, 451–457.

Allgaier, H., Poralla, K. & Jung, G. (1985) ω-Cycloheptyl-α-hydroxyundecanoic acid, a new fatty acid from a thermoacidophilic *Bacillus* species. *Liebigs Annalen der Chemie* 1985, 378–382.

Andersson, M., Laukkanen, M., Nurmiaho-Lassila, E.-L., Rainey, F.A., Niemelä, S.I. & Salkinoja-Salonen, M. (1995) *Bacillus thermosphaericus* sp. nov., a new thermophilic ureolytic *Bacillus* isolated from air. *Systematic and Applied Microbiology* 18, 203–220.

Ash, C., Farrow, J.A.E., Wallbanks, S. & Collins, M.D. (1991) Phylogenetic heterogeneity of the genus *Bacillus* revealed by comparative analysis of small subunit ribosomal RNA sequences. *Letters in Applied Microbiology* 13, 202–206.

Ash, C., Priest, F.G. & Collins, M.D. (1993) Molecular identification of rRNA group 3 bacilli (Ash, Farrow, Wallbanks, and Collins) using a PCR probe test. Proposal for the creation of a new genus *Paenibacillus*. *Antonie van Leeuwenhoek* 64, 253–260.

Chung, Y.R., Kim, C.H., Hwang, I. & Chun, J. (2000) *Paenibacillus koreensis* sp. nov., a new species that produces an iturin-like antifungal compound. *International Journal of Systematic and Evolutionary Microbiology* 50, 1495–1500.

Claus, D. & Berkeley, R.C.W. (1986) Genus *Bacillus* Cohn 1872. In: *Bergey's Manual of Systematic Bacteriology* (eds P.H.A. Sneath et al.), Vol. 2, pp. 1105–1140. Williams & Wilkins, Baltimore, MD.

Clejan, S., Krulwich, T.A., Mondrus, K.R. & Seto-Young, D. (1986) Membrane lipid composition of obligately alkalophilic strains of *Bacillus* spp. *Journal of Bacteriology* 168, 334–340.

Darland, G. & Brock, T.D. (1971) *Bacillus acidocaldarius* sp. nov., an acidophilic thermophilic spore-forming bacterium. *Journal of General Microbiology* 67, 9–15.

Daron, H.H. (1973) Nutritional alteration of the fatty acid composition of a thermophilic *Bacillus* species. *Journal of Bacteriology* 116, 1096–1099.

De Rosa, M., Gambacorta, A., Minale, L. & Bulock, J.D. (1971) Cyclohexane fatty acids from a thermophilic bacterium. *Chemical Communications* 1971, 1334.

Deinhard, G., Blanz, P., Poralla, K. & Altan, E. (1987a) *Bacillus acidoterrestris* sp. nov., a new thermotolerant acidophile isolated from different soils. *Systematic and Applied Microbiology* 10, 47–53.

Deinhard, G., Saar, J., Krischke, W. & Poralla, K. (1987b) *Bacillus cycloheptanicus* sp. nov., a new thermoacidophile containing ω-cycloheptane fatty acids. *Systematic and Applied Microbiology* **10**, 68–73.

Duffner, F.M., Reinscheid, U.M., Bauer, M.P., Mutzel, A. & Müller, R. (1997) Strain differentiation and taxonomic characterization of a thermophilic group of phenol-degrading bacilli. *Systematic and Applied Microbiology* **20**, 602–611.

Dufresne, S., Bouscluet, J., Boissinot, M. & Guay, R. (1996) *Sulfobacillus disulfidooxidans* sp. nov., a new acidophilic, disulfide-oxidizing, Gram-positive, spore-forming bacterium, *International Journal of Systematic Bacteriology* **46**, 1056–1064.

Durand, P. (1996) Primary structure of the 16S rRNA gene of *Sulfobacillus thermosulfidooxidans* by direct sequencing of PCR amplified gene and its similarity with that of other moderately thermophilic chemolithotrophic bacteria. *Systematic and Applied Microbiology* **19**, 360–364.

Elo, S., Suominen, I., Kämpfer, P. *et al.* (2001) *Paenibacillus borealis* sp. nov., a nitrogen-fixing species isolated from spruce forest humus in Finland. *International Journal of Systematic and Evolutionary Microbiology* **51**, 535–545.

Fortina, M.G., Pukall, R., Schumann, P. *et al.* (2001) *Ureibacillus* gen. nov., a new genus to accommodate *Bacillus thermosphaericus* (Andersson *et al.* 1995), emendation of *Ureibacillus thermosphaericus* and description of *Ureibacillus terrenus* sp. nov. *International Journal of Systematic and Evolutionary Microbiology* **51**, 447–455.

Fritze, D. (1996) *Bacillus haloalkaliphilus* sp. nov. *International Journal of Systematic Bacteriology* **46**, 98–101.

Garabito, M.J., Arahal, D.R., Mellado, E., Márquez, M.C. & Ventosa, A. (1997) *Bacillus salexigens* sp. nov., a new moderately halophilic *Bacillus* species. *International Journal of Systematic Bacteriology* **47**, 735–741.

Golovacheva, R.S. & Karavaiko, G.I. (1978) A new genus of thermophilic spore-forming bacteria, *Sulfobacillus*. *Microbiology* **47**, 658–665.

Goodfellow, M. & Minnikin, D. (1981) Lipids in the classification of *Bacillus* and related taxa. In: *The Aerobic Endospore-Forming Bacteria* (eds R.C.W. Berkeley & M. Goodfellow), pp. 59–90. Academic Press, London.

Heyndrickx, M., Vandemeulebroecke, K., Scheldeman, P. *et al.* (1995) *Paenibacillus* (formerly *Bacillus*) *gordonae* (Pichinoty *et al.* 1986) Ash *et al.* 1994 is a later synonym of *Paenibacillus* (formerly *Bacillus*) *validus* (Nakamura 1984) Ash *et al.* 1994: emended description of *P. validus*. *International Journal of Systematic Bacteriology* **45**, 661–669.

Heyndrickx, M., Vandemeulebroecke, K., Hoste, B. *et al.* (1996a) Reclassification of *Paenibacillus* (formerly *Bacillus*) *pulvifaciens* (Nakamura 1984) Ash *et al.* 1994, a later synonym of *Paenibacillus* (formerly *Bacillus*) *larvae* (White 1906) Ash *et al.* 1994, as a subspecies of *P. larvae*, with descriptions of *P. larvae* as *P. larvae* subsp. *larvae* and *P. larvae* subsp. *pulvifaciens*. *International Journal of Systematic Bacteriology* **46**, 270–279.

Heyndrickx, M., Vandemeulebroecke, K., Scheldeman, P. *et al.* (1996b) A polyphasic reassessment of the genus *Paenibacillus*, reclassification of *Bacillus lautus* (Nakamura 1984) as *Paenibacillus lautus* comb. nov. and *Bacillus peoriae* (Montefusco *et al.* 1993) as *Paenibacillus peoriae* n nov., and emended descriptions of *P. lautus* and of *P. peoriae*. *International Journal of Systematic Bacteriology* **46**, 988–1003.

Heyndrickx, M., Lebbe, L., Vancanneyt, M. *et al.* (1997) A polyphasic reassessment of the genus *Aneurinibacillus*, reclassification of *Bacillus thermoaerophilus* (Meier-Stauffer *et al.* 1996) as *Aneurinibacillus thermoaerophilus* comb. nov. and emended descriptions of *A. aneurinilyticus* corrig., *A. migulanus*, and *A. thermoaerophilus*. *International Journal of Systematic Bacteriology* **47**, 808–817.

Heyndrickx, M., Lebbe, L., Kersters, K., De Vos, P., Forsyth, G. & Logan, N.A. (1998) *Virgibacillus*: a new genus to accommodate *Bacillus pantothenticus* (Proom and Knight 1950). Emended description of *Virgibacillus pantothenticus*. *International Journal of Systematic Bacteriology* **48**, 99–106.

Heyndrickx, M., Lebbe, L., Hoste, B. *et al.* (1999) Proposal of *Virgibacillus proomii* sp. nov. and emended description of *Virgibacillus pantotheniticus* (Proom and Knight 1950) Heyndrickx *et al.* 1998. *International Journal of Systematic Bacteriology* **49**, 1083–1090.

James, A.T. & Martin, A.J.P. (1952) Gas-liquid partition chromatography: the separation and microestimation of volatile fatty acids from formic acid to dodecanoic acid. *Biochemical Journal* **50**, 679–690.

Kämpfer, P. (1994) Limits and possibilities of total fatty acid analysis for classification and identification of *Bacillus* species. *Systematic and Applied Microbiology* **17**, 86–96.

Kaneda, T. (1966) Biosynthesis of branched chain fatty acids. IV. Factors affecting relative abundance of fatty acids produced by *Bacillus subtilis*. *Canadian Journal of Microbiology* **12**, 510–514.

Kaneda, T. (1977) Fatty acids of the genus *Bacillus*: an example of branched-chain preferences. *Bacteriological Reviews* **41**, 391–418.

Kaneda, T., Smith, E.J. & Naik, D.N. (1983) Fatty acid composition and primer specifity of *de novo* fatty acid synthetase in *Bacillus globisporus, Bacillus insolitus,* and *Bacillus psychrophilus. Canadian Journal of Microbiology* 29, 1634–1641.

Kunigh, T., Borst, E.-M., Breunig, A. *et al.* (1995) *Bacillus oleronius* sp. nov., a member of the hind-gut flora of the termite *Reticulitermes santonensis* (Feytaud). *Canadian Journal of Microbiology* 41, 699–706.

Kuroshima, K.-I., Sakane, T., Takata, R. & Yokota, A. (1996) *Bacillus ehimensis* sp. nov. and *Bacillus chitinolyticus* sp. nov., new chitinolytic members of the gnus *Bacillus. International Journal of Systematic Bacteriology* 46, 76–80.

Lawrence, D., Heitefuss, S. & Seifert, H.S. (1991) Differentiation of *Bacillus anthracis* from *Bacillus cereus* by gas chromatographic whole-cell fatty acid analysis. *Journal of Clinical Microbiology* 29, 1508–1512.

Logan, N.A., Lebbe, L., Hoste, B. *et al.* (2000) Aerobic endospore-forming bacteria from geothermal environments in northern Victoria Land, Antarctica, and Candlemas Island, South Sandwich archipelago, with the proposal of *Bacillus fumarioli* sp. nov. *International Journal of Systematic and Evolutionary Microbiology* 50, 1741–1753.

Manachini, P.L., Fortina, M.G., Parini, C. & Craveri, R. (1985) *Bacillus thermoruber* sp. nov., nom. rev., a red-pigmented thermophilic bacterium. *International Journal of Systematic Bacteriology* 35, 493–496.

Meehan, C., Bjourson, A.J. & McMullan, G. (2001) *Paenibacillus azoreducens* sp. nov., a synthetic azo dye decolorizing bacterium from industrial wastewater. *International Journal of Systematic and Evolutionary Microbiology* 51, 1681–1685.

Meier-Stauffer, K., Busse, H.-J., Rainey, F.A. *et al.* (1996) Description of *Bacillus thermoaerophilus* sp. nov., to include sugar beet isolates and *Bacillus brevis* ATCC 12990. *International Journal of Systematic Bacteriology* 46, 532–541.

Nakamura, L.K. (1984) *Bacillus amylolyticus* sp. nov., nom. rev., *Bacillus lautus* sp. nov., nom. rev., *Bacillus pabuli* sp. nov., nom. rev., and *Bacillus validus* sp. nov., nom. rev. *International Journal of Systematic Bacteriology* 34, 224–226.

Nakamura, L.K. (1991) *Bacillus brevis* Migula 1900 taxonomy: reassociation and base composition of DNA. *International Journal of Systematic Bacteriology* 41, 510–515.

Nakamura, L.K. (1993) DNA relatedness of *Bacillus brevis* Migula 1900 strains and proposal of *Bacillus agri* sp. nov., nom. rev., and *Bacillus centrosporus* sp. nov., nom. rev. *International Journal of Systematic Bacteriology* 43, 20–25.

Nakamura, L.K. (1996) *Paenibacillus apiarius* sp. nov. *International Journal of Systematic Bacteriology* 46, 688–693.

Nakamura, L.K. (1998) *Bacillus pseudomycoides* sp. nov. *International Journal of Systematic Bacteriology* 48, 1031–1035.

Nakamura, L.K. & Jackson, M.A. (1995) Clarification of the taxonomy of *Bacillus mycoides. International Journal of Systematic Bacteriology* 45, 46–49.

Nazina, T.N., Tourova, T.P., Poltaraus, A.B. *et al.* (2001) Taxonomic study of aerobic thermophilic bacilli: descriptions of *Geobacillus subterraneus* gen. nov., sp. nov. and *Geobacillus uzenensis* sp. nov. from petroleum reservoirs and transfer of *Bacillus stearothermophilus, Bacillus thermocatenulatus, Bacillus thermoleovorans, Bacillus kaustophilus, Bacillus thermoglucosidasius* and *Bacillus thermodenitrificans* to *Geobacillus* as the new combinations *G. stearothermophilus, G. thermocatenulatus, G. thermoleovorans, G. kaustophilus, G. thermoglucosidasius* and *G. thermodenitrificans. International Journal of Systematic and Evolutionary Microbiology* 51, 433–446.

Nielsen, P., Rainey, F.A., Outtrup, F.A., Priest, F.G. & Fritze, D. (1994) Comparative 16S rDNA sequence analysis of some alkaliphilic bacilli and the establishment of a sixth rRNA group within the genus *Bacillus.* FEMS *Microbiology Letters* 117, 61–66.

Nielsen, P., Fritze, D. & Priest, F.G. (1995) Phenetic diversity of alkaliphilic *Bacillus* strains: proposal for nine new species. *Microbiology* 141, 1745–1761.

Niimura, Y., Koh, E., Yanagida, F., Suzuki, K., Komagata, K. &. Kozaki, M. (1990) *Amphibacillus xylanus* gen. nov., sp. nov., a facultatively anaerobic sporeforming xylan-digesting bacterium which lacks cytochrome, quinone, and catalase. *International Journal of Systematic Bacteriology* 40, 297–301.

Norris, J.R., Berkeley, R.C.W., Logan, N.A. & O'Donnell, A.G. (1981). The genera *Bacillus* and *Sporolactobacillus.* In: *The Prokaryotes* (eds M.P. Starr *et al.*), pp. 1711–1742. Springer, Berlin.

Oshima, M. & Ariga, T. (1975) ω-Cyclohexyl fatty acids in acidophilic thermophilic bacteria. *Journal of Biological Chemistry* 250, 6963–6968.

Palmisano, M.M., Nakamura, L.K., Duncan, K.E., Istock, C.A. & Cohan, F.M. (2001) *Bacillus sonorensis* sp. nov., a close relative of *Bacillus licheniformis,* isolated from soil in the Sonoran Desert, Arizona. *International Journal of Systematic and Evolutionary Microbiology* 51, 1671–1679.

Petterson, B., Rippere, K.E., Yousten, A.A. & Priest, F.G. (1999) Transfer of *Bacillus lentimorbus* and *Bacillus popilliae* to the genus *Paenibacillus* with emended description of *Paenibacillus lentimorbus* comb. nov. and *Paenibacillus popilliae* comb. nov. *International Journal of Systematic Bacteriology* **49**, 531–540.

Rainey, F.A., Fritze, D. & Stackebrandt, E. (1994) The genetic diversity of thermophilic members of the genus *Bacillus* as revealed by 16S rDNA analysis. *FEMS Microbiology Letters* **115**, 205–211.

Rheims, H., Frühling, A., Schumann, P., Rohde, M. & Stackebrandt, E. (1999) *Bacillus silvestris* sp. nov., a new member of the genus *Bacillus* that contains lysine in its cell wall. *International Journal of Systematic Bacteriology* **49**, 795–802.

Rippere, K.E., Johnson, J.L. & Yousten, A.A. (1997) DNA similarities among mosquito-pathogenic and nonpathogenic strains of *Bacillus sphaericus*. *International Journal of Systematic Bacteriology* **47**, 214–216.

Roberts, M.S., Nakamura, L.K. & Cohan, F.M. (1996) *Bacillus vallismortis* sp. nov., a close relative to *Bacillus subtilis*, isolated from soil in Death Valley, California. *International Journal of Systematic Bacteriology* **46**, 470–475.

Rössler, D., Ludwig, W., Schleifer, K.-H. *et al.* (1991) Phylogenetic diversity in the genus *Bacillus* as seen by 16S rRNA sequencing studies. *Systematic and Applied Microbiology* **14**, 266–269.

Saito, K. (1960) Chromatographic studies on bacterial fatty acids. *Journal of Biochemistry* **47**, 699–719

Shida, O., Takagi, H., Kadowaki, K. *et al.* (1994a) *Bacillus aneurinolyticus* sp. nov., nom. rev. *International Journal of Systematic Bacteriology* **44**, 143–150.

Shida, O., Takagi, H., Kadowaki, K., Udaka, S. & Komagata, K. (1994b) *Bacillus galactophilus* is a later subjective synonym of *Bacillus agri*. *International Journal of Systematic Bacteriology* **44**, 172–173.

Shida, O., Takagi, H., Kadowaki, K., Udaka, S., Nakamura, L.K. & Komagata, K. (1995) Proposal of *Bacillus reuszeri* sp. nov., *Bacillus formosus* sp. nov., nom. rev., and *Bacillus borstelensis* sp. nov., nom. rev. *International Journal of Systematic Bacteriology* **45**, 93–100.

Shida, O., Takagi, H., Kadowaki, K. & Komagata, K. (1996) Proposal for two new genera, *Brevibacillus* gen. nov. and *Aneurinibacillus* gen. nov. *International Journal of Systematic Bacteriology* **46**, 939–946.

Shida, O., Takagi, H., Kadowaki, K., Nakamura, L.K. & Komagata, K. (1997a) Transfer of *Bacillus alginolyticus*, *Bacillus chondroitinus*, *Bacillus curdlanolyticus*, *Bacillus glucanolyticus*, *Bacillus kobensis*, and *Bacillus thiaminolyticus* to the genus *Paenibacillus* and emended description of the genus *Paenibacillus*. *International Journal of Systematic Bacteriology* **47**, 289–298.

Shida, O., Takagi, H., Kadowaki, K., Nakamura, L.K. & Komagata, K. (1997b) Emended description of *Paenibacillus amylolyticus* and description of *Paenibacillus illinoisensis* sp. nov. and *Paenibacillus chibensis* sp. nov. *International Journal of Systematic Bacteriology* **47**, 299–306.

Spanka, R. & Fritze, D. (1993) *Bacillus cohnii* sp. nov., a new, obligately alkaliphilic, oval-spore-forming *Bacillus* species with ornithine and aspartic acid instead of diaminopimelic acid in the cell wall. *International Journal of Systematic Bacteriology* **43**, 150–156.

Spring, S., Ludwig, W., Marquez, M.C., Ventosa, A. & Schleifer, K.-H. (1996) *Halobacillus* gen. nov., with descriptions of *Halobacillus litoralis* sp. nov. and *Halobacillus trueperi* sp. nov., and transfer of *Sporosarcina halophila* to *Halobacillus halophilus* comb. nov. *International Journal of Systematic Bacteriology* **43**, 150–156.

Suzuki, K., Goodfellow, M. & O'Donnell, A.G. (1993) Cell envelopes and classification. In: *Handbook of New Bacterial Systematics* (eds M. Goodfellow & A.G. O'Donnell), pp. 195–238. Academic Press, London.

Takagi, H., Shida, O., Kadowaki, K., Komagata, K. & Udaka, S. (1993) Characterization of *Bacillus brevis*, with descriptions of *Bacillus migulanus* sp. nov., *Bacillus choshinensis* sp. nov., *Bacillus parabrevis* sp. nov., and *Bacillus galactophilus* sp. nov. *International Journal of Systematic Bacteriology* **43**, 221–231.

Tcherpakow, M., Ben-Jacob, E. & Gutnick, D.L. (1999) *Paenibacillus dendritiformis* sp. nov. proposal for a new pattern-forming species and its localization within a phylogenetic cluster. *International Journal of Systematic Bacteriology* **49**, 239–246.

Touzel, J.P., O'Donohue, M., Debeire, P., Samain, E. & Breton, C. (2000) *Thermobacillus xylanilyticus* gen. nov. sp. nov., a new aerobic thermophilic xylan-degrading bacterium isolated from farm soil. *International Journal of Systematic and Evolutionary Microbiology* **50**, 315–320.

Vandamme, P., Pot, B., Gillis, M., De Vos, P., Kersters, K. & Swings, J. (1996) Polyphasic taxonomy, a consensus approach to bacterial systematics. *Microbiological Reviews* **60**, 407–438.

Wainö, M., Tindall, B.J., Schumann, P. & Ingvorsen, K. (1999) *Gracilibacillus* gen. nov., with description of *Gracilibacillus halotolerans* gen. nov. sp. nov.; transfer of *Bacillus dipsosauri* to *Gracili-*

bacillus dipsosauri comb. nov., and *Bacillus salexigens* to the genus *Salibacillus salexigens* comb. nov. *International Journal of Systematic Bacteriology* 49, 821–831.

Weerkamp, A. & Heinen, W. (1972) Effect of temperature on the fatty acid composition of the extreme thermophiles *Bacillus caldolyticus* and *Bacillus caldotenax*. *Journal of Bacteriology* 109, 443–446.

Welch, D.F. (1991) Applications of cellular fatty acid analysis. *Clinical Microbiology Reviews* 4, 422–438.

Willecke, K. & Pardee, A.B. (1971) Fatty acid requiring mutant of *Bacillus subtilis* defective in brached chain α-keto acid dehydrogenase. *Journal of Biological Chemistry* 246, 5264–5272.

Wisotzkey, J.D., Jurtshuk, P., Jr, Fox, G.E., Deinhard, G. & Poralla, K. (1992) Comparative sequence analysis on the 16S rRNA (rDNA) of *Bacillus acidocaldarius*, *Bacillus acidoterrestris*, and *Bacillus cycloheptanicus* and proposal for creation of a new genus, *Alicyclobacillus* gen. nov. *International Journal of Systematic Bacteriology* 42, 263–269.

Yoon, J.-H., Yim, D.K., Lee, J.-S. *et al.* (1998) *Paenibacillus campinasensis* sp. nov., a cyclodexrin-producing bacterium isolated in Brazil. *International Journal of Systematic Bacteriology* 48, 833–837.

Yoon, J.-H., Lee, K.-C., Weiss, N., Kho, Y.-H., Kang, K.H. & Park, Y.-H. (2001a) *Sporosarcina aquimarina* sp. nov., a bacterium isolated from seawater in Korea, and transfer of *Bacillus globisporus* (Larkin and Stokes 1967), *Bacillus psychrophilus* (Nakamura 1984) and *Bacillus pasteurii* (Chester 1898) to the genus *Sporosarcina* as *Sporosarcina globispora* comb. nov., *Sporosarcina psychrophila* comb. nov. and *Sporosarcina pasteurii* comb. nov., and emended description of the genus *Sporosarcina*. *International Journal of Systematic and Evolutionary Microbiology* 51, 1079–1086.

Yoon, J.-H., Weiss, N., Lee, K.-C., Lee, I.-S., Kang, K.H. & Park, Y.-H. (2001b) *Jeotgalibacillus alimentarius* gen. nov., sp. nov., a novel bacterium isolated from jeotgal with L-lysine in the cell wall, and reclassification of *Bacillus marinus* Rüger 1983 as *Marinibacillus marinus* gen. nov., comb. nov. *International Journal of Systematic and Evolutionary Microbiology* 51, 2087–2093.

Yoon, J.-H., Kang, S.-S., Lee, K.-C. *et al.* (2001c) *Bacillus jeotgali* sp. nov., isolated from jeotgal, Korean traditional fermented seafood. *International Journal of Systematic and Evolutionary Microbiology* 51, 1087–1092.

Yumoto, I., Yamazaki, K., Sawabe, T. *et al.* (1998) *Bacillus horti* sp. nov., a new gram-negative alkaliphilic bacillus. *International Journal of Systematic Bacteriology* 48, 565–571.

Chapter 19

Some Concluding Observations

J.R. Norris

Introduction

When planning the conference on which this book is based, the organizers chose to focus on two rather different themes: systematics and applications. This proved to be a wise decision. Both sections covered a great deal of ground, making this publication an important reference point in the developing story of the genus; equally valuable for research workers studying the aerobic endospore-formers and for students entering the field.

The organizers intended that the two themes would interact and cross-feed one another to their mutual advantage. This certainly happened, as will be evident to the reader. Techniques and concepts crossed the border between the two themes at numerous points, emphasizing the importance of systematics for the rational development of many applications and, at the same time, enabling the more applied topics to provide new perspectives for the systematist.

At the end of the conference there was a clear hope that it would be followed in perhaps three years' time by a further meeting. The present chapter therefore comments on some of the points raised in the book and offers a few suggestions about possible topics for such a future meeting.

Classification

Seen from the outside, the taxonomy of the aerobic endospore-formers today is complex, not a little confusing, and very exciting. The complexity and confusion stem from the interaction of three approaches: traditional characterization based on phenotypic analysis, the more recent methods of characterization at the molecular level (especially those aimed at describing the structures of DNA and RNA), and the diagnostician's urgent need in some cases to identify new isolates rapidly and unequivocally.

The excitement arises from the rapid development of our understanding of the genetic structure of bacteria, bringing with it new perspectives on systematic relationships and the opportunity to study the phylogenies of the taxa. With gene sequences nearing completion for *Bacillus anthracis*, *B. cereus* and *B. thuringiensis*, it is not surprising that there was a feeling at the meeting of keen anticipation of major developments, as reflected by the chapters of Stackebrandt and Swiderski, Turnbull *et al.*, Fritze (Chapters 2, 3 and 8), and several others.

As the reader tries to weigh the strengths and weaknesses of different approaches to characterization, it is generally hoped that the modern phylogenetic relationships should not completely overturn those based on the more traditional phenotypic methods. Of course, it is comforting and reassuring when the conclusions of these two approaches agree, and a reflection of much painstaking and extensive work, but there is no reason why they should always be in agreement.

The very nature of the organisms with which we work usually dictates that:

1 we study thoroughly only those we can isolate and grow in the laboratory;

2 we only study them under conditions we can easily produce in the laboratory; and

3 we only study those aspects of their structure and behaviour that we can easily test in the laboratory.

Inevitably we are restricted to studying only a part – possibly quite a small part – of the phenotypic expression of the whole genome. Why should we expect the results of traditional characterization studies to mirror a pattern of relationships based on information derived at the genetic level and reflecting phylogeny?

This is a basic question in microbial systematics at present. It is posed clearly by Stackebrandt and Swiderski in Chapter 2, and underlies several of the discussions elsewhere. It leads me to ask whether we still pay too much attention to taxonomic structures based on phenotypic characterization. Within a few years we will have sufficient information about the structure of the genomes of these bacteria to know how they are related to each other at the genetic level and in an evolutionary sense. In the meantime our main requirement is for simple, reliable methods enabling us to identify a particular isolate, often for a specific reason – whether it is toxic, or liable to cause food spoilage, or survive a critical control point in a HACCP procedure for example. The point is well made by Turnbull and by Logan in Chapters 3 and 9.

Some people cannot get very excited about the details of taxonomy. To them it does not seem particularly important to define genus, species, variety, biotype and so forth. If the protein crystal is removed from a *B. thuringiensis*, does it really matter if the organism is renamed *B. cereus*?

The reality of the situation today is that we are able to take a *B. cereus* or *B. thuringiensis*, strip out undesirable characteristics and then introduce carefully selected toxins and other properties into the shell that is left. In this way it is possible to make a new, effective, insecticide tailor-made for a specific application. What does this do to the taxonomy of the group? What should the resulting organism be called, given that a name should be predictive of an organism's properties?

We can see several examples of this kind of situation. De Vos clearly sets out some of the key issues in Chapter 10. Outtrup, Jørgensen and Chanway (Chapters 14 and 15) give us fascinating insights into the development of strains for industrial and agricultural applications – and captivating glimpses of the future. Although it might be difficult to give names to some of the specialized industrial organisms that have been developed, names remain essential for communication and for patenting.

Identification

Techniques for analysing and interpreting DNA sequences are probably unlikely to become generally available to the majority of routine microbiological laboratories in the near future. We should not, however, fall into the trap of underestimating the rate of technological progress and the effects it can have on the cost and ease of processes that may today seem difficult and expensive. When a digital watch fails it may be thrown away because it is not cost-effective to have it repaired, and yet it would represent more than the entire computing capacity that existed in the world in 1950, the year this author went to university.

Nevertheless, for the next few years – or decades perhaps – the needs of microbiologists whose main concern is the rapid identification of bacteria will be met by techniques of the kind described by Magee and Goodacre, and Kämpfer (Chapters 17 and 18). The problems of identification are well described by Fritze and Logan (Chapters 8 and 9). The various different approaches now being developed for whole-cell analysis at the molecular level are beginning to offer help to those concerned with rapid and accurate identification. They generate new kinds of information but, of course, create further headaches for classical taxonomists. The sophisticated statistical methods that are used to analyse the information generated by these techniques provide new perspectives on the handling of characterization data, such as full DNA sequences. Data handling and analysis might be a topic suitable for a future conference.

Applications

The chapters concerned with applications give an excellent overview of some very important subjects, and provide a valuable reference point for students and research workers alike. Especially interesting is Chanway's review of *Bacillus* species and relatives as plant growth-promoters (Chapter 15). Granum (Chapter 4) provides an excellent summary of *B. cereus* food poisoning, and Heyndrickx and Scheldeman (Chapter 6) give us a valuable overview of the food spoilage potential of the aerobic endospore-formers. A future meeting might usefully see an expansion of these applied topics to cover such themes as anthrax, opportunistic pathogenicity, and human and animal pathogenicity in general. Also of interest is the problem of endospore-formers as contaminants in industrial processes, and the use of *Bacillus* species as indicator organisms in the monitoring of sterilization procedures.

The work on thermophiles described by Nicolaus *et al.* (Chapter 5) and that on halophiles presented by Arahal and Ventosa (Chapter 7), both rely heavily on the application of efficient sampling techniques in the field. A session devoted to modern sampling and recovery methods for aerobic endospore-formers could well provide an interesting and valuable addition to a future meeting.

Insect pathogens

The insect pathogens, and the biological insecticides derived from them, represent an important theme in any discussion of the aerobic endospore-formers nowadays. The field is well covered by Bishop, Van Rie and Priest (Chapters 11, 12 and 13). One can only marvel at the subtlety of the mechanisms evolved by *B. thuringiensis* and *B. sphaericus* to enable them to colonize the strange ecological niche represented by the insect gut. Together these chapters provide a fascinating section of the book. This is not only because of the elegance of the work described and the beauty of the molecular mechanisms involved. It is also valuable to have available such timely reviews and balanced accounts when *B. thuringiensis* (*Bt*) has achieved some notoriety as a target for those opposed to genetic modification (GM) of foods.

With the development of transgenic crop technology, *Bt* has transformed in the minds of the public from being 'organic' to being 'GM'; changing overnight from the Fairy Godmother of the environmentalists to become the Wicked Witch of the anti-GM lobby. This situation serves to emphasize the communication gap that exists between scientists and the public.

Questions about the potential of *Bt* strains to cause infections in humans were raised at several points in the meeting. They remained unanswered, however, and this must be a serious concern. Two million kilograms of *Bt* spores have been sprayed worldwide. Clearly there have been no obvious major problems associated with this heavy inoculation of the Earth's surface, but there are suggestions that the organism may be involved in food poisoning outbreaks and perhaps in other conditions such as skin and eye infections in man. Bishop's review of this topic (Chapter 11) is most welcome and timely. Given that infections with *Bt* might easily be diagnosed as being caused by *B. cereus*, it is clearly time that possible *Bt* infections were subject to proper examination by laboratories competent to make the necessary diagnoses. This is an important issue, deserving urgent attention. Discussion of a report of such a study would make a valuable focal point for a future conference.

Interaction with the environment

The first person actually to see *B. thuringiensis* was probably Louis Pasteur. In 1849 he reported seeing rod-shaped bacteria in the body fluid of silkworms suffering from a disease called flachérie, whose symptoms suggest *Bt* infection. When in 1902 the organism turned up again, in silkworms in Japan, Shigatane Ishiwata had at his disposal all the basic techniques to isolate, purify, characterize and study the organism. He was able to provide a good description of *Bacillus sotto* and to show that it was the cause of flachérie.

Ishiwata's success owed much to the work of Robert Koch, who introduced new techniques and concepts into microbiology at a very early stage in the development of the subject. Koch handed to the early microbiologists a tremendously powerful set of tools, and virtually set the pattern for the development of microbiology for the next 100 years. He did, however, also do the subject a disservice by focusing attention, in those early formative years, on the pure culture – a man-made, laboratory artefact virtually unknown in the natural world. It is only relatively recently that we have come to recognize the importance of the complex web of interactions that take place between microorganisms and other life forms in the environment, although, as Chanway (Chapter 15) shows, such ideas are not really new.

Throughout the book there are many references to the apparent ease with which bacteria are able to acquire genetic information from other organisms. Mahillon pays particular attention to this subject (Chapter 16), which also underlies many other comments both in the book and during discussion at the conference. Should we perhaps think rather less about bacteria as individual entities, and rather more about their interactions with the environments in which they live? Should we be asking more penetrating questions about the ease with which bacteria can exchange information with that gene pool?

Our extensive knowledge of the aerobic endospore-formers and their behaviour provides us with a set of well-characterized markers. These may prove to be useful tools with which to carry out some imaginative investigations into the interactions of bacteria with their environments. The rewards of such a study might well prove to be highly significant for our understanding of microbial interactions in general.

Index